Experimental and computational mathematics:
Selected writings

Jonathan Borwein
and
Peter Borwein

PSIpress

Perfectly Scientific Press
www.perfscipress.com

Perfectly Scientific Press
3754 SE Knight St.
Portland, OR 97202

Copyright © 2010 by Perfectly Scientific Press.

All Rights Reserved. No part of this book may be reproduced, used, scanned, or distributed in any printed or electronic form or in any manner without written permission except for brief quotations embodied in critical articles and reviews.

Second Perfectly Scientific Press paperback edition: December 2010.
Perfectly Scientific Press paperback ISBN: 978-1-935638-05-6

Cover design by Julia Canright.

Cover background by Armin Straub, based on the complex-plane plot of a random walk's moment $W_4(s)$, obtained via Meijer-G functions and numerical quadrature.

Visit our website at www.perfscipress.com

Printed in the United States of America.
9 8 7 6 5 4 3 2 1

This book was printed on 15% post-consumer waste paper.

Foreword

Great things often begin in humble ways.

My 25-year-long collaboration with Jonathan and Peter Borwein began in 1985, when I read their paper on fast methods for computing elementary constants and functions, which article is included as the first chapter of this volume. After reading this engaging and intriguing article, I was inspired to try to implement some of these algorithms on the computer, using a high-precision arithmetic software package that I wrote specifically for this task. Indeed, the Borwein algorithms worked as advertised, yielding many thousands of correct digits in just a few seconds on a state-of-the-art computer at the time. I was particularly intrigued by the Borwein algorithm for pi. I then contacted the Borweins to tell of my interest in their work, and, as they say, the rest is history.

As can be readily seen in the articles here, the Borweins are inarguably the world's leading exponents of utilizing state-of-the-art computer technology to discover and prove new and fundamental mathematical results. They employ raw numerical computation, symbolic processing and advanced visualization facilities in their work. The Borweins have spread this gospel in countless fascinating and cogent lectures, as well as in numerous books and over 200 published papers. As a direct result of the Borweins' influence, hundreds of researchers worldwide are now engaged in "experimental" and computationally-assisted mathematics, and the pace of mathematical discovery has measurably quickened.

In reading through these papers, I am struck that the Borweins have also made an important contribution to a more fundamental problem in the field: How can researchers, laboring at the state of the art in the very difficult and demanding arena of modern mathematics, communicate the excitement of their work to young minds who potentially will form the next generation of mathematicians?

The Borweins have found the answer: (a) Bring computers into all aspects of the mathematical research arena, thereby attracting thousands of young, computer-savvy students into the fold, and letting them experience first-hand the excitement of discovering heretofore unknown facts of mathematics; and (b) Highlight the numerous intriguing connections of this work to other fields of mathematics, computer science and modern scientific philosophy.

These articles exemplify the exploratory spirit of truly pioneering work. They are destined to be read over and over again for decades to come. What's more, in most cases these papers can be read and comprehended even by persons of modest mathematical training. Enjoy!

David H. Bailey
Lawrence Berkeley National Laboratory
September 2010

Preface

This is a representative collection of most of our writings about the nature of mathematics over the past twenty-five years. Many are jointly authored, others are by one or the other of us with various coauthors—often with our longtime collaborator and friend David Bailey who has generously written a foreword to the volume. We have included more technical papers only when they form the basis for discussion in the more general papers collected.

For each article in this selection, we have written a few paragraphs of introduction situating the given paper in the collection and where appropriate bringing it up to date. For the most part, we let the articles speak for themselves and suggest the reader consult the index which we have added to improve navigation between selections. We would like to thank Joshua Borwein for writing that index.

Jonathan Borwein
Peter Borwein
September 2010

Acknowledgments

For our families, many members of which were not born when we started collaborating. And particularly for our wives Jennifer and Judith who have sustained and supported us for all these years!

Contents

	Page
Foreword	i
Preface	iii
Acknowledgments	v
1 The arithmetic-geometric mean and fast computation of elementary functions	1
2 On the complexity of familiar functions and numbers	19
3 Ramanujan and pi	35
4 Ramanujan, modular equations, and approximations to pi or how to compute a billion digits of pi	45
5 Strange series and high precision fraud	67
6 The amazing number π	87
7 Experimental mathematics: Recent developments and future outlook	93
8 Visible structures in number theory	113
9 The experimental mathematician: The pleasure of discovery and the role of proof	129
10 Experimental mathematics: Examples, methods and implications	165
11 Ten problems in experimental mathematics	179
12 Implications of experimental mathematics for the philosophy of mathematics	211
13 Exploratory experimentation and computation	241
14 Closed forms: What they are and why they matter	265
Index	295

1. The arithmetic-geometric mean and fast computation of elementary functions

Discussion

This article was written just after we had first digested the requisite classical analysis and number theory to understand why fast computation requires elliptic integrals. It owes much to the great 19th century analysts—especially Gauss, Legendre, Jacobi and Cayley. We found their accounting of the theory to be both more accessible and more computationally helpful than later, more abstract writings on the subject. The article has an equal debt to still active researchers, such as Richard Brent, Gene Salamin and Bill Gosper.

By great good luck, the first elliptic integral computation of π was announced just as the article was being completed. Thus, the theory and the practice were successfully married.

This is further detailed in the next two articles.

The objects described herein are revisited throughout the collection, thereby substantiating Freeman Dyson's elegant observation in his review of *Nature's Numbers* by Ian Stewart (Basic Books 1995). Dyson writes:

> I see some parallels between the shifts of fashion in mathematics and in music. In music, the popular new styles of jazz and rock became fashionable a little earlier than the new mathematical styles of chaos and complexity theory. Jazz and rock were long despised by classical musicians, but have emerged as art-forms more accessible than classical music to a wide section of the public. Jazz and rock are no longer to be despised as passing fads. Neither are chaos and complexity theory. But still, classical music and classical mathematics are not dead. Mozart lives, and so does Euler. When the wheel of fashion turns once more, quantum mechanics and hard analysis will once again be in style.[1]

Source

J.M. Borwein and P.B. Borwein, "The arithmetic-geometric mean and fast computation of elementary functions," *SIAM Review*, **26** (1984), 351–366.

---

[1] *American Mathematical Monthly*, August-September 1996, 612.

THE ARITHMETIC-GEOMETRIC MEAN AND FAST COMPUTATION OF ELEMENTARY FUNCTIONS*

J. M. BORWEIN† AND P. B. BORWEIN†

Abstract. We produce a self contained account of the relationship between the Gaussian arithmetic-geometric mean iteration and the fast computation of elementary functions. A particularly pleasant algorithm for π is one of the by-products.

Introduction. It is possible to calculate 2^n decimal places of π using only n iterations of a (fairly) simple three-term recursion. This remarkable fact seems to have first been explicitly noted by Salamin in 1976 [16]. Recently the Japanese workers Y. Tamura and Y. Kanada have used Salamin's algorithm to calculate π to 2^{23} decimal places in 6.8 hours. Subsequently 2^{24} places were obtained ([18] and private communication). Even more remarkable is the fact that all the elementary functions can be calculated with similar dispatch. This was proved (and implemented) by Brent in 1976 [5]. These extraordinarily rapid algorithms rely on a body of material from the theory of elliptic functions, all of which was known to Gauss. It is an interesting synthesis of classical mathematics with contemporary computational concerns that has provided us with these methods. Brent's analysis requires a number of results on elliptic functions that are no longer particularly familiar to most mathematicians. Newman in 1981 stripped this analysis to its bare essentials and derived related, though somewhat less computationally satisfactory, methods for computing π and log. This concise and attractive treatment may be found in [15].

Our intention is to provide a mathematically intermediate perspective and some bits of the history. We shall derive implementable (essentially) quadratic methods for computing π and all the elementary functions. The treatment is entirely self-contained and uses only a minimum of elliptic function theory.

1. 3.14159265358979323846264338327950288419697. The calculation of π to great accuracy has had a mathematical import that goes far beyond the dictates of utility. It requires a mere 39 digits of π in order to compute the circumference of a circle of radius 2×10^{25} meters (an upper bound on the distance travelled by a particle moving at the speed of light for 20 billion years, and as such an upper bound on the radius of the universe) with an error of less than 10^{-12} meters (a lower bound for the radius of a hydrogen atom).

Such a calculation was in principle possible for Archimedes, who was the first person to develop methods capable of generating arbitrarily many digits of π. He considered circumscribed and inscribed regular n-gons in a circle of radius 1. Using $n = 96$ he obtained

$$3.1405\cdots = \frac{6336}{2017.25} < \pi < \frac{14688}{4673.5} = 3.1428.$$

If $1/A_n$ denotes the area of an inscribed regular 2^n-gon and $1/B_n$ denotes the area of a circumscribed regular 2^n-gon about a circle of radius 1 then

$$(1.1) \qquad A_{n+1} = \sqrt{A_n B_n}, \qquad B_{n+1} = \frac{A_{n+1} + B_n}{2}.$$

---

*Received by the editors February 8, 1983, and in revised form November 21, 1983. This research was partially sponsored by the Natural Sciences and Engineering Research Council of Canada.

†Department of Mathematics, Dalhousie University, Halifax, Nova Scotia, Canada B3H 4H8.

This two-term iteration, starting with $A_2 := \frac{1}{2}$ and $B_2 := \frac{1}{4}$, can obviously be used to calculate π. (See Edwards [9, p. 34].) A_{15}^{-1}, for example, is 3.14159266 which is correct through the first seven digits. In the early sixteen hundreds Ludolph von Ceulen actually computed π to 35 places by Archimedes' method [2].

Observe that $A_n := 2^{-n} \operatorname{cosec}(\theta/2^n)$ and $B_n := 2^{-n-1} \operatorname{cotan}(\theta/2^{n+1})$ satisfy the above recursion. So do $A_n := 2^{-n} \operatorname{cosech}(\theta/2^n)$ and $B_n := 2^{-n-1} \operatorname{cotanh}(\theta/2^{n+1})$. Since in both cases the common limit is $1/\theta$, the iteration can be used to calculate the standard inverse trigonometric and inverse hyperbolic functions. (This is often called Borchardt's algorithm [6], [19].)

If we observe that

$$A_{n+1} - B_{n+1} = \frac{1}{2(\sqrt{A_n}/\sqrt{B_n} + 1)} (A_n - B_n)$$

we see that the error is decreased by a factor of approximately four with each iteration. This is linear convergence. To compute n decimal digits of π, or for that matter arcsin, arcsinh or log, requires $O(n)$ iterations.

We can, of course, compute π from arctan or arcsin using the Taylor expansion of these functions. John Machin (1680–1752) observed that

$$\pi = 16 \arctan\left(\frac{1}{5}\right) - 4 \arctan\left(\frac{1}{239}\right)$$

and used this to compute π to 100 places. William Shanks in 1873 used the same formula for his celebrated 707 digit calculation. A similar formula was employed by Leonhard Euler (1707–1783):

$$\pi = 20 \arctan\left(\frac{1}{7}\right) + 8 \arctan\left(\frac{3}{79}\right).$$

This, with the expansion

$$\arctan(x) = \frac{y}{x}\left(1 + \frac{2}{3} y + \frac{2 \cdot 4}{3 \cdot 5} y^2 + \cdots\right)$$

where $y = x^2/(1 + x^2)$, was used by Euler to compute π to 20 decimal places in an hour. (See Beckman [2] or Wrench [21] for a comprehensive discussion of these matters.) In 1844 Johann Dase (1824–1861) computed π correctly to 200 places using the formula

$$\frac{\pi}{4} = \arctan\left(\frac{1}{2}\right) + \arctan\left(\frac{1}{5}\right) + \arctan\left(\frac{1}{8}\right).$$

Dase, an "idiot savant" and a calculating prodigy, performed this "stupendous task" in "just under two months." (The quotes are from Beckman, pp. 105 and 107.)

A similar identity:

$$\pi = 24 \arctan\left(\frac{1}{8}\right) + 8 \arctan\left(\frac{1}{57}\right) + 4 \arctan\left(\frac{1}{239}\right)$$

was employed, in 1962, to compute 100,000 decimals of π. A more reliable "idiot savant", the IBM 7090, performed this calculation in a mere 8 hrs. 43 mins. [17].

There are, of course, many series, products and continued fractions for π. However, all the usual ones, even cleverly evaluated, require $O(\sqrt{n})$ operations ($+$, $\times$, $\div$, $\sqrt{\ }$) to arrive at n digits of π. Most of them, in fact, employ $O(n)$ operations for n digits, which is essentially linear convergence. Here we consider only full precision operations. For a time complexity analysis and a discussion of time efficient algorithms based on binary splitting see [4].

The algorithm employed in [17] requires about 1,000,000 operations to compute 1,000,000 digits of π. We shall present an algorithm that reduces this to about 200 operations. The algorithm, like Salamin's and Newman's requires some very elementary elliptic function theory. The circle of ideas surrounding the algorithm for π also provides algorithms for all the elementary functions.

2. Extraordinarily rapid algorithms for algebraic functions. We need the following two measures of speed of convergence of a sequence (a_n) with limit L. If there is a constant C_1 so that

$$|a_{n+1} - L| \leq C_1 |a_n - L|^2$$

for all n, then we say that (a_n) converges to L *quadratically*, or with *second order*. If there is a constant $C_2 > 1$ so that, for all n,

$$|a_n - L| \leq C_2^{-2^n}$$

then we say that (a_n) converges to L *exponentially*. These two notions are closely related; quadratic convergence implies exponential convergence and both types of convergence guarantee that a_n and L will "agree" through the first $O(2^n)$ digits (provided we adopt the convention that .9999...9 and 1.000...0 agree through the required number of digits).

Newton's method is perhaps the best known second order iterative method. Newton's method computes a zero of $f(x) - y$ by

$$(2.1) \qquad x_{n+1} := x_n - \frac{f(x_n) - y}{f'(x_n)}$$

and hence, can be used to compute f^{-1} quadratically from f, at least locally. For our purposes, finding suitable starting values poses little difficulty. Division can be performed by inverting $(1/x) - y$. The following iteration computes $1/y$:

$$(2.2) \qquad x_{n+1} := 2x_n - x_n^2 y.$$

Square root extraction ($\sqrt{y}$) is performed by

$$(2.3) \qquad x_{n+1} := \frac{1}{2}\left(x_n + \frac{y}{x_n}\right).$$

This ancient iteration can be traced back at least as far as the Babylonians. From (2.2) and (2.3) we can deduce that division and square root extraction are of the same order of complexity as multiplication (see [5]). Let $M(n)$ be the "amount of work" required to multiply two n digit numbers together and let $D(n)$ and $S(n)$ be, respectively, the "amount of work" required to invert an n digit number and compute its square root, to n digit accuracy. Then

$$D(n) = O(M(n))$$

and

$$S(n) = O(M(n)).$$

We are not bothering to specify precisely what we mean by work. We could for example count the number of single digit multiplications. The basic point is as follows. It requires

$O(\log n)$ iterations of Newton's method (2.2) to compute $1/y$. However, at the ith iteration, one need only work with accuracy $O(2^i)$. In this sense, Newton's method is self-correcting. Thus,

$$D(n) = O\left(\sum_{i=1}^{\log n} M(2^i)\right) = O(M(n))$$

provided $M(2^i) \geq 2M(2^{i-1})$. The constants concealed beneath the order symbol are not even particularly large. Finally, using a fast multiplication, see [12], it is possible to multiply two n digits numbers in $O(n \log(n) \log \log(n))$ single digit operations.

What we have indicated is that, for the purposes of asymptotics, it is reasonable to consider multiplication, division and root extraction as equally complicated and to consider each of these as only marginally more complicated than addition. Thus, when we refer to operations we shall be allowing addition, multiplication, division and root extraction.

Algebraic functions, that is roots of polynomials whose coefficients are rational functions, can be approximated (calculated) exponentially using Newton's method. By this we mean that the iterations converge exponentially and that each iterate is itself suitably calculable. (See [13].)

The difficult trick is to find a method to exponentially approximate just one elementary transcendental function. It will then transpire that the other elementary functions can also be exponentially calculated from it by composition, inversion and so on.

For this Newton's method cannot suffice since, if f is algebraic in (2.1) then the limit is also algebraic.

The only familiar iterative procedure that converges quadratically to a transcendental function is the arithmetic-geometric mean iteration of Gauss and Legendre for computing complete elliptic integrals. This is where we now turn. We must emphasize that it is difficult to exaggerate Gauss' mastery of this material and most of the next section is to be found in one form or another in [10].

3. The real AGM iteration. Let two positive numbers a and b with $a > b$ be given. Let $a_0 := a$, $b_0 := b$ and define

$$(3.1) \qquad a_{n+1} := \frac{1}{2}(a_n + b_n), \qquad b_{n+1} := \sqrt{a_n b_n}$$

for n in $\mathbb{N}$.

One observes, as a consequence of the arithmetic-geometric mean inequality, that $a_n \geq a_{n+1} \geq b_{n+1} \geq b_n$ for all n. It follows easily that (a_n) and (b_n) converge to a common limit L which we sometimes denote by $AG(a, b)$. Let us now set

$$(3.2) \qquad c_n := \sqrt{a_n^2 - b_n^2} \quad \text{for } n \in \mathbb{N}.$$

It is apparent that

$$(3.3) \qquad c_{n+1} = \frac{1}{2}(a_n - b_n) = \frac{c_n^2}{4a_{n+1}} \leq \frac{c_n^2}{4L},$$

which shows that (c_n) converges quadratically to zero. We also observe that

$$(3.4) \qquad a_n = a_{n+1} + c_{n+1} \quad \text{and} \quad b_n = a_{n+1} - c_{n+1}$$

which allows us to define a_n, b_n and c_n for negative n. These negative terms can also be generated by the *conjugate scale* in which one starts with $a_0' := a_0$ and $b_0' := c_0$ and defines

(a'_n) and (b'_n) by (3.1). A simple induction shows that for any integer n

$$(3.5) \quad a'_n = 2^{-n} a_{-n}, \quad b'_n = 2^{-n} c_{-n}, \quad c'_n = 2^{-n} b_{-n}.$$

Thus, backward iteration can be avoided simply by altering the starting values. For future use we define the quadratic *conjugate* $k' := \sqrt{1 - k^2}$ for any k between 0 and 1.

The limit of (a_n) can be expressed in terms of a *complete elliptic integral of the first kind*,

$$(3.6) \quad I(a, b) := \int_0^{\pi/2} \frac{d\theta}{\sqrt{a^2 \cos^2 \theta + b^2 \sin^2 \theta}}.$$

In fact

$$(3.7) \quad I(a, b) = \frac{1}{2} \int_{-\infty}^{\infty} \frac{dt}{\sqrt{(a^2 + t^2)(b^2 + t^2)}}$$

as the substitution $t := a \tan \theta$ shows. Now the substitution of $u := \tfrac{1}{2}(t - (ab/t))$ and some careful but straightforward work [15] show that

$$(3.8) \quad I(a, b) = I\left(\left(\frac{a + b}{2}\right), \sqrt{ab}\right).$$

It follows that $I(a_n, b_n)$ is independent of n and that, on interchanging limit and integral,

$$I(a_0, b_0) = \lim_{n \to \infty} I(a_n, b_n) = I(L, L).$$

Since the last integral is a simple arctan (or directly from (3.6)) we see that

$$(3.9) \quad I(a_0, b_0) = \frac{\pi}{2} AG(a_0, b_0).$$

Gauss, of course, had to derive rather than merely verify this remarkable formula. We note in passing that $AG(\cdot, \cdot)$ is positively homogeneous.

We are now ready to establish the underlying limit formula.

PROPOSITION 1.

$$(3.10) \quad \lim_{k \to 0^+} \left[\log\left(\frac{4}{k}\right) - I(1, k) \right] = 0.$$

Proof. Let

$$A(k) := \int_0^{\pi/2} \frac{k' \sin \theta \, d\theta}{\sqrt{k^2 + (k')^2 \cos^2 \theta}}$$

and

$$B(k) := \int_0^{\pi/2} \sqrt{\frac{1 - k' \sin \theta}{1 + k' \sin \theta}} \, d\theta.$$

Since $1 - (k' \sin \theta)^2 = \cos^2 \theta + (k \sin \theta)^2 = (k' \cos \theta)^2 + k^2$, we can check that

$$I(1, k) = A(k) + B(k).$$

Moreover, the substitution $u := k' \cos \theta$ allows one to evaluate

$$(3.11) \quad A(k) := \int_0^{k'} \frac{du}{\sqrt{u^2 + k^2}} = \log\left(\frac{1 + k'}{k}\right).$$

356   J. M. BORWEIN AND P. B. BORWEIN

Finally, a uniformity argument justifies

$$\lim_{k \to 0^+} B(k) = B(0) = \int_0^{\pi/2} \frac{\cos \theta \, d\theta}{1 + \sin \theta} = \log 2, \tag{3.12}$$

and (3.11) and (3.12) combine to show (3.10). □

It is possible to give various asymptotics in (3.10), by estimating the convergence rate in (3.12).

PROPOSITION 2. *For $k \in (0, 1]$*

$$\left| \log\left(\frac{4}{k}\right) - I(1, k) \right| \leq 4k^2 I(1, k) \leq 4k^2 (8 + |\log k|). \tag{3.13}$$

Proof. Let

$$\Delta(k) := \log\left(\frac{4}{k}\right) - I(1, k).$$

As in Proposition 1, for $k \in (0, 1]$,

$$|\Delta(k)| \leq \left| \log\left(\frac{2}{k}\right) - \log\left(\frac{1 + k'}{k}\right) \right| + \left| \int_0^{\pi/2} \left[\sqrt{\frac{1 - k' \sin \theta}{1 + k' \sin \theta}} - \sqrt{\frac{1 - \sin \theta}{1 + \sin \theta}} \right] d\theta \right|. \tag{3.14}$$

We observe that, since $1 - k' = 1 - \sqrt{1 - k^2} < k^2$,

$$\left| \log\left(\frac{2}{k}\right) - \log\left(\frac{1 + k'}{k}\right) \right| = \left| \log\left(\frac{1 + k'}{2}\right) \right| \leq 1 - k' \leq k^2. \tag{3.15}$$

Also, by the mean value theorem, for each θ there is a $\gamma \in [0, k]$, so that

$$0 \leq \left[\sqrt{\frac{1 - k' \sin \theta}{1 + k' \sin \theta}} - \sqrt{\frac{1 - \sin \theta}{1 + \sin \theta}} \right]$$

$$\leq \left[\sqrt{\frac{1 - (1 - k^2) \sin \theta}{1 + (1 - k^2) \sin \theta}} - \sqrt{\frac{1 - \sin \theta}{1 + \sin \theta}} \right]$$

$$= \left[\frac{\sqrt{1 + (1 - \gamma^2) \sin \theta}}{\sqrt{1 - (1 - \gamma^2) \sin \theta}} \cdot \frac{2\gamma \sin \theta}{(1 + (1 - \gamma^2) \sin \theta)^2} \right] k$$

$$\leq \frac{2\gamma k}{\sqrt{1 - (1 - \gamma^2) \sin \theta}} \leq \frac{2k^2}{\sqrt{1 - (1 - k^2) \sin \theta}}.$$

This yields

$$\left| \int_0^{\pi/2} \left[\sqrt{\frac{1 - k' \sin \theta}{1 + k' \sin \theta}} - \sqrt{\frac{1 - \sin \theta}{1 + \sin \theta}} \right] d\theta \right| \leq 2k^2 \int_0^{\pi/2} \frac{d\theta}{\sqrt{1 - k' \sin \theta}} \leq 2\sqrt{2} \, k^2 I(1, k)$$

which combines with (3.14) and (3.15) to show that

$$|\Delta(k)| \leq (1 + 2\sqrt{2})k^2 I(1, k) \leq 4k^2 I(1, k).$$

We finish by observing that

$$kI(1, k) \leq \frac{\pi}{2}$$

ARITHMETIC-GEOMETRIC MEAN AND FAST COMPUTATION

allows us to deduce that

$$I(1, k) \leq 2\pi k + \log\left(\frac{4}{k}\right).$$

□

Similar considerations allow one to deduce that

(3.16) $$|\Delta(k) - \Delta(h)| \leq 2\pi |k - h|$$

for $0 < k, h < 1/\sqrt{2}$.

The next proposition gives all the information necessary for computing the elementary functions from the AGM.

PROPOSITION 3. *The AGM satisfies the following identity (for all initial values):*

(3.17) $$\lim_{n \to \infty} 2^{-n} \frac{a'_n}{a_n} \log\left(\frac{4a_n}{c_n}\right) = \frac{\pi}{2}.$$

Proof. One verifies that

$$\frac{\pi}{2} = \lim_{n \to \infty} a'_n I(a'_0, b'_0) \qquad \text{(by (3.9))}$$

$$= \lim_{n \to \infty} a'_n I(a'_{-n}, b'_{-n}) \qquad \text{(by (3.8))}$$

$$= \lim_{n \to \infty} a'_n I(2^n a_n, 2^n c_n) \qquad \text{(by 3.5))}.$$

Now the homogeneity properties of $I(\cdot, \cdot)$ show that

$$I(2^n a_n, 2^n c_n) = \frac{2^{-n}}{a_n} I\left(1, \frac{c_n}{a_n}\right).$$

Thus

$$\frac{\pi}{2} = \lim_{n \to \infty} 2^{-n} \frac{a'_n}{a_n} I\left(1, \frac{c_n}{a_n}\right),$$

and the result follows from Proposition 1. □

From now on we fix $a_0 := a'_0 := 1$ and consider the iteration as a function of $b_0 := k$ and $c_0 := k'$. Let P_n and Q_n be defined by

(3.18) $$P_n(k) := \left(\frac{4a_n}{c_n}\right)^{2^{1-n}}, \qquad Q_n(k) := \frac{a_n}{a'_n},$$

and let $P(k) := \lim_{n \to \infty} P_n(k)$, $Q(k) := \lim_{n \to \infty} Q_n(k)$. Similarly let $a := a(k) := \lim_{n \to \infty} a_n$ and $a' := a'(k) := \lim_{n \to \infty} a'_n$.

THEOREM 1. *For $0 < k < 1$ one has*:

(a) $P(k) = \exp(\pi Q(k))$,

(3.19) (b) $0 \leq P_n(k) - P(k) \leq \frac{16}{1 - k^2}\left(\frac{a_n - a}{a}\right),$

(c) $|Q_n(k) - Q(k)| \leq \frac{a'|a - a_n| + a|a' - a'_n|}{(a')^2}.$

Proof. (a) is an immediate rephrasing of Proposition 3, while (c) is straightforward.

To see (b) we observe that

$$P_{n+1} = P_n \cdot \left(\frac{a_{n+1}}{a_n}\right)^{2^{1-n}} \tag{3.20}$$

because $4a_{n+1} c_{n+1} = c_n^2$, as in (3.3). Since $a_{n+1} \leq a_n$ we see that

$$O \leq P_n - P_{n+1} \leq \left[1 - \left(\frac{a_{n+1}}{a_n}\right)^{2^{1-n}}\right] P_n \leq \left(1 - \frac{a_{n+1}}{a_n}\right) P_0,$$

$$P_n - P_{n+1} \leq \left(\frac{a_n - a_{n+1}}{a}\right) P_0 \tag{3.21}$$

since a_n decreases to a. The result now follows from (3.21) on summation. □

Thus, the theorem shows that both P and Q can be computed exponentially since (a_n) can be so calculated. In the following sections we will use this theorem to give implementable exponential algorithms for π and then for all the elementary functions.

We conclude this section by rephrasing (3.19a). By using (3.20) repeatedly we derive that

$$P = \frac{16}{1 - k^2} \prod_{n=0}^{\infty} \left(\frac{a_{n+1}}{a_n}\right)^{2^{1-n}}. \tag{3.22}$$

Let us note that

$$\frac{a_{n+1}}{a_n} = \frac{a_n + b_n}{2a_n} = \frac{1}{2}\left(1 + \frac{b_n}{a_n}\right),$$

and $x_n := b_n/a_n$ satisfies the one-term recursion used by Legendre [14]

$$x_{n+1} := \frac{2\sqrt{x_n}}{x_n + 1} \qquad x_0 := k. \tag{3.23}$$

Thus, also

$$P_{n+1}(k) = \frac{16}{1 - k^2} \prod_{j=0}^{n+1} \left(\frac{1 + x_j}{2}\right)^{2^{1-j}} = \left(\frac{1 + x_n}{1 - x_n}\right)^{2^{-n}}. \tag{3.24}$$

When $k := 2^{-1/2}$, $k = k'$ and one can explicitly deduce that $P(2^{-1/2}) = e^{\pi}$. When $k = 2^{-1/2}$ (3.22) is also given in [16].

4. Some interrelationships. A centerpiece of this exposition is the formula (3.17) of Proposition 3.

$$\lim_{n \to \infty} \frac{1}{2^n} \log\left(\frac{4a_n}{c_n}\right) = \frac{\pi}{2} \lim_{n \to \infty} \frac{a_n}{a'_n}, \tag{4.1}$$

coupled with the observation that both sides converge exponentially. To approximate $\log x$ exponentially, for example, we first find a starting value for which

$$\left(\frac{4a_n}{c_n}\right)^{1/2^n} \to x.$$

This we can do to any required accuracy quadratically by Newton's method. Then we compute the right limit, also quadratically, by the AGM iteration. We can compute exp analogously and since, as we will show, (4.1) holds for complex initial values we can also get the trigonometric functions.

There are details, of course, some of which we will discuss later. An obvious detail is that we require π to desired accuracy. The next section will provide an exponentially converging algorithm for π also based only on (4.1). The principle for it is very simple. If we differentiate both sides of (4.1) we lose the logarithm but keep the π!

Formula (3.10), of Proposition 1, is of some interest. It appears in King [11, pp. 13, 38] often without the "4" in the log term. For our purposes the "4" is crucial since without it (4.1) will only converge linearly (like $(\log 4)/2^n$). King's 1924 monograph contains a wealth of material on the various iterative methods related to computing elliptic integrals. He comments [11, p. 14]:

"The limit [(4.1) without the "4"] does not appear to be generally known, although an equivalent formula is given by Legendre (*Fonctions éliptiques,* t. I, pp. 94–101)."

King adds that while Gauss did not explicitly state (4.1) he derived a closely related series expansion and that none of this "appears to have been noticed by Jacobi or by subsequent writers on elliptic functions." This series [10, p. 377] gives (4.1) almost directly.

Proposition 1 may be found in Bowman [3]. Of course, almost all the basic work is to be found in the works of Abel, Gauss and Legendre [1], [10] and [14]. (See also [7].) As was noted by both Brent and Salamin, Proposition 2 can be used to estimate log given π. We know from (3.13) that, for $0 < k \leq 10^{-3}$,

$$\left| \log\left(\frac{4}{k}\right) - I(1, k) \right| < 10k^2 |\log k|.$$

By subtraction, for $0 < x < 1$, and $n \geq 3$,

(4.2) $\qquad |\log(x) - [I(1, 10^{-n}) - I(1, 10^{-n}x)]| < n\, 10^{-2(n-1)}$

and we can compute log exponentially from the AGM approximations of the elliptic integrals in the above formula. This is in the spirit of Newman's presentation [15]. Formula (4.2) works rather well numerically but has the minor computational drawback that it requires computing the AGM for small initial values. This leads to some linear steps (roughly $\log(n)$) before quadratic convergence takes over.

We can use (3.16) or (4.2) to show directly that π is exponentially computable. With $k := 10^{-n}$ and $h := 10^{-2n} + 10^{-n}$ (3.16) yields with (3.9) that, for $n \geq 1$,

$$\left| \log(10^{-n} + 1) - \frac{\pi}{2}\left[\frac{1}{AG(1, 10^{-n})} - \frac{1}{AG(1, 10^{-n} + 10^{-2n})} \right] \right| \leq 10^{1-2n}.$$

Since $|\log(x+1)/x - 1| \leq x/2$ for $0 < x < 1$, we derive that

(4.3) $\qquad \left| \frac{2}{\pi} - \left[\frac{10^n}{AG(1, 10^{-n})} - \frac{10^n}{AG(1, 10^{-n} + 10^{-2n})} \right] \right| \leq 10^{1-n}.$

Newman [15] gives (4.3) with a rougher order estimate and without proof. This analytically beautiful formula has the serious computational drawback that obtaining n digit accuracy for π demands that certain of the operations be done to twice that precision.

Both Brent's and Salamin's approaches require *Legendre's relation*: for $0 < k < 1$

(4.4) $\qquad I(1, k)J(1, k') + I(1, k')J(1, k) - I(1, k)I(1, k') = \dfrac{\pi}{2}$

where $J(a, b)$ is the *complete elliptic integral of the second kind* defined by

$$J(a, b) := \int_0^{\pi/2} \sqrt{a^2 \cos^2 \theta + b^2 \sin^2 \theta}\, d\theta.$$

The elliptic integrals of the first and second kind are related by

$$(4.5) \qquad J(a_0, b_0) = \left(a_0^2 - \frac{1}{2}\sum_{n=0}^{\infty} 2^n c_n^2\right) I(a_0, b_0)$$

where, as before, $c_n^2 = a_n^2 - b_n^2$ and a_n and b_n are computed from the AGM iteration.

Legendre's proof of (4.4) can be found in [3] and [8]. His elegant elementary argument is to differentiate (4.4) and show the derivative to be constant. He then evaluates the constant, essentially by Proposition 1. Strangely enough, Legendre had some difficulty in evaluating the constant since he had problems in showing that $k^2 \log(k)$ tends to zero with k [8, p. 150].

Relation (4.5) uses properties of the ascending Landen transformation and is derived by King in [11].

From (4.4) and (4.5), noting that if k equals $2^{-1/2}$ then so does k', it is immediate that

$$(4.6) \qquad \pi = \frac{[2AG(1, 2^{-1/2})]^2}{1 - \sum_{n=1}^{\infty} 2^{n+1} c_n^2}.$$

This concise and surprising exponentially converging formula for π is used by both Salamin and Brent. As Salamin points out, by 1819 Gauss was in possession of the AGM iteration for computing elliptic integrals of the first kind and also formula (4.5) for computing elliptic integrals of second kind. Legendre had derived his relation (4.4) by 1811, and as Watson puts it [20, p. 14] "in the hands of Legendre, the transformation [(3.23)] became a most powerful method for computing elliptic integrals." (See also [10], [14] and the footnotes of [11].) King [11, p. 39] derives (4.6) which he attributes, in an equivalent form, to Gauss. It is perhaps surprising that (4.6) was not suggested as a practical means of calculating π to great accuracy until recently.

It is worth emphasizing the extraordinary similarity between (1.1) which leads to linearly convergent algorithms for all the elementary functions, and (3.1) which leads to exponentially convergent algorithms.

Brent's algorithms for the elementary functions require a discussion of incomplete elliptic integrals and the Landen transform, matters we will not pursue except to mention that some of the contributions of Landen and Fagnano are entertainingly laid out in an article by G.N. Watson entitled "The Marquis [Fagnano] and the Land Agent [Landen]" [20]. We note that Proposition 1 is also central to Brent's development though he derives it somewhat tangentially. He also derives Theorem 1(a) in different variables via the Landen transform.

5. An algorithm for π. We now present the details of our exponentially converging algorithm for calculating the digits of π. Twenty iterations will provide over two million digits. Each iteration requires about ten operations. The algorithm is very stable with all the operations being performed on numbers between $\frac{1}{2}$ and 7. The eighth iteration, for example, gives π correctly to 694 digits.

THEOREM 2. *Consider the three-term iteration with initial values*

$$\alpha_0 := \sqrt{2}, \quad \beta_0 := 0, \quad \pi_0 := 2 + \sqrt{2}$$

given by

$$\text{(i)} \quad \alpha_{n+1} := \frac{1}{2}(\alpha_n^{1/2} + \alpha_n^{-1/2}),$$

ARITHMETIC-GEOMETRIC MEAN AND FAST COMPUTATION

(ii) $\beta_{n+1} := \alpha_n^{1/2}\left(\dfrac{\beta_n + 1}{\beta_n + \alpha_n}\right)$,

(iii) $\pi_{n+1} := \pi_n \beta_{n+1}\left(\dfrac{1 + \alpha_{n+1}}{1 + \beta_{n+1}}\right)$.

Then π_n converges exponentially to π and

$$|\pi_n - \pi| \leq \frac{1}{10^{2^n}}.$$

Proof. Consider the formula

(5.1) $$\frac{1}{2^n}\log\left(4\frac{a_n}{c_n}\right) - \frac{\pi}{2}\frac{a_n}{a'_n}$$

which, as we will see later, converges exponentially at a uniform rate to zero in some (complex) neighbourhood of $1/\sqrt{2}$. (We are considering each of $a_n, b_n, c_n, a'_n, b'_n, c'_n$ as being functions of a complex initial value k, i.e. $b_0 = k$, $b'_0 = \sqrt{1 - k^2}$, $a_0 = a'_0 = 1$.)

Differentiating (5.1) with respect to k yields

(5.2) $$\frac{1}{2^n}\left(\frac{\dot{a}_n}{a_n} - \frac{\dot{c}_n}{c_n}\right) - \frac{\pi}{2}\frac{a_n}{a'_n}\left(\frac{\dot{a}_n}{a_n} - \frac{\dot{a}'_n}{a'_n}\right)$$

which also converges uniformly exponentially to zero in some neighbourhood of $1/\sqrt{2}$. (This general principle for exponential convergence of differentiated sequences of analytic functions is a trivial consequence of the Cauchy integral formula.) We can compute $\dot{a}_n, \dot{b}_n$ and $\dot{c}_n$ from the recursions

$$\dot{a}_{n+1} := \frac{\dot{a}_n + \dot{b}_n}{2},$$

(5.3) $$\dot{b}_{n+1} := \frac{1}{2}\left(\dot{a}_n\sqrt{\frac{b_n}{a_n}} + \dot{b}_n\sqrt{\frac{a_n}{b_n}}\right),$$

$$\dot{c}_{n+1} := \frac{1}{2}(\dot{a}_n - \dot{b}_n),$$

where $\dot{a}_0 := 0$, $\dot{b}_0 := 1$, $a_0 := 1$ and $b_0 := k$.

We note that a_n and b_n map $\{z \mid Re(z) > 0\}$ into itself and that $\dot{a}_n$ and $\dot{b}_n$ (for sufficiently large n) do likewise.

It is convenient to set

(5.4) $$\alpha_n := \frac{a_n}{b_n} \quad \text{and} \quad \beta_n := \frac{\dot{a}_n}{\dot{b}_n}$$

with

$$\alpha_0 := \frac{1}{k} \quad \text{and} \quad \beta_0 := 0.$$

We can derive the following formulae in a completely elementary fashion from the basic relationships for a_n, b_n and c_n and (5.3):

(5.5) $$\dot{a}_{n+1} - \dot{b}_{n+1} = \frac{1}{2}(\sqrt{a_n} - \sqrt{b_n})\left(\frac{\dot{a}_n}{\sqrt{a_n}} - \frac{\dot{b}_n}{\sqrt{b_n}}\right),$$

$$(5.6) \quad 1 - \frac{a_{n+1}\,\dot{c}_{n+1}}{\dot{a}_{n+1}\,c_{n+1}} = \frac{2(\alpha_n - \beta_n)}{(\alpha_n - 1)(\beta_n + 1)},$$

$$(5.7) \quad \alpha_{n+1} = \frac{1}{2}(\alpha_n^{1/2} + \alpha_n^{-1/2}),$$

$$(5.8) \quad \beta_{n+1} = \alpha_n^{1/2}\left(\frac{\beta_n + 1}{\beta_n + \alpha_n}\right),$$

$$(5.9) \quad \alpha_{n+1} - 1 = \frac{1}{2\alpha_n^{1/2}}(\alpha_n^{1/2} - 1)^2,$$

$$(5.10) \quad \alpha_{n+1} - \beta_{n+1} = \frac{\alpha_n^{1/2}}{2} \frac{(1 - \alpha_n)(\beta_n - \alpha_n)}{\alpha_n(\beta_n + \alpha_n)},$$

$$(5.11) \quad \frac{\alpha_{n+1} - \beta_{n+1}}{\alpha_{n+1} - 1} = \frac{(1 + \alpha_n^{1/2})^2}{(\beta_n + \alpha_n)} \cdot \frac{(\alpha_n - \beta_n)}{(\alpha_n - 1)}.$$

From (5.7) and (5.9) we deduce that $\alpha_n \to 1$ uniformly with second order in compact subsets of the open right half-plane. Likewise, we see from (5.8) and (5.10) that $\beta_n \to 1$ uniformly and exponentially. Finally, we set

$$(5.12) \quad \gamma_n := \frac{1}{2^n}\left(\frac{\alpha_n - \beta_n}{\alpha_n - 1}\right).$$

We see from (5.11) that

$$(5.13) \quad \gamma_{n+1} = \frac{(1 + \alpha_n^{1/2})}{2(\beta_n + \alpha_n)} \gamma_n$$

and also from (5.6) that

$$(5.14) \quad \frac{\gamma_n}{1 + \beta_n} = \frac{1}{2^{n+1}}\left(1 - \frac{a_{n+1}\,\dot{c}_{n+1}}{\dot{a}_{n+1}\,c_{n+1}}\right).$$

Without any knowledge of the convergence of (5.1) one can, from the preceding relationships, easily and directly deduce the exponential convergence of (5.2), in $\{z \mid |z - \frac{1}{2}| \le c < \frac{1}{2}\}$. We need the information from (5.1) only to see that (5.2) converges to zero.

The algorithm for π comes from multiplying (5.2) by $a_n/\dot{a}_n$ and starting the iteration at $k := 2^{-1/2}$. For this value of k $a'_n = a_n$, $(\dot{a}'_n) = -\dot{a}_n$ and

$$\frac{1}{2^{n+1}}\left(1 - \frac{a_{n+1}\,\dot{c}_{n+1}}{\dot{a}_{n+1}\,c_{n+1}}\right) \to \pi$$

which by (5.14) shows that

$$\pi_n := \frac{\gamma_n}{1 + \beta_n} \to \pi.$$

Some manipulation of (5.7), (5.8) and (5.13) now produces (iii). The starting values for α_n, β_n and γ_n are computed from (5.4). Other values of k will also lead to similar, but slightly more complicated, iterations for π.

To analyse the error one considers

$$\frac{\gamma_{n+1}}{1 + \beta_{n+1}} - \frac{\gamma_n}{1 + \beta_n} = \left[\frac{(1 + \alpha_n^{1/2})^2}{2(\beta_n + \alpha_n)(1 + \beta_{n+1})} - \frac{1}{(1 + \beta_n)}\right]\gamma_n$$

and notes that, from (5.9) and (5.10), for $n \geq 4$,

$$|\alpha_n - 1| \leq \frac{1}{10^{2^n+2}} \quad \text{and} \quad |\beta_n - 1| \leq \frac{1}{10^{2^n+2}}.$$

(One computes that the above holds for $n = 4$.) Hence,

$$\left|\frac{\gamma_{n+1}}{1 + \beta_{n+1}} - \frac{\gamma_n}{1 + \beta_n}\right| \leq \left|\frac{1}{10^{2^n+1}}\right| |\gamma_n|$$

and

$$\left|\frac{\gamma_n}{1 + \beta_n} - \pi\right| \leq \frac{1}{10^{2^n}}. \qquad \square$$

In fact one can show that the error is of order $2^n e^{-\pi 2^{n+1}}$.

If we choose integers in $[\delta, \delta^{-1}]$, $0 < \delta < \frac{1}{2}$ and perform n operations $(+, -, \times, \div, \sqrt{\ })$ then the result is always less than or equal to δ^{2^n}. Thus, if $\gamma > \delta$, it is not possible, using the above operations and integral starting values in $[\delta, \delta^{-1}]$, for every n to compute π with an accuracy of $O(\gamma^{-2^n})$ in n steps. In particular, convergence very much faster than that provided by Theorem 2 is not possible.

The analysis in this section allows one to derive the Gauss-Salamin formula (4.6) without using Legendre's formula or second integrals. This can be done by combining our results with problems 15 and 18 in [11]. Indeed, the results of this section make quantitative sense of problems 16 and 17 in [11]. King also observes that Legendre's formula is actually equivalent to the Gauss–Salamin formula and that each may be derived from the other using only properties of the AGM which we have developed and equation (4.5).

This algorithm, like the algorithms of §4, is not self correcting in the way that Newton's method is. Thus, while a certain amount of time may be saved by observing that some of the calculations need not be performed to full precision it seems intrinsic (though not proven) that $O(\log n)$ full precision operations must be executed to calculate π to n digits. In fact, showing that π is intrinsically more complicated from a time complexity point of view than multiplication would prove that π is transcendental [5].

6. The complex AGM iteration. The AGM iteration

$$a_{n+1} := \frac{1}{2}(a_n + b_n), \qquad b_{n+1} := \sqrt{a_n b_n}$$

is well defined as a complex iteration starting with $a_0 := 1$, $b_0 := z$. Provided that z does not lie on the negative real axis, the iteration will converge (to what then must be an analytic limit). One can see this geometrically. For initial z in the right half-plane the limit is given by (3.9). It is also easy to see geometrically that a_n and b_n are always nonzero.

The iteration for $x_n := b_n/a_n$ given in the form (3.23) as $x_{n+1} := 2\sqrt{x_n}/x_{n+1}$ satisfies

(6.1) $$(x_{n+1} - 1) = \frac{(1 - \sqrt{x_n})^2}{1 + x_n}.$$

This also converges in the cut plane $\mathbb{C} - (-\infty, 0]$. In fact, the convergence is uniformly exponential on compact subsets (see Fig. 1). With each iteration the angle θ_n between x_n and 1 is at least halved and the real parts converge uniformly to 1.

It is now apparent from (6.1) and (3.24) that

(6.2) $$P_n(k) := \left(\frac{4a_n}{c_n}\right)^{2^{1-n}} = \left(\frac{1 + x_n}{1 - x_n}\right)^{2^{-n}}$$

364
J. M. BORWEIN AND P. B. BORWEIN

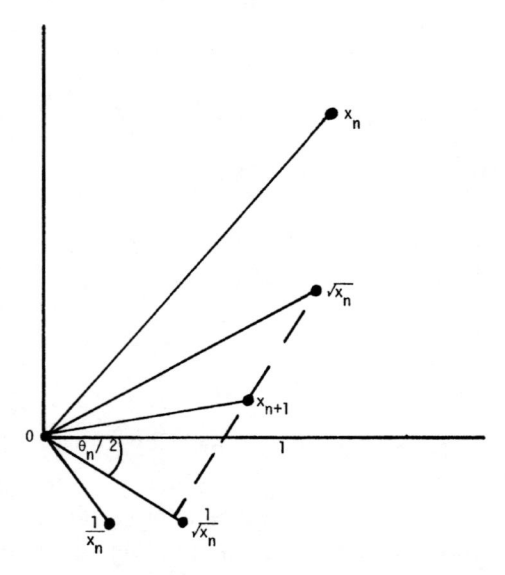

FIG. 1.

and also,

$$Q_n(k) := \frac{a_n}{a'_n}$$

converge exponentially to analytic limits on compact subsets of the complex plane that avoid

$$D := \{z \in \mathbb{C} \mid z \notin (-\infty, 0] \cup [1, \infty)\}.$$

Again we denote the limits by P and Q. By standard analytic reasoning it must be that (3.19a) still holds for k in D.

Thus one can compute the complex exponential—and so also cos and sin—exponentially using (3.19). More precisely, one uses Newton's method to approximately solve $Q(k) = z$ for k and then computes $P_n(k)$. The outcome is e^z. One can still perform the root extractions using Newton's method. Some care must be taken to extract the correct root and to determine an appropriate starting value for the Newton inversion. For example $k := 0.02876158$ yields $Q(k) = 1$ and $P_4(k) = e$ to 8 significant places. If one now uses k as an initial estimate for the Newton inversions one can compute $e^{1+i\theta}$ for $|\theta| \leq \pi/8$. Since, as we have observed, e is also exponentially computable we have produced a sufficient range of values to painlessly compute $\cos \theta + i \sin \theta$ with no recourse to any auxiliary computations (other than π and e, which can be computed once and stored). By contrast Brent's trigonometric algorithm needs to compute a different logarithm each time.

The most stable way to compute P_n is to use the fact that one may update c_n by

(6.3) $$c_{n+1} = \frac{c_n^2}{4a_{n+1}}.$$

One then computes a_n, b_n and c_n to desired accuracy and returns

$$\left(\frac{4a_n}{c_n}\right)^{1/2^n} \quad \text{or} \quad \left(\frac{2(a_n + b_n)}{c_n}\right)^{1/2^n}.$$

This provides a feasible computation of P_n, and so of exp or log.

In an entirely analogous fashion, formula (4.2) for log is valid in the cut complex plane. The given error estimate fails but the convergence is still exponential. Thus (4.2) may also be used to compute all the elementary functions.

7. Concluding remarks and numerical data. We have presented a development of the AGM and its uses for rapidly computing elementary functions which is, we hope, almost entirely self-contained and which produces workable algorithms. The algorithm for π is particularly robust and attractive. We hope that we have given something of the flavour of this beautiful collection of ideas, with its surprising mixture of the classical and the modern. An open question remains. Can one entirely divorce the central discussion from elliptic integral concerns? That is, can one derive exponential iterations for the elementary functions without recourse to some nonelementary transcendental functions? It would be particularly nice to produce a direct iteration for e of the sort we have for π which does not rely either on Newton inversions or on binary splitting.

The algorithm for π has been run in an arbitrary precision integer arithmetic. (The algorithm can be easily scaled to be integral.) The errors were as follows:

Iterate	Digits correct	Iterate	Digits correct
1	3	6	170
2	8	7	345
3	19	8	694
4	41	9	1392
5	83	10	2788

Formula (4.2) was then used to compute 2 log (2) and log (4), using π estimated as above and the same integer package. Up to 500 digits were computed this way. It is worth noting that the error estimate in (4.2) is of the right order.

The iteration implicit in (3.22) was used to compute e^π in a double precision Fortran. Beginning with $k := 2^{-1/2}$ produced the following data:

Iterate	$P_n - e^\pi$	$a_n/b_n - 1$
1	1.6×10^{-1}	1.5×10^{-2}
2	2.8×10^{-9}	2.8×10^{-5}
3	1.7×10^{-20}	9.7×10^{-11}
4	$< 10^{-40}$	1.2×10^{-21}

Identical results were obtained from (6.3). In this case $y_n := 4a_n/c_n$ was computed by the two term recursion which uses x_n, given by (3.23), and

$$(7.1) \qquad y_0^2 := \frac{16}{1 - k^2}, \qquad y_{n+1} = \left(\frac{1 + x_n}{2}\right)^2 y_n^2.$$

One observes from (7.1) that the calculation of y_n is very stable.

We conclude by observing that the high precision root extraction required in the AGM [18], was actually calculated by inverting $y = 1/x^2$. This leads to the iteration

$$(7.2) \qquad x_{n+1} = \frac{3x_n - x_n^3 y}{2}$$

for computing $y^{-1/2}$. One now multiplies by y to recapture $\sqrt{y}$. This was preferred because it avoided division.

REFERENCES

[1] N. H. ABEL, *Oeuvres complètes,* Grondahl and Son, Christiana, 1881.
[2] P. BECKMAN, *A History of Pi,* Golem Press, Ed. 4, 1977.
[3] F. BOWMAN, *Introduction to Elliptic Functions,* English Univ. Press, London, 1953.
[4] R. P. BRENT, *Multiple-precision zero-finding and the complexity of elementary function evaluation,* in Analytic Computational Complexity, J. F. Traub, ed., Academic Press, New York, 1975, pp. 151–176.
[5] ———, *Fast multiple-precision evaluation of elementary functions,* J. Assoc. Comput. Mach., 23 (1976), pp. 242–251.
[6] B. C. CARLSON, *Algorithms involving arithmetic and geometric means,* Amer. Math. Monthly, 78(1971), pp. 496–505.
[7] A. CAYLEY, *An Elementary Treatise on Elliptic Functions,* Bell and Sons, 1895, republished by Dover, New York, 1961.
[8] A. EAGLE, *The Elliptic Functions as They Should Be,* Galloway and Porter, Cambridge, 1958.
[9] C. H. EDWARDS, Jr. *The Historical Development of the Calculus,* Springer-Verlag, New York, 1979.
[10] K. F. GAUSS, *Werke,* Bd 3, Gottingen, 1866, 361–403.
[11] L. V. KING, *On the direct numerical calculation of elliptic functions and integrals,* Cambridge Univ. Press, Cambridge, 1924.
[12] D. KNUTH, *The Art of Computer Programming. Vol. 2: Seminumerical Algorithms,* Addison-Wesley, Reading, MA, 1969.
[13] H. T. KUNG AND J. F. TRAUB, *All algebraic functions can be computed fast,* J. Assoc. Comput. Mach., 25(1978), 245–260.
[14] A. M. LEGENDRE, *Exercices de calcul integral,* Vol. 1, Paris, 1811.
[15] D. J. NEUMAN, *Rational approximation versus fast computer methods,* in Lectures on Approximation and Value Distribution, Presses de l'Université de Montréal, Montreal, 1982, pp. 149–174.
[16] E. SALAMIN, *Computation of π using arithmetic-geometric mean,* Math. Comput. 135(1976), pp. 565–570.
[17] D. SHANKS AND J. W. WRENCH, Jr. *Calculation of π to 100,000 decimals,* Math. Comput., 16 (1962), pp. 76–79.
[18] Y. TAMURA AND Y. KANADA, *Calculation of π to 4,196,293 decimals based on Gauss-Legendre algorithm,* preprint.
[19] J. TODD, *Basic Numerical Mathematics,* Vol. 1, Academic Press, New York, 1979.
[20] G. N. WATSON, *The marquis and the land agent,* Math. Gazette, 17 (1933), pp. 5–17.
[21] J. W. WRENCH, Jr. *The evolution of extended decimal approximations to π.* The Mathematics Teacher, 53(1960), pp. 644–650.

2. On the complexity of familiar functions and numbers

Discussion

One of the startling observations of the latter third of the 20th century is that many, perhaps most, mathematical objects are easy to compute. In fact, these objects can be computed far more easily than one might have expected or one might have guessed in pre-computer days. A related observation is that the usual computational method is rarely the best, and the best known method is rarely 'natural'.

This is both of interest mathematically and practically. The mathematical technology that allows us to multiply large numbers (say 1 billion digit examples) and the mathematical technology that allows sophisticated medical scanners to process images in real time are fundamentally the same. Fast Fourier Transforms and their relatives are now vital in many computational contexts but they were a curiosity when observed by people like James Cooley and John Tukey in the 1960s.

The questions that are raised in this paper are fundamental. They concern the efficient computation of familiar functions and our knowledge has not changed much since we first posed these questions. As Larry Nazareth wrote:

> "A real number complexity model appropriate for this context [continuous optimization] is given in the recent landmark work of Blum, Cucker, Shub, and Smale.[1] In discussing their motivation for seeking a suitable theoretical foundation for modern scientific computing, where most of the algorithms are 'real number algorithms' the authors of this work quote the following illuminating remarks of John von Neumann, made in 1948:
>
>> "There exists today a very elaborate system of formal logic, and specifically, of logic applied to mathematics. This is a discipline with many good sides but also serious weaknesses... Everybody who has worked in formal logic will confirm that it is one of the technically most refractory parts of mathematics. The reason for this is that it deals with rigid, all-or-none concepts, and has very little contact with the continuous concept of the real or the complex number, that is with mathematical analysis. Yet analysis is the technically most successful and best-elaborated part of mathematics. Thus formal logic, by the nature of its approach, is cut off from the best cultivated portions of mathematics, and forced onto the most difficult mathematical terrain, into combinatorics.
>>
>> The theory of automata, of the digital, all-or-none type as discussed up to now, is certainly a chapter in formal logic. It would, therefore, seem that it will have to share this unattractive property of formal logic. It will have to be, from the mathematical point of view, combinatorial rather than analytical."

Source

J.M. Borwein and P.B. Borwein, "On the complexity of familiar functions and numbers," *SIAM Review*, **30** (1988), 589–601.

---

[1] L. Blum, P. Cucker, M. Shub and S. Smale, *Complexity and Real Computation*, Springer-Verlag, New York, 2008

ON THE COMPLEXITY OF FAMILIAR FUNCTIONS AND NUMBERS*

J. M. BORWEIN[†] AND P. B. BORWEIN[†]

Abstract. This paper examines low-complexity approximations to familiar functions and numbers. The intent is to suggest that it is possible to base a taxonomy of such functions and numbers on their computational complexity. A central theme is that traditional methods of approximation are often very far from optimal, while good or optimal methods are often very far from obvious. For most functions, provably optimal methods are not known; however the gap between what is known and what is possible is often small. A considerable number of open problems are posed and a number of related examples are presented.

Key words. elementary functions, pi, low-complexity approximation, reduced-complexity approximation, rational approximation, algebraic approximation, computation of digits, open problems

AMS(MOS) subject classifications. 68C25, 41A30, 10A30

1. Introduction. We examine various methods for evaluating familiar functions and numbers to high precision. Primarily, we are interested in the asymptotic behavior of these methods. The kinds of questions we pose are:

(1) How much work (by various types of computational or approximation measures) is required to evaluate n digits of a given function or number?

(2) How do analytic properties of a function relate to the efficacy with which it can be approximated?

(3) To what extent are analytically simple numbers or functions also easy to compute?

(4) To what extent is it easy to compute analytically simple functions?

Even partial answers to these questions are likely to be very difficult. Some, perhaps easier, specializations of the above are:

(5) Why is the function $\sqrt{x}$ easier to compute than exp? Why is it only marginally easier?

(6) Why is the Taylor series often the wrong way to compute familiar functions?

(7) Why is the number $\sqrt{2}$ easier to compute than e or π? Why is it only marginally easier?

(8) Why is the number $.1234567891011\cdots$ computationally easier than π or e?

(9) Why is computing just the nth digit of $\exp(x)$ really no easier than computing all the first n digits?

(10) Why is computing just the nth digit of π really no easier than computing all the first n digits?

Answers to (7) and (10) are almost certainly far beyond the scope of current number-theoretic techniques. Partial answers to some of the remaining questions are available.

The traditional way to compute elementary functions, such as exp or log, is to use a partial sum of the Taylor series or a related polynomial or rational approximation. These are analytically tractable approximations, and over the class of such

---

* Received by the editors February 2, 1987; accepted for publication (in revised form) September 11, 1987. The second author's research was supported in part by the Natural Sciences and Engineering Research Council of Canada.

[†] Department of Mathematics, Statistics, and Computing Science, Dalhousie University, Halifax, Nova Scotia, Canada B3H 3J5.

approximations are often optimal or near optimal. For example, the nth partial sums to exp are asymptotically the best polynomial approximations in the uniform norm on the unit disc in the complex plane, in the sense that if s_n is the nth partial sum of the Taylor expansion and p_n is any polynomial of degree n, then for large n,

$$\|\exp(z) - s_n(z)\|_D < \left[1 + \frac{4}{n}\right] \|\exp(z) - p_n(z)\|_D.$$

Here $\|\ \|_D$ denotes the supremum norm over the unit disc in the complex plane (see [5]). If the measure of the amount of work is the degree of the approximation, as it has been from a conventional point of view, then the story for exp might end here.

Questions (1)–(3) above have a very elegant answer for polynomial approximation in the form of the Bernstein–Jackson theorems [11]. These, for example, tell us that a function is entire if and only if the error in best uniform polynomial approximation of degree n on an interval tends to zero faster than geometrically, with a similar exact differentiability classification of a function in terms of the rate of polynomial approximation.

If we wish to compute n digits of $\log(x)$ using a Taylor polynomial then we employ a polynomial of degree n and perform $O(n)$ rational operations, while for $\exp(x)$ we require $O(n/\log n)$ rational operations to compute n digits. The slight improvement for exp reflects the faster convergence rate of the Taylor series. Padé approximants, best rational approximants and best polynomial approximants all behave in roughly the same fashion, except that the constants implicit in the order symbol change [5], [8].

A startling observation is that there exist rational functions that give n digits of log, exp, or any elementary function but require only $O((\log n)^k)$ rational operations to evaluate. These approximants are of degree $O(n)$ but can be evaluated in $O((\log n)^k)$ infinite-precision arithmetic operations. The simplest example of such a function is x^n which can be evaluated in $O(\log n)$ arithmetic operations by repeated squaring. While we cannot very explicitly construct these low-rational-complexity approximations to exp or log, it is clear that much of their simplicity results from squarings of intermediate terms. The moral is that it is appropriate and useful to view x^n as having the complexity of a general polynomial of degree $\log n$, not of degree n.

The existence of such approximants is a consequence of the construction of low-bit-complexity algorithms for log and π resting on the Arithmetic-Geometric Mean (AGM) iteration of Gauss, Lagrange, and Legendre (see §2 for definitions). These algorithms were discovered and examined by Beeler, Gosper, and Schroeppel [3], Brent [9], and Salamin [21] in the 1970s. A complete exposition is available in [5]. These remarkable algorithms are both theoretically and practically faster than any of the traditional methods for extended precision evaluation of elementary functions. The exact point at which they start to outperform the usual series expansions depends critically on implementation; the switchover comes somewhere in the 100- to 1000-digit range.

The main purpose of this paper is to catalogue the known results on complexity of familiar functions. We now appear to know enough structure to at least speculate on the existence of a reasonable taxonomy of functions based on their computational complexity. Here we have in mind something that relates computational properties of functions to their analytic or algebraic properties, something vaguely resembling the Bernstein–Jackson theorems in the polynomial case.

Likewise we would like to suggest the possibility of a taxonomy of numbers based on their computational nature. Here, we are looking for something that resembles

Mahler's classification of transcendendentals in terms of their rate of algebraic approximation [15].

It is not our intention to provide a taxonomy; this must await further progress in the field. We do hope, however, to present enough examples and pose enough interesting questions to persuade the reader that it is fruitful to pursue such an end.

2. Definitions. We consider four notions of complexity.

(1) *Rational complexity.* We say that a function f has *rational complexity* $O_{rat}(s(n))$ on a set A if there exists a sequence of rational functions R_n so that

(a) $|R_n(x) - f(x)| < 10^{-n}$ for all $x \in A$;

(b) asymptotically, R_n can be evaluated using no more than $O(s(n))$ rational operations (i.e., infinite-precision additions, subtractions, multiplications, and divisions).

That exp has rational complexity $O_{rat}(\log^3 n)$ means that there is a sequence of rational functions, the nth being evaluable in roughly $\log^3 n$ arithmetic operations, giving an n-digit approximation to exp. The subscript on the order symbol is for emphasis.

We will sometimes use Ω and Ω_{rat} as the lower bound order symbols. Whenever we talk about "n-digit precision" or "computing n digits" we mean computing to an accuracy of 10^{-n}.

(2) *Algebraic complexity.* We say that a function f has *algebraic complexity* $O_{alg}(s(n))$ on a set A if there exists a sequence of algebraic functions A_n so that

(a) $|A_n(x) - f(x)| < 10^{-n}$ for all $x \in A$;

(b) asymptotically, all the A_n can be evaluated using no more than $O(s(n))$ algebraic operations (i.e., infinite-precision solutions of a fixed number of prespecified algebraic equations).

This algebraic complexity measure allows us, for example, to use square root extractions in the calculation of the approximants and to count them on an equal footing with the rational operations. This is often appropriate because, from a bit-complexity point of view, root extraction is equivalent to multiplication (see §4). Note that we allow only a finite number of additional algebraic operations—so while we might allow for computing square roots, cube roots, and seventeenth roots, we would not allow an infinite number of different orders of roots.

Neither of the above measures takes account of the fact that low-precision operations are easier than high-precision operations.

(3) *Bit complexity.* We say that a function f has *bit complexity* $O_{bit}(s(n))$ on a set A if there exists a sequence of approximations B_n so that

(a) $|B_n(x) - f(x)| < 10^{-n}$ for all $x \in A$;

(b) B_n is the output of an algorithm (given input n and x) that evaluates the B_n to n-digit accuracy using $O(s(n))$ *single-digit* operations $(+, -, \times)$.

This is the appropriate measure of time complexity on a serial machine. (See [1] for more formal definitions.)

We wish to capture in the next definition the notion of how complex it is to compute only the nth digit of a function.

(4) *Digit complexity.* We say that a function f has *digit complexity* $O_{dig}(s(n))$ on a set A if there exists a sequence of approximations D_n so that

(a) $D_n(x)$ gives the nth digit of $f(x)$. By this we mean that $D_n(x)$ differs from the n through $(n+k)$th digits of $f(x)$ by at most 10^{-k} for any preassigned fixed k;

(b) D_n is the output of an algorithm (given input n and x) that evaluates the D_n to k digits using $O(s(n))$ *single-digit* operations.

This definition of agreement of nth digits takes account of the fact that sequences of repeated nines can occur. We really want to say that $.19999\cdots$ and $.2000\cdots$ agree in the first digit. As it stands, the definition above exactly computes only the nth digit to a probability dependent on k.

It is also assumed that accessing the kth through nth digit of input of x is an $O_{\text{bit}}(\max(n-k,\log k))$ operation, so that accessing the first n digits is $O_{\text{bit}}(n)$ while accessing just the nth bit is $O_{\text{bit}}(\log n)$.

Addition is $O_{\text{rat}}(1)$, $O_{\text{alg}}(1)$, $O_{\text{bit}}(n)$, and $O_{\text{dig}}(\log n)$. Here we take the set A, where we seek a uniform algorithm, to be the unit square in $\mathbb{R}^2$. The usual addition algorithm gives the upper bounds shown above. Addition is one of the very few cases where we know the exact result. Trivial uniqueness considerations show that addition is $\Omega_{\text{bit}}(n)$, and hence all the above orders are exact.

It comes as a major surprise of this side of theoretical computer science that the usual way of multiplying is far from optimal from a bit-complexity point of view. The usual multiplication algorithm has bit complexity $\Omega_{\text{bit}}(n^2)$. However, it is possible to construct a multiplication which is $O_{\text{bit}}(n \log n \log \log n)$. This is based on the Fast Fourier Transform and is due to Schönhage and Strassen (see [1], [16]). The extent to which the log terms are necessary is not known. Given a standard model of computation the best known lower bound is the trivial one, $\Omega_{\text{bit}}(n)$. We will denote the *bit complexity of multiplication* by $M(n)$.

3. A table of results. The state of our current knowledge is contained in Table 1. The orders of the various measures of complexities for computing n digits (or in the final case the nth digit) compose the columns. In each case, except addition, the only upper bound we know for the digit complexity is the same as the bit-complexity bound. When we deal with functions, we assume that we are on a compact region of the domain of the given function that is bounded away from any singularities and that contains an interval. Numbers may be considered as functions whose domain is a singleton.

For our purposes *hypergeometric functions* are functions of the form

$$f(x) := \sum a_n x^n \quad \text{where } a_n/a_{n-1} = R(n)$$

and R is a fixed rational function (with coefficients in $\mathbb{Q}$).

TABLE 1

Type of function	O_{rat}	O_{alg}	O_{bit}	Ω_{dig}
(1) Addition	1	1	n	$\log n$
(2) Multiplication	1	1	$n \log n \log \log n$	n
(3) Algebraic (nonlinear)	$\log n$	1	$M(n)$	n
(4) log (complete elliptic integrals)	$\log^2 n$	$\log n$	$(\log n)M(n)$	n
(5) exp	$\log^3 n$	$\log^2 n$	$(\log n)M(n)$	n
(6) Elementary (nonlinear)	$\log^k n$	$\log^k n$	$(\log n)M(n)$	n
(7) Hypergeometric (over $\mathbb{Q}$)	$n^{1/2+}$	$n^{1/2+}$	$(\log^2 n)M(n)$	n
(8) Gamma and zeta	$n^{1/2+}$	$n^{1/2+}$	$n^{1/2+}M(n)$	n
(9) Gamma and zeta on $\mathbb{Q}$	$n^{1/2+}$	$n^{1/2+}$	$(\log^2 n)M(n)$	$\log n$
(10) pi, log (2), $\Gamma(\tfrac{1}{3})$	$\log^2 n$	$\log n$	$(\log n)M(n)$	$\log n$
(11) Euler's constant (Catalan's constant)	$n^{1/2+}$	$n^{1/2+}$	$(\log^2 n)M(n)$	$\log n$

Elementary functions are functions built from rational functions (with rational coefficients) exp and log by any number of additions, multiplications, compositions, and solutions of algebraic equations.

A number of techniques are employed in deriving Table 1. Our intention is to indicate the most useful of these without going into too much detail. The next four sections outline the derivations of most of the bounds.

4. Newton's method. The calculation of algebraic functions, given that we have algorithms for addition and multiplication, is entirely an exercise in applying Newton's method to solving equations of the form $f(x) - y = 0$. Newton's method for $1/x - y = 0$ gives the iteration

(a) $x_{n+1} := 2x_n - yx_n^2$,

while for $x^2 - y = 0$ the iteration is

(b) $x_{n+1} := (x_n + y/x_n)/2$.

These two iterations converge quadratically. Thus $O(\log n)$ iterations give n digits of $1/y$ and $\sqrt{y}$, respectively, and we have given an $O_{\text{rat}}(\log n)$ algorithm for square root extraction.

The quadratic rate of convergence is only half the story. Because Newton's method is *self-correcting*, in the sense that a small perturbation in x_n does not change the limit, it is possible to start with a single-digit estimate and double the precision with each iteration. Thus the bit complexity of root extraction is

$$O(M(1) + M(2) + M(4) + \cdots + M(n)) = O(M(n)).$$

This leads to $O_{\text{bit}}(M(n))$ algorithms for root extraction and division, and a similar analysis works for any algebraic function. This explains most of (1)–(3) in Table 1. We also have the interesting result that the computation of digits of any algebraic number is asymptotically no more complicated than multiplication. (These results on the complexity of algebraic functions may be found in [5] and [9].)

The approximation in (a), x_n, is in fact the $(2^n - 1)$st Taylor polynomial to $1/y$ at 1. In (b), x_n is in fact the $(2^n, 2^n - 1)$st Padé approximant to $\sqrt{y}$ at 1. (See [5] or [11] for further material on Padé approximants.) This is one of the very few cases where Newton's method generates familiar approximants.

Newton's method is also useful for inverting functions. The inverse of f is computed from the iteration

$$x_{n+1} := x_n - [f(x_n) - y]/f'(x_n).$$

For any reasonable f this gives the same bit complexity estimate for f^{-1} as for f. Inverting by Newton's method multiplies the rational and algebraic complexities by $\log n$.

5. The AGM. The two-term iteration with starting values $a_0 := x \in (0, 1]$ and $b_0 := 1$ given by

$$a_{n+1} := (a_n + b_n)/2, \quad b_{n+1} := \sqrt{(a_n \cdot b_n)}$$

converges quadratically to $m(1, x)$, where

$$\frac{1}{m(1,x)} = \frac{2}{\pi} \int_0^{\pi/2} \frac{dt}{\sqrt{1-(1-x^2)\sin^2 t}}.$$

This is the arithmetic-geometric mean iteration of Gauss, Lagrange, and Legendre. This latter complete elliptic integral is $2K'(x)/\pi$ and is a nonelementary transcendental function with complexity

$$O_{\text{alg}}(\log(n)), \quad O_{\text{rat}}(\log^2(n)), \quad O_{\text{bit}}(\log(n)M(n)).$$

It is also essentially the only identifiable nonelementary limit of a quadratically converging fixed iteration and as such is of central importance [5].

One way to get a low complexity algorithm for log is to use the logarithmic asymptote of K' at 0. This gives the estimate

$$|(2/\pi)\log x - 1/m(1, 10^{-n}) + 1/m(1, x10^{-n})| < n10^{-2(n-1)}, \quad n > 3, \quad x \in [.5, 1].$$

Up to computing π, this allows for the derivation of algorithms with the complexity of entry (4) in Table 1. Algorithms for π can be derived from the same kinds of considerations (see [4], [5], [9], [18], [21]). Probably the fastest known algorithm for π is the quartic example given below [5], [2].

ALGORITHM. Let $\alpha_0 := 6 - 4\sqrt{2}$ and $y_0 := \sqrt{2} - 1$. Let

$$y_{n+1} := [1 - (1 - y_n^4)^{1/4}]/[1 + (1 - y_n^4)^{1/4}]$$

and

$$\alpha_{n+1} := (1 + y_{n+1})^4 \alpha_n - 2^{2n+3} y_{n+1}(1 + y_{n+1} + y_{n+1}^2).$$

Then $1/\alpha_n$ tends to π quartically and

$$0 < \alpha_n - \frac{1}{\pi} < 16 \cdot 4^n \exp(-2 \cdot 4^n \pi).$$

The exponential function may be derived from log by inverting using Newton's method. This continues to work for appropriate complex values. The elementary functions are now built from log and exp and the solution of algebraic equations in these quantities. The constant k in the rational- and bit-complexity estimates depends on the number of these equations that require solution. This explains entries (5), (6), and (10) in Table 1, except for $\Gamma(\frac{1}{3})$. (This and a few other values of Γ arise as algebraic combinations of complete elliptic integrals and pi.) (Substantial additional material on this section is to be found in [5].)

6. FFT methods. The Fast Fourier Transform (FFT) is a way of solving the following two problems:

(a) Given the coefficients of a polynomial of degree $n - 1$, evaluate the polynomial at all n of the nth roots of unity.

(b) Given the values of a polynomial of degree $n - 1$ at the nth roots of unity, compute the coefficients of the polynomial.

These two problems are actually equivalent (see [1], [5], [16]). The important observation made by Cooley and Tukey in the 1960s is that both of these problems are solvable with rational complexity $O_{\text{rat}}(n \log n)$, rather than the complexity of $\Omega_{\text{rat}}(n^2)$ that the usual methods require (i.e., Horner's method). This is an enormously useful algorithm.

We can multiply two polynomials of degree n with complexity $O_{\text{rat}}(n \log n)$ by using the FFT three times. First we compute the values of the two polynomials at $2n + 1$ roots of unity. Then we work out the coefficients of the polynomial of degree $2n$ that agrees with the product at these roots.

Variations on this technique allow for the evaluation of a rational function of degree n at n points in $O_{\text{rat}}(n \log^2 n)$ and $O_{\text{bit}}(n \log^2 n M(k))$, where k is the precision to which we are working [5].

Fast multiplications are constructed by observing that multiplication of numbers is much like multiplication of polynomials whose coefficients are the digits, the additional complication being the "carries."

How does this give reduced-complexity algorithms? We illustrate with $\log(1-x)$. Let

$$s_{n^2}(x) := \sum_{k=1}^{n^2} \frac{x^k}{k}$$

and write

$$s_{n^2}(x) = \sum_{k=0}^{n-1} x^k p(kn) \quad \text{where } p(y) := \sum_{j=1}^{n} \frac{x^j}{j+y}.$$

Now evaluate $p(0)$, $p(n)$, $\cdots$, $pn(n-1)$ using FFT methods, and then evaluate s_{n^2}. This gives an $O_{\text{rat}}(n^{1/2}(\log n)^2)$ and $O_{\text{bit}}(n^{1/2}(\log n)^2 M(n))$ algorithm for log. At any fixed rational value r, we get an $O_{\text{bit}}((\log n)^2 M(n))$ for $\log r$. For this final estimate we must take advantage of the reduced precision possible for intermediate calculations.

This is not as good an estimate as the AGM estimates for log. It is, however, a much more generally applicable method. We can orchestrate the calculation, much as above, for any hypergeometric function. This is how the estimates in line (7) in Table 1 are deduced. Schroeppel [3], [22] shows how a similar circle of ideas can be used to give $O_{\text{bit}}(\log^k n\, M(n))$ algorithms for the solutions of linear differential equations whose coefficients are rational functions with coefficients in $\mathbb{Q}$.[1]

The gamma function, Γ, can be computed from the estimate

$$\left| \Gamma(x) - N^x \sum_{k=0}^{6N} \frac{(-1)^k N^k}{k!(x+k)} \right| < 2Ne^{-N}, \quad x \in [1,2]$$

(see [5] for details). The zeta function, ζ, is then computable from Riemann's integral [24]:

$$\zeta(x)\Gamma\left(\frac{x}{2}\right)\pi^{-x/2} - \frac{1}{x(x-1)} = \int_1^\infty \frac{t^{(1-x)/2} + t^{x/2}}{t} \sum_{n=1}^\infty e^{-n^2 \pi t}\, dt.$$

We truncate both the integral and the sum. These two formulae explain lines (8) and (9) of Table 1.

Catalan's constant

$$G := \sum_{n=0}^\infty \frac{(-1)^n}{(2n+1)^2}$$

can be computed from Ramanujan's sum

$$\frac{8}{3}G = \frac{\pi}{3}\log(2+\sqrt{3}) + \sum_{m=0}^\infty \frac{m!\,m!}{(2m+1)^2(2m)!},$$

while Euler's constant, γ, can be computed from the asymptotic expansion

$$\gamma = -\log x - \sum_{k=1}^\infty \frac{(-x)^k}{k \cdot k!} + O(\exp(-x)), \quad x > 1.$$

---

[1] Chudnovsky and Chudnovsky [26] provide a low-bit complexity approach to solutions of linear differential equations in [26].

This gives line (11) of Table 1. Some of the details may be found in [5] and [6].

A variation of the above method for computing γ has been used by Brent and McMillan [10] to compute over 29,000 partial quotients of the continued fraction of γ. From this computation it follows that if γ is rational its denominator exceeds $10^{15,000}$.

7. Digit complexity. The aim of this section is to explain the last column in Table 1. The main observation is that the digit complexity of computing the mth digit ($m \leq n$) of the product of two n-digit numbers is $\Omega_{\text{dig}}(m)$. This is essentially just a uniqueness argument the details of which may be pursued in [7].

Now suppose that f is analytic around zero (C^3 suffices). Then

$$f(x) = a + bx + cx^2 + O(x^3)$$

or equivalently

$$cx^2 = f(x) - a - bx + O(x^3).$$

If f is of low-digit complexity then, as above, truncating after one term gives a low-complexity algorithm for $a + bx$. Recall that addition is $O_{\text{dig}}(\log n)$. This in turn gives a low-digit complexity evaluation of cx^2 in a neighborhood of zero, but evaluation of cx^2 is essentially equivalent to multiplication. Once again, the details are available in [7]. Thus, if f is any nonlinear C^3 function it is $\Omega_{\text{dig}}(n)$, or we would have too good an algorithm for calculating the mth digit of multiplication.

We now have the following type of theorem.

THEOREM. *If f is a nonlinear elementary function (on an interval) then f is*

$$O_{\text{bit}}(n(\log n)^k) \quad \text{and} \quad \Omega_{\text{dig}}(n).$$

This is now close to an exact result. Actually we can say considerably more. For example, we have the following theorem.

THEOREM. *If f is a nonlinear C^3 function (on an interval) then the set of x for which the digit complexity of $f(x)$ is $o(n)$ by any algorithm is of the first Baire category.*

A set of first Baire category is small in a topological sense (see [25]).

We define the class of *sublinear numbers* by calling a number x sublinear if the digit complexity of x is $O_{\text{dig}}(n^{1-})$. Call α a *sublinear multiplier* if the function αx is sublinear for all $x \in [0, 1]$ (given both α and x as input).

THEOREM. *The set of sublinear multipliers is a nonempty set of the first Baire category.*

Two more definitions are useful in relation to numbers of very low digit complexity. We say that x is *sparse* if x has digit complexity $O_{\text{dig}}(n^\delta)$ for all $\delta > 0$, and we say that α is a *sparse multiplier* if αx is sparse for all $x \in [0, 1]$. Sparse multipliers have sparse digits. Indeed, let $S := \{x \mid \#(\text{nonzero digits of } x \text{ among the first } n \text{ digits}) = O(n^\delta) \text{ for all } \delta > 0\}$.

THEOREM. *The set of sparse multipliers is exactly the set S.*

Thus there are uncountably many sparse multipliers and hence also uncountably many sublinear multipliers.

These are base-dependent notions. The previous theorem shows that $\frac{1}{2}$ is a sparse multiplier base 2 but not base 3. We can prove directly that irrational sparse multipliers must be transcendental. Various questions concerning these matters will be raised in the next sections.

8. Questions on the complexity of functions. The hardest problems associated with Table 1 of §3 concern the almost complete lack of nontrivial lower bound

COMPLEXITY OF FAMILIAR FUNCTIONS AND NUMBERS

estimates. This reflects the current state of affairs in theoretical computer science. Not only is the question of whether $P = NP$ still open, it is still not resolved that any NP problems are nonlinear. Friedman [13], for example, shows that we can take maxima over the class of polynomially computable functions if and only if $P = NP$ and that we can integrate over this class if and only if $P = \#P$. While these notions are somewhat tangential to our concerns they do indicate that some of our problems are likely to be hard.

One of the reasons for looking at the rational complexity is that it is likely to be a little more amenable to analysis. We can show that exp and log *cannot* have rational complexity $o(\log n)$. This is a consequence of the known estimates in approximating exp and log by rational functions of degree n [5], [8]. Note that n rational operations can generate a rational function of at most degree 2^n. Thus there is only a small gap between the known and best possible rational complexity estimates for log.

Question 1. Does log have rational complexity $O_{\text{rat}}(\log n)$?

The extra power of log in the rational complexity of exp over that of log is almost certainly an artifact of the method. So at least one power of log ought to be removable.

Question 2. Show that exp has rational complexity $O_{\text{rat}}(\log^2 n)$. Does exp have rational complexity $O_{\text{rat}}(\log n)$?

The low-complexity approximants to exp and log are constructed indirectly. It would be valuable to have a direct construction.

Question 3. Construct, as explicitly as possible, approximants to exp and log with complexity $O_{\text{rat}}(\log^k n)$.

There is a big difference in the rational complexity of exp and of Γ. It is tempting to speculate that this is artificial.

Question 4. Does Γ have rational complexity $O_{\text{rat}}(\log^k n)$?

Ideally we would like to identify those functions with this complexity.

Question 5. Classify (analytic) functions with rational complexity $O_{\text{rat}}(\log^k n)$.

This last question is almost certainly very hard.

We would expect there to be little difference between rational complexity and algebraic complexity.

Question 6. Does any of exp, log, or K have rational complexity essentially slower than its algebraic complexity?

In the case of bit complexity, there are no nontrivial lower bounds. At best we can say that the bit complexity is always at least that of multiplication. Thus a crucial first step is the content of the next question.

Question 7. Show that exp, log, or any of the functions we have considered is *not* $O_{\text{bit}}(M(n))$.

It is easy to construct entire functions with very low bit complexity; we simply use very rapidly converging power series. Thus there exist nonalgebraic analytic functions with bit complexity $O_{\text{bit}}(a_n M(n))$, where a_n is any sequence tending to infinity. However, the following question appears to be open.

Question 8. Does there exist a nonalgebraic analytic function with bit complexity $O_{\text{bit}}(M(n))$?

A negative answer to this question would also resolve the question preceding it.

A very natural class to examine is the class of functions that satisfy algebraic differential equations (not necessarily linear). Almost all familiar functions arise in this context. Even an unlikely example like the theta function

$$\theta_3(q) := \sum_{n \in \mathbb{Z}} q^{n^2},$$

satisfies a nonlinear algebraic differential equation, as Jacobi showed (see [20]).

Question 9. How do solutions of algebraic differential equations fit into the complexity table?

We end with a question on digit complexity.

Question 10. Does there exist an analytic function whose digit complexity is essentially faster than its bit complexity? Does there exist an analytic function with digit complexity $O_{\text{dig}}(n)$?

There exist functions of the form

$$\sum a_n |x - b_n|$$

with low-digit complexity, where a_n and b_n are low-digit complexity numbers. Possibly we can construct nowhere differentiable functions that are sublinear, in the sense of digit complexity.

9. Questions on the complexity of numbers. Questions concerning the transcendence of functions tend to be easier than questions on the transcendence of individual numbers. In much the same way, questions on the complexity of functions tend to be easier than those on the complexity of specific numbers. The intent of this section is to pose various problems that suggest the link between complexity and transcendence. Such questions, while raised before, tend to have been concerned just with the notion of computability rather than also considering the rate of the computation (see [14]).

The class of sublinear numbers, defined in §7, contains all rational numbers; it also contains known transcendents such as

$$\alpha := .12345678910111213\cdots.$$

However, while the rationals are in this class in a base-independent fashion, it is not at all clear that the above number α is sublinear in bases relatively prime to 10. The 10^{10}th digit, base 10, is 1. What is it in base 2?

Question 11. Are there any irrational numbers that are sublinear in every base?

It is easy to generate numbers that are sublinear in particular bases. Numbers such as

$$a := .d_1 d_2 \cdots \qquad \begin{aligned} d_i &:= 1 \quad \text{if } i \text{ is a square,} \\ d_i &:= 0 \quad \text{otherwise,} \end{aligned}$$

or

$$b := .d_1 d_2 \cdots \qquad \begin{aligned} d_i &:= 1 \quad \text{if } i \text{ is a power of 2,} \\ d_i &:= 0 \quad \text{otherwise,} \end{aligned}$$

are sublinear in whatever base is specified. It is tempting to conjecture that the next question has a positive answer.

Question 12. Must an irrational number that is sublinear (in all bases) be transcendental?

Loxton and van der Poorten [17] show that a particular very special class of sublinear numbers, namely those generated by finite automata, are either rational or transcendental. These are numbers for which computation of the nth digit essentially requires no memory of the preceding digits. The base dependence of these numbers is discussed in [12].

Question 13. Is either of π or e sublinear (in any base)?

Almost certainly the answer to this question is no. There is an interesting observation relating to this. Consider the series

$$\frac{1}{\pi} = \sum_{n=0}^{\infty} \binom{2n}{n}^3 \frac{42n+5}{2^{12n+4}}.$$

This series due to Ramanujan [5], [19] has numerators that grow roughly, e.g., 2^{6n}, while the denominators are powers of 2. Thus, as has been observed, we can compute the second length n block of binary digits of $1/\pi$ without computing the first block. Likewise, in base 10, we can compute the second block of length n of decimal digits of $\sqrt{(\tfrac{5}{3})}$ from the series

$$\frac{1}{\sqrt{1-4x}} = \sum_{n=0}^{\infty} \binom{2n}{n} x^n.$$

In neither case, however, is there any reduction in the order of complexity.

It seems likely that computing the nth digit of π is an $\Omega_{\text{dig}}(n)$ calculation. Thus, we might make the strong conjecture that no one will ever compute the 10^{1000}th digit of π. This number arises from an (over)estimate of the number of electrons in the known universe and as such almost certainly overestimates the amount of storage that will ever be available for such a calculation.

The set of sparse multipliers is a subset of the sublinear numbers that can be shown directly to contain no irrational algebraics. We do not know this about sparse numbers, though we strongly suspect it to be true.

Recall that a sparse multiplier has mostly zero digits and observe that a nonintegral rational cannot possess a terminating expansion in two relatively prime bases. This suggests the following question.

Question 14. Do there exist irrationals that are sparse multipliers in two relatively prime bases? Do there exist irrationals whose digits are asymptotically mostly zeros in two relatively prime bases?

Many of these questions are at least partly related to questions on normality [23]. Virtually nothing is known about the normality of familiar numbers. The following is a somewhat related question by Mahler.

Question 15 (Mahler [15]). Does there exist a nonrational function

$$f(x) := \sum_{n=0}^{\infty} a_n x^n$$

where the a_n are a bounded sequence of positive integers, that maps algebraic numbers in the unit disc to algebraic numbers?

Suppose that such an example exists, and suppose the a_n are bounded by 9. Then

$$f(1/1000) = a_0.00a_1 00a_2 \cdots$$

is a thoroughly nonnormal irrational algebraic. Thus, in some sense, Mahler's question is a very weak conjecture concerning normality. Note also that, if in such an example the a_n were sublinearly computable, we would have produced sublinearly computable algebraic irrationalities.

Perhaps we will be able to distinguish rational numbers by their digit complexity. What can we hope to say about algebraic numbers? A natural class to look at is the class of numbers that are *linear (in multiplication)*, that is, numbers with bit complexity $O_{\text{bit}}(M(n))$. This class contains all algebraic numbers in a base-independent fashion. It also contains numbers such as

$$a := \sum_{n=0}^{\infty} \frac{1}{3^{3^n}} \quad \text{and} \quad b := \prod_{n=0}^{\infty} (1 + 3^{-3^n}),$$

also in a base-independent fashion.

Question 16. Can we identify the class of numbers that are linear in multiplication?

This is almost certainly hard. As is the following question.

Question 17. Are either e or π linear in multiplication?

A negative answer to the above would include a proof of the transcendence of π. The place to start might be with the following.

Question 18. Can we construct any natural nonlinear number?

Our current state of knowledge is that γ and G have bit complexity $O_{\text{bit}}(\log^2 nM(n))$.

Question 19. Are γ and G both $O_{\text{bit}}(\log nM(n))$?

We might expect that elementary functions cannot take sublinear numbers to sublinear numbers.

Question 20. Does there exist a number $a \neq 0$ so that both a and $\exp(a)$ are sublinear (in some base)? Can a and $\exp(a)$ both be linear in multiplication?

It seems likely that the answer is no. Question 20 should also be asked about other elementary transcendental functions.

For simple nonelementary functions Question 20 has a positive answer. Consider the function $F := (2/\pi)K$, where K is the complete elliptic integral of the first kind. Then F satisfies a linear differential equation of order 2 and is a nonelementary transcendental function. However, if

$$k(q) := q^{1/2} \left(\sum_{n \in \mathbb{Z}} q^{n^2+n} \right)^2 \bigg/ \left(\sum_{n \in \mathbb{Z}} q^{n^2} \right)^2,$$

then

$$F(k(q)) = \left(\sum_{n \in \mathbb{Z}} q^{n^2} \right)^2,$$

and when $q := 1/10^{2k}$ both $F(k(q))$ and $k(q)$ are linear in multiplication, at least in base 10. (This is because the series above have particularly low complexity for $q := 1/10^{2k}$.)

Note also that the function

$$\theta_3(q) := \sum_{n \in \mathbb{Z}} q^{n^2},$$

which satisfies a nonlinear algebraic differential equation, takes sublinear numbers of the form $q := 1/10^n$ to sublinear numbers (base 10).

10. Conclusion. Many issues have not been touched upon at all. One such issue is the overhead costs of these low-complexity algorithms. This amounts to a discussion of the constants buried in the asymptotic estimates. Sometimes the theoretically low-complexity algorithms are also of low complexity practically. This is the case for AGM-related algorithms for complete elliptic integrals. These are probably the algorithms of choice in any precision. The AGM-related algorithms for log and exp will certainly not outperform more traditional methods in the usual ranges in which we compute (less than 100 digits). Some of the FFT-related algorithms are probably of

COMPLEXITY OF FAMILIAR FUNCTIONS AND NUMBERS

only theoretical interest, even for computing millions of digits, because the overhead constants are so large. In other cases, such as multiplication or the computation of π, an FFT-related method is vital for very high precision computations.

We have not succeeded in completely answering any of the questions in the Introduction. In large part, this is because we have virtually no methods for handling lower bounds for such problems. The questions raised in this paper seem to be fundamental. The partial answers have provided a number of substantial surprises. For these reasons we believe these questions are deserving of study.

REFERENCES

[1] A. V. AHO, J. E. HOPCROFT, AND J. D. ULLMAN, *The Design and Analysis of Computer Algorithms*, Addison–Wesley, Reading, MA, 1974.

[2] D. H. BAILEY, *The computation of π to 29,360,000 decimal digits using Borweins' quartically convergent algorithm*, Math. Comp., 50 (1988), pp. 283–296.

[3] M. BEELER, R. W. GOSPER, AND R. SCHROEPPEL, *Hakmem*, MIT Artificial Intelligence Lab, Massachusetts Institute of Technology, Cambridge, MA, 1972.

[4] J. M. BORWEIN AND P. B. BORWEIN, *The arithmetic-geometric mean and fast computation of elementary functions*, SIAM Rev., 26 (1984), pp. 351–365.

[5] ———, *Pi and the AGM—A Study in Analytic Number Theory and Computational Complexity*, John Wiley, New York, 1987.

[6] P. B. BORWEIN, *Reduced complexity evaluation of hypergeometric functions*, J. Approx. Theory, 50 (1987), pp. 193–199.

[7] ———, *Digit complexity*, in preparation.

[8] D. BRAESS, *Nonlinear Approximation Theory*, Springer-Verlag, Berlin, 1986.

[9] R. P. BRENT, *Fast multiple-precision evaluation of elementary functions*, J. Assoc. Comput. Mach., 23 (1976), pp. 242–251.

[10] R. P. BRENT AND E. M. MCMILLAN, *Some new algorithms for high-precision calculation of Euler's constant*, Math. Comput., 34 (1980), pp. 305–312.

[11] E. W. CHENEY, *Introduction to Approximation Theory*, McGraw-Hill, New York, 1966.

[12] A. COBHAM, *On the base-dependence of sets of numbers recognizable by finite automata*, Math. Systems Theory, 3 (1969), pp. 186–192.

[13] H. FRIEDMAN, *The computational complexity of maximization and integration*, Adv. in Math., 53 (1984), pp. 80–98.

[14] J. HARTMANIS AND R. E. STEARNS, *On the computational complexity of algorithms*, Trans. Amer. Math. Soc., 117 (1965), pp. 265–306.

[15] K. MAHLER, *Lectures on Transcendental Numbers*, Lecture Notes in Math. 546, Springer-Verlag, Berlin, New York, 1976.

[16] D. KNUTH, *The Art of Computer Programming*, Vol. 2: *Seminumerical Algorithms*, Addison–Wesley, Reading, MA, 1981.

[17] J. H. LOXTON AND A. J. VAN DER POORTEN, *Arithmetic properties of the solution of a class of functional equations*, J. Reine Angew. Math., 330 (1982), pp. 159–172.

[18] D. J. NEWMAN, *Rational approximation versus fast computer methods*, in Lectures on Approximation and Value Distribution, Presses de l'Université de Montreal, Montreal, Canada, 1982, pp. 149–174.

[19] S. RAMANUJAN, *Modular equations and approximations to π*, Quart. J. Math., 45 (1914), 350–372.

[20] L. A. RUBEL, *Some research problems about algebraic differential equations*, Trans. Amer. Math. Soc., 280 (1983), pp. 43–52.

[21] E. SALAMIN, *Computation of π using arithmetic-geometric mean*, Math. Comput. 30 (1976), pp. 565–570.

[22] R. SCHROEPPEL, unpublished manuscript.

[23] S. WAGON, *Is π normal?* Math. Intelligencer, 7 (1985), pp. 65–67.

[24] E. T. WHITTAKER AND G. N. WATSON, *A Course of Modern Analysis*, Fourth edition, Cambridge University Press, London, 1927.

[25] A. WILANSKY, *Modern Methods in Topological Vector Spaces*, McGraw-Hill, New York, 1978.

[26] D. V. CHUDNOVSKY AND G. V. CHUDNOVSKY, *Approximations and complex multiplication according to Ramanujan*, in Ramanujan Revisited, G. Andrews, R. Ashey, B. Berndt, K. Ramanathan, R. Rankin, eds., Academic Press, San Diego, CA, 1988.

3. Ramanujan and pi

Discussion

This article and the next describe Ramanujan, his work on Pi and our own related contributions. It is gratifying to see its diverse reprintings, which reflect the continuing general interest in both Pi and in Ramanujan. As the computational capacity to do concrete mathematics has expanded and its popularity has grown, the sense of Ramanujan's strangeness has abated. At the same time, the completion of a fully edited edition *Ramanujan's Notebooks*, by Bruce Berndt, and of the *Lost Notebook*, by Berndt and George Andrews, has made the magnitude of Ramanujan's particular genius even more apparent. In his two-part biography of Ramanujan, Allyn Jackson wrote:

> In a mathematical conversation, someone suggested to Grothendieck that they should consider a particular prime number. "You mean an actual number?" Grothendieck asked. The other person replied, yes, an actual prime number. Grothendieck suggested, "All right, take 57."
>
> But Grothendieck must have known that 57 is not prime, right? Absolutely not, said David Mumford of Brown University. "He doesn't think concretely." Consider by contrast the Indian mathematician Ramanujan, who was intimately familiar with properties of many numbers, some of them huge. That way of thinking represents a world antipodal to that of Grothendieck. "He really never worked on examples," Mumford observed. "I only understand things through examples and then gradually make them more abstract. I don't think it helped Grothendieck in the least to look at an example. He really got control of the situation by thinking of it in absolutely the most abstract possible way. It's just very strange. That's the way his mind worked."—Allyn Jackson[1]

Source

J.M. Borwein and P.B. Borwein, "Ramanujan and Pi," *Scientific American*, February 1988, 112–117. Japanese ed. April, Russian ed. April, German ed. May. Reprinted as the following:

1. pp. 647–659 of WORLD TREASURY OF PHYSICS, ASTRONOMY, AND MATHEMATICS, T. Ferris Ed., Little, Brown and Co, 1991.

2. pp. 60–68 of *Moderne Mathematik*, G. Faltings Ed., Spektrum Verlag, 1996 (with a computational update).

3. pp. 187–199 of *Ramanujan: Essays and Surveys,* Bruce C. Berndt and Robert A. Rankin Eds., AMS-LMS History of Mathematics, volume 22, 2001.

---

[1] From a two-part biography in 2004 in the *Notices of the AMS*.

Ramanujan and Pi

Some 75 years ago an Indian mathematical genius developed ways of calculating pi with extraordinary efficiency. His approach is now incorporated in computer algorithms yielding millions of digits of pi

by Jonathan M. Borwein and Peter B. Borwein

Pi, the ratio of any circle's circumference to its diameter, was computed in 1987 to an unprecedented level of accuracy: more than 100 million decimal places. Last year also marked the centenary of the birth of Srinivasa Ramanujan, an enigmatic Indian mathematical genius who spent much of his short life in isolation and poor health. The two events are in fact closely linked, because the basic approach underlying the most recent computations of pi was anticipated by Ramanujan, although its implementation had to await the formulation of efficient algorithms (by various workers including us), modern supercomputers and new ways to multiply numbers.

Aside from providing an arena in which to set records of a kind, the quest to calculate the number to millions of decimal places may seem rather pointless. Thirty-nine places of pi suffice for computing the circumference of a circle girdling the known universe with an error no greater than the radius of a hydrogen atom. It is hard to imagine physical situations requiring more digits. Why are mathematicians and computer scientists not satisfied with, say, the first 50 digits of pi?

Several answers can be given. One is that the calculation of pi has become something of a benchmark computation: it serves as a measure of the sophistication and reliability of the computers that carry it out. In addition, the pursuit of ever more accurate values of pi leads mathematicians to intriguing and unexpected niches of number theory. Another and more ingenuous motivation is simply "because it's there." In fact, pi has been a fixture of mathematical culture for more than two and a half millenniums.

Furthermore, there is always the chance that such computations will shed light on some of the riddles surrounding pi, a universal constant that is not particularly well understood, in spite of its relatively elementary nature. For example, although it has been proved that pi cannot ever be exactly evaluated by subjecting positive integers to any combination of adding, subtracting, multiplying, dividing or extracting roots, no one has succeeded in proving that the digits of pi follow a random distribution (such that each number from 0 to 9 appears with equal frequency). It is possible, albeit highly unlikely, that after a while all the remaining digits of pi are 0's and 1's or exhibit some other regularity. Moreover, pi turns up in all kinds of unexpected places that have nothing to do with circles. If a number is picked at random from the set of integers, for instance, the probability that it will have no repeated prime divisors is six divided by the square of pi. No different from other eminent mathematicians, Ramanujan was prey to the fascinations of the number.

The ingredients of the recent approaches to calculating pi are among the mathematical treasures unearthed by renewed interest in Ramanujan's work. Much of what he did, however, is still inaccessible to investigators. The body of his work is contained in his "Notebooks," which are personal records written in his own nomenclature. To make matters more frustrating for mathematicians who have studied the "Notebooks," Ramanujan generally did not include formal proofs for his theorems. The task of deciphering and editing the "Notebooks" is only now nearing completion, by Bruce C. Berndt of the University of Illinois at Urbana-Champaign.

To our knowledge no mathematical redaction of this scope or difficulty has ever been attempted. The effort is certainly worthwhile. Ramanujan's legacy in the "Notebooks" promises not only to enrich pure mathematics but also to find application in various fields of mathematical physics. Rodney J. Baxter of the Australian National University, for example, acknowledges that Ramanujan's findings helped him to solve such problems in statistical mechanics as the so-called hard-hexagon model, which considers the behavior of a system of interacting particles laid out on a honeycomblike grid. Similarly, Carlos J. Moreno of the City University of New York and Freeman J. Dyson of the Institute for Advanced Study have pointed out that Ramanujan's work is beginning to be applied by physicists in superstring theory.

Ramanujan's stature as a mathematician is all the more astonishing when one considers his limited formal education. He was born on December 22, 1887, into a somewhat impoverished family of the Brahmin caste in the town of Erode in southern India and grew up in Kumbakonam, where his father was an accountant to a clothier. His mathematical precocity was recognized early, and at the age of seven he was given a scholarship to the Kumbakonam Town High School. He is said to have recited mathematical formulas to his schoolmates—including the value of pi to many places.

When he was 12, Ramanujan mastered the contents of S. L. Loney's rather comprehensive *Plane Trigonometry,* including its discussion of the sum and products of infinite sequences, which later were to figure prominently in his work. (An infinite sequence is an unending string of terms, often generated by a simple formula. In this context the interesting sequences are those whose terms can be added or multiplied to yield an identifiable, finite value. If the terms are added, the resulting expression is called a series; if they are multiplied, it is called a product.) Three years later he borrowed the *Synopsis of Elementary Results in Pure Mathematics,* a listing of some 6,000 theorems (most of them given without proof) compiled by G. S. Carr, a tutor at the University of Cambridge. Those two books were the basis of Ramanujan's mathematical training.

In 1903 Ramanujan was admitted to a local government college. Yet total absorption in his own mathematical diversions at the expense of everything else caused him to fail his examinations, a pattern repeated four years later at another college in Madras. Ramanujan did set his avocation aside—if only temporarily—to look for a job after his marriage in 1909. Fortunately in 1910 R. Ramachandra Rao, a well-to-do patron of mathematics, gave him a monthly stipend largely on the strength of favorable recommendations from various sympathetic Indian mathematicians and the findings he already had jotted down in the "Notebooks."

In 1912, wanting more conventional work, he took a clerical position in the Madras Port Trust, where the chairman was a British engineer, Sir Francis Spring, and the manager was V. Ramaswami Aiyar, the founder of the Indian Mathematical Society. They encouraged Ramanujan to communicate his results to three prominent British mathematicians. Two apparently did not respond; the one who did was G. H. Hardy of Cambridge, now regarded as the foremost British mathematician of the period.

Hardy, accustomed to receiving crank mail, was inclined to disregard Ramanujan's letter at first glance the day it arrived, January 16, 1913. But after dinner that night Hardy and a close colleague, John E. Littlewood, sat down to puzzle through a list of 120 formulas and theorems Ramanujan had appended to his letter. Some hours later they had reached a verdict: they were seeing the work of a genius and not a crackpot. (According to his own "pure-talent scale" of mathematicians, Hardy was later to rate Ramanujan a 100, Littlewood a 30 and himself a 25. The German mathematician David Hilbert, the most influential figure of the time, merited only an 80.) Hardy described the revelation and its consequences as the one romantic incident in his life. He wrote that some of Ramanujan's formulas defeated him completely, and yet "they must be true, because if they were not true, no one would have had the imagination to invent them."

Hardy immediately invited Ramanujan to come to Cambridge. In spite of his mother's strong objections as well as his own reservations, Ramanujan set out for England in March of 1914. During the next five years Hardy and Ramanujan worked together at Trinity College. The blend of Hardy's technical expertise and Ramanujan's raw brilliance produced an unequaled collaboration. They published a series of seminal papers on the properties of various arithmetic functions, laying the groundwork for the answer to such questions as: How many prime divisors is a given number likely to have? How many ways can one express a number as a sum of smaller positive integers?

In 1917 Ramanujan was made a Fellow of the Royal Society of London and a Fellow of Trinity College—the first Indian to be awarded either honor. Yet as his prominence grew his health deteriorated sharply, a decline perhaps accelerated by the difficulty of maintaining a strict vegetarian diet in war-rationed England. Although Ramanujan was in and out of sanatoriums, he continued to pour forth new results. In 1919, when peace made travel abroad safe again, Ramanujan returned to India. Already an icon for young Indian intellectuals, the 32-year-old Ramanujan died on April 26, 1920, of what was then diagnosed as tuberculosis but now is thought to have been a severe vitamin deficiency. True to mathematics until the end, Ramanujan did not slow down during his last, pain-racked months, producing the re-

SRINIVASA RAMANUJAN, born in 1887 in India, managed in spite of limited formal education to reconstruct almost single-handedly much of the edifice of number theory and to go on to derive original theorems and formulas. Like many illustrious mathematicians before him, Ramanujan was fascinated by pi: the ratio of any circle's circumference to its diameter. Based on his investigation of modular equations (*see box on page 114*), he formulated exact expressions for pi and derived from them approximate values. As a result of the work of various investigators (including the authors), Ramanujan's methods are now better understood and have been implemented as algorithms.

markable work recorded in his so-called "Lost Notebook."

Ramanujan's work on pi grew in large part out of his investigation of modular equations, perhaps the most thoroughly treated subject in the "Notebooks." Roughly speaking, a modular equation is an algebraic relation between a function expressed in terms of a variable x—in mathematical notation, $f(x)$—and the same function expressed in terms of x raised to an integral power, for example $f(x^2)$, $f(x^3)$ or $f(x^4)$. The "order" of the modular equation is given by the integral power. The simplest modular equation is the second-order one: $f(x) = 2\sqrt{f(x^2)}/[1 + f(x^2)]$. Of course, not every function will satisfy a modular equation, but there is a class of functions, called modular functions, that do. These functions have various surprising symmetries that give them a special place in mathematics.

Ramanujan was unparalleled in his ability to come up with solutions to modular equations that also satisfy other conditions. Such solutions are called singular values. It turns out that solving for singular values in certain cases yields numbers whose natural logarithms coincide with pi (times a constant) to a surprising number of places [*see box on page 114*]. Applying this general approach with extraordinary virtuosity, Ramanujan produced many remarkable infinite series as well as single-term approximations for pi. Some of them are given in Ramanujan's one formal paper on the subject, *Modular Equations and Approximations to* π, published in 1914.

Ramanujan's attempts to approximate pi are part of a venerable tradition. The earliest Indo-European civilizations were aware that the area of a circle is proportional to the square of its radius and that the circumference of a circle is directly proportional to its diameter. Less clear, however, is when it was first realized that the ratio of any circle's circumference to its diameter and the ratio of any circle's area to the square of its radius are in fact the same constant, which today is designated by the symbol π. (The symbol, which gives the constant its name, is a latecomer in the history of mathematics, having been introduced in 1706 by the English mathematical writer William Jones and popularized by the Swiss mathematician Leonhard Euler in the 18th century.)

Archimedes of Syracuse, the greatest mathematician of antiquity, rigorously established the equivalence of the two ratios in his treatise *Measurement of a Circle*. He also calculated a value for pi based on mathematical principles rather than on direct measurement of a circle's circumference, area and diameter. What Archimedes did was to inscribe and circumscribe regular polygons (polygons whose sides are all the same length) on a circle assumed to have a diameter of one unit and to consider

ARCHIMEDES' METHOD for estimating pi relied on inscribed and circumscribed regular polygons (polygons with sides of equal length) on a circle having a diameter of one unit (or a radius of half a unit). The perimeters of the inscribed and circumscribed polygons served respectively as lower and upper bounds for the value of pi. The sine and tangent functions can be used to calculate the polygons' perimeters, as is shown here, but Archimedes had to develop equivalent relations based on geometric constructions. Using 96-sided polygons, he determined that pi is greater than $3^{10}/_{71}$ and less than $3^{1}/_{7}$.

Chapter 3

the polygons' respective perimeters as lower and upper bounds for possible values of the circumference of the circle, which is numerically equal to pi [*see illustration on opposite page*].

This method of approaching a value for pi was not novel: inscribing polygons of ever more sides in a circle had been proposed earlier by Antiphon, and Antiphon's contemporary, Bryson of Heraclea, had added circumscribed polygons to the procedure. What was novel was Archimedes' correct determination of the effect of doubling the number of sides on both the circumscribed and the inscribed polygons. He thereby developed a procedure that, when repeated enough times, enables one in principle to calculate pi to any number of digits. (It should be pointed out that the perimeter of a regular polygon can be readily calculated by means of simple trigonometric functions: the sine, cosine and tangent functions. But in Archimedes' time, the third century B.C., such functions were only partly understood. Archimedes therefore had to rely mainly on geometric constructions, which made the calculations considerably more demanding than they might appear today.)

Archimedes began with inscribed and circumscribed hexagons, which yield the inequality $3 < \pi < 2\sqrt{3}$. By doubling the number of sides four times, to 96, he narrowed the range of pi to between $3\frac{10}{71}$ and $3\frac{1}{7}$, obtaining the estimate $\pi \approx 3.14$. There is some evidence that the extant text of *Measurement of a Circle* is only a fragment of a larger work in which Archimedes described how, starting with decagons and doubling them six times, he got a five-digit estimate: $\pi \approx 3.1416$.

Archimedes' method is conceptually simple, but in the absence of a ready way to calculate trigonometric functions it requires the extraction of roots, which is rather time-consuming when done by hand. Moreover, the estimates converge slowly to pi: their error decreases by about a factor of four per iteration. Nevertheless, all European attempts to calculate pi before the mid-17th century relied in one way or another on the method. The 16th-century Dutch mathematician Ludolph van Ceulen dedicated much of his career to a computation of pi. Near the end of his life he obtained a 32-digit estimate by calculating the perimeter of inscribed and circumscribed polygons having 2^{62} (some 10^{18}) sides. His value for pi, called the Ludolphian number in parts of Europe, is said to have served as his epitaph.

The development of calculus, largely by Isaac Newton and Gottfried Wilhelm Leibniz, made it possible to calculate pi much more expeditiously. Calculus provides efficient techniques for computing a function's derivative (the rate of change in the function's value as its variables change) and its integral (the sum of the function's values over a range of variables). Applying the techniques, one can demonstrate that inverse trigonometric functions are given by integrals of quadratic functions that describe the curve of a circle. (The inverse of a trigonometric function gives the angle that corresponds to a particular value of the function. For example, the inverse tangent of 1 is 45 degrees or, equivalently, $\pi/4$ radians.)

(The underlying connection between trigonometric functions and algebraic expressions can be appreciated by considering a circle that has a radius of one unit and its center at the origin of a Cartesian x-y plane. The equation for the circle—whose area is numerically equal to pi—is $x^2 + y^2 = 1$, which is a restatement of the Pythagorean theorem for a right triangle with a hypotenuse equal to 1. Moreover, the sine and cosine of the angle between the positive x axis and any point on the circle are equal respectively to the point's coordinates, y and x; the angle's tangent is simply y/x.)

Of more importance for the purposes of calculating pi, however, is the fact that an inverse trigonometric function can be "expanded" as a series, the terms of which are computable from the derivatives of the function. Newton himself calculated pi to 15 places by adding the first few terms of a series that can be derived

---

WALLIS' PRODUCT (1665)

$$\frac{\pi}{2} = \frac{2 \times 2}{1 \times 3} \times \frac{4 \times 4}{3 \times 5} \times \frac{6 \times 6}{5 \times 7} \times \frac{8 \times 8}{7 \times 9} \times \cdots = \prod_{n=1}^{\infty} \frac{4n^2}{4n^2 - 1}$$

GREGORY'S SERIES (1671)

$$\frac{\pi}{4} = 1 - \frac{1}{3} + \frac{1}{5} - \frac{1}{7} + \cdots = \sum_{n=0}^{\infty} \frac{(-1)^n}{2n+1}$$

MACHIN'S FORMULA (1706)

$$\frac{\pi}{4} = 4 \arctan(1/5) - \arctan(1/239), \quad \text{where } \arctan X = X - \frac{X^3}{3} + \frac{X^5}{5} - \frac{X^7}{7} + \cdots = \sum_{n=0}^{\infty} (-1)^n \frac{X^{(2n+1)}}{2n+1}$$

RAMANUJAN (1914)

$$\frac{1}{\pi} = \frac{\sqrt{8}}{9,801} \sum_{n=0}^{\infty} \frac{(4n)![1,103 + 26,390n]}{(n!)^4 396^{4n}}, \quad \text{where } n! = n \times (n-1) \times (n-2) \times \cdots \times 1 \text{ and } 0! = 1$$

BORWEIN AND BORWEIN (1987)

$$\frac{1}{\pi} = 12 \sum_{n=0}^{\infty} \frac{(-1)^n (6n)![212,175,710,912\sqrt{61} + 1,657,145,277,365 + n(13,773,980,892,672\sqrt{61} + 107,578,229,802,750)]}{(n!)^3(3n)![5,280(236,674 + 30,303\sqrt{61})]^{(3n-3/2)}}$$

TERMS OF MATHEMATICAL SEQUENCES can be summed or multiplied to yield values for pi (divided by a constant) or its reciprocal. The first two sequences, discovered respectively by the mathematicians John Wallis and James Gregory, are probably among the best-known, but they are practically useless for computational purposes. Not even 100 years of computing on a supercomputer programmed to add or multiply the terms of either sequence would yield 100 digits of pi. The formula discovered by John Machin made the calculation of pi feasible, since calculus allows the inverse tangent (arc tangent) of a number, x, to be expressed in terms of a sequence whose sum converges more rapidly to the value of the arc tangent the smaller x is. Virtually all calculations for pi from the beginning of the 18th century until the early 1970's have relied on variations of Machin's formula. The sum of Ramanujan's sequence converges to the true value of $1/\pi$ much faster: each successive term in the sequence adds roughly eight more correct digits. The last sequence, formulated by the authors, adds about 25 digits per term; the first term (for which n is 0) yields a number that agrees with pi to 24 digits.

as an expression for the inverse of the sine function. He later confessed to a colleague: "I am ashamed to tell you to how many figures I carried these calculations, having no other business at the time."

In 1674 Leibniz derived the formula $1 - 1/3 + 1/5 - 1/7\ldots = \pi/4$, which is the inverse tangent of 1. (The general inverse-tangent series was originally discovered in 1671 by the Scottish mathematician James Gregory. Indeed, similar expressions appear to have been developed independently several centuries earlier in India.) The error of the approximation, defined as the difference between the sum of n terms and the exact value of $\pi/4$, is roughly equal to the $n+1$th term in the series. Since the denominator of each successive term increases by only 2, one must add approximately 50 terms to get two-digit accuracy, 500 terms for three-digit accuracy and so on. Summing the terms of the series to calculate a value for pi more than a few digits long is clearly prohibitive.

An observation made by John Machin, however, made it practicable to calculate pi by means of a series expansion for the inverse-tangent function. He noted that pi divided by 4 is equal to 4 times the inverse tangent of 1/5 minus the inverse tangent of 1/239. Because the inverse-tangent series for a given value converges more quickly the smaller the value is, Machin's formula greatly simplified the calculation. Coupling his formula with the series expansion for the inverse tangent, Machin computed 100 digits of pi in 1706. Indeed, his technique proved to be so powerful that all extended calculations of pi from the beginning of the 18th century until recently relied on variants of the method.

Two 19th-century calculations deserve special mention. In 1844 Johann Dase computed 205 digits of pi in a matter of months by calculating the values of three inverse tangents in a Machin-like formula. Dase was a calculating prodigy who could multiply 100-digit numbers entirely in his head—a feat that took him roughly eight hours. (He was perhaps the closest precursor of the modern supercomputer, at least in terms of memory capacity.) In 1853 William Shanks outdid Dase by publishing his computation of pi to 607 places, although the digits that followed the 527th place were wrong. Shank's task took years and was a rather routine, albeit laborious, application of Machin's formula. (In what must itself be some kind of record, 92 years passed before Shank's error was detected, in a comparison between his value and a 530-place approximation produced by D. F. Ferguson with the aid of a mechanical calculator.)

The advent of the digital computer saw a renewal of efforts to calculate ever more digits of pi, since the machine was ideally suited for lengthy, repetitive "number crunching." ENIAC, one of the first digital computers, was applied to the task in June, 1949, by John von Neumann and his colleagues. ENIAC produced 2,037 digits in 70 hours. In 1957 G. E. Felton attempted to compute 10,000 digits of pi, but owing to a machine error only the first 7,480 digits were correct. The 10,000-digit goal was reached by F. Genuys the following year on an IBM 704 computer. In 1961 Daniel Shanks and John W. Wrench, Jr., calculated 100,000 digits of pi in less than nine hours on an IBM 7090. The million-digit mark was passed in 1973 by Jean Guilloud and M. Bouyer, a feat that took just under a day of computation on a CDC 7600. (The computations done by Shanks and Wrench and by Guilloud and Bouyer were in fact carried out twice using different inverse-tangent identities for pi. Given the history of both human and machine error in these calculations, it is only after such verification that modern "digit hunters" consider a record officially set.)

Although an increase in the speed of computers was a major reason ever more accurate calculations for pi could be performed, it soon became clear that there were inescapable limits. Doubling the number of digits lengthens computing time by at least a factor of four, if one applies the traditional methods of performing arithmetic in computers. Hence even allowing for a hundredfold increase in computational speed, Guilloud and Bouyer's program would have required at least a quarter century to produce a billion-digit value for pi. From the perspective of the early 1970's such a computation did not seem realistically practicable.

Yet the task is now feasible, thanks

MODULAR FUNCTIONS AND APPROXIMATIONS TO PI

A modular function is a function, $\lambda(q)$, that can be related through an algebraic expression called a modular equation to the same function expressed in terms of the same variable, q, raised to an integral power: $\lambda(q^p)$. The integral power, p, determines the "order" of the modular equation. An example of a modular function is

$$\lambda(q) = 16q \prod_{n=1}^{\infty} \left(\frac{1+q^{2n}}{1+q^{2n-1}} \right)^8.$$

Its associated seventh-order modular equation, which relates $\lambda(q)$ to $\lambda(q^7)$, is given by

$$\sqrt[8]{\lambda(q)\lambda(q^7)} + \sqrt[8]{[1-\lambda(q)][1-\lambda(q^7)]} = 1.$$

Singular values are solutions of modular equations that must also satisfy additional conditions. One class of singular values corresponds to computing a sequence of values, k_p, where

$$k_p = \sqrt{\lambda(e^{-\pi\sqrt{p}})}$$

and p takes integer values. These values have the curious property that the logarithmic expression

$$\frac{-2}{\sqrt{p}} \log\left(\frac{k_p}{4}\right)$$

coincides with many of the first digits of pi. The number of digits the expression has in common with pi increases with larger values of p.

Ramanujan was unparalleled in his ability to calculate these singular values. One of his most famous is the value when p equals 210, which was included in his original letter to G. H. Hardy. It is

$$k_{210} = (\sqrt{2}-1)^2(2-\sqrt{3})(\sqrt{7}-\sqrt{6})^2(8-3\sqrt{7})(\sqrt{10}-3)^2(\sqrt{15}-\sqrt{14})(4-\sqrt{15})^2(6-\sqrt{35}).$$

This number, when plugged into the logarithmic expression, agrees with pi through the first 20 decimal places. In comparison, k_{240} yields a number that agrees with pi through more than one million digits.

Applying this general approach, Ramanujan constructed a number of remarkable series for pi, including the one shown in the illustration on the preceding page. The general approach also underlies the two-step, iterative algorithms in the top illustration on the opposite page. In each iteration the first step (calculating y_n) corresponds to computing one of a sequence of singular values by solving a modular equation of the appropriate order; the second step (calculating α_n) is tantamount to taking the logarithm of the singular value.

Chapter 3

not only to faster computers but also to new, efficient methods for multiplying large numbers in computers. A third development was also crucial: the advent of iterative algorithms that quickly converge to pi. (An iterative algorithm can be expressed as a computer program that repeatedly performs the same arithmetic operations, taking the output of one cycle as the input for the next.) These algorithms, some of which we constructed, were in many respects anticipated by Ramanujan, although he knew nothing of computer programming. Indeed, computers not only have made it possible to apply Ramanujan's work but also have helped to unravel it. Sophisticated algebraic-manipulation software has allowed further exploration of the road Ramanujan traveled alone and unaided 75 years ago.

One of the interesting lessons of theoretical computer science is that many familiar algorithms, such as the way children are taught to multiply in grade school, are far from optimal. Computer scientists gauge the efficiency of an algorithm by determining its bit complexity: the number of times individual digits are added or multiplied in carrying out an algorithm. By this measure, adding two n-digit numbers in the normal way has a bit complexity that increases in step with n; multiplying two n-digit numbers in the normal way has a bit complexity that increases as n^2. By traditional methods, multiplication is much "harder" than addition in that it is much more time-consuming.

Yet, as was shown in 1971 by A. Schönhage and V. Strassen, the multiplication of two numbers can in theory have a bit complexity only a little greater than addition. One way to achieve this potential reduction in bit complexity is to implement so-called fast Fourier transforms (FFT's). FFT-based multiplication of two large numbers allows the intermediary computations among individual digits to be carefully orchestrated so that redundancy is avoided. Because division and root extraction can be reduced to a sequence of multiplications, they too can have a bit complexity just slightly greater than that of addition. The result is a tremendous saving in bit complexity and hence in computation time. For this reason all recent efforts to calculate pi rely on some variation of the FFT technique for multiplication.

Yet for hundreds of millions of dig-

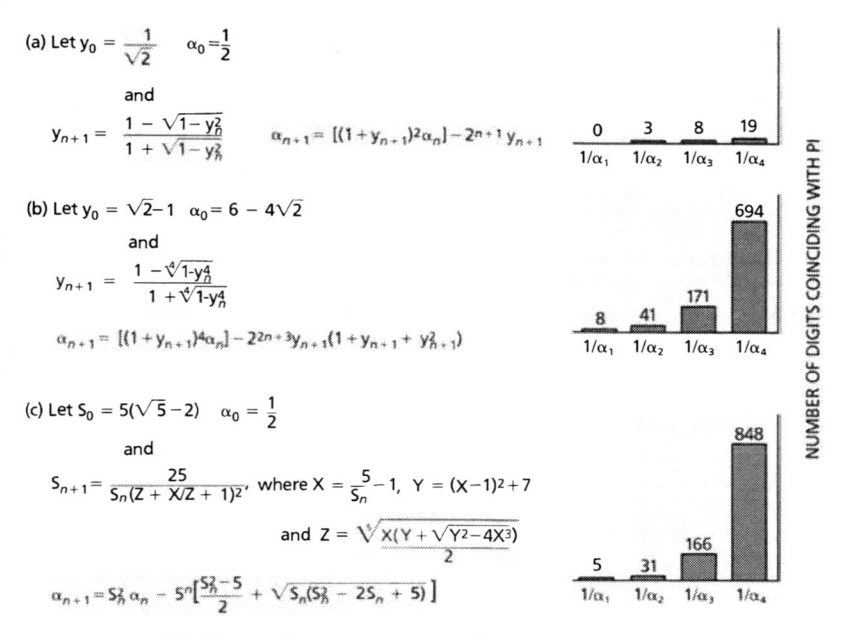

(a) Let $y_0 = \frac{1}{\sqrt{2}}$ $\alpha_0 = \frac{1}{2}$

and

$y_{n+1} = \frac{1 - \sqrt{1-y_n^2}}{1 + \sqrt{1-y_n^2}}$ $\alpha_{n+1} = [(1+y_{n+1})^2 \alpha_n] - 2^{n+1} y_{n+1}$

(b) Let $y_0 = \sqrt{2}-1$ $\alpha_0 = 6 - 4\sqrt{2}$

and

$y_{n+1} = \frac{1 - \sqrt[4]{1-y_n^4}}{1 + \sqrt[4]{1-y_n^4}}$

$\alpha_{n+1} = [(1+y_{n+1})^4 \alpha_n] - 2^{2n+3} y_{n+1}(1+y_{n+1}+y_{n+1}^2)$

(c) Let $S_0 = 5(\sqrt{5}-2)$ $\alpha_0 = \frac{1}{2}$

and

$S_{n+1} = \frac{25}{S_n(Z + X/Z + 1)^2}$, where $X = \frac{5}{S_n} - 1$, $Y = (X-1)^2 + 7$

and $Z = \sqrt[5]{\frac{X(Y + \sqrt{Y^2 - 4X^3})}{2}}$

$\alpha_{n+1} = S_n^2 \alpha_n - 5^n\left[\frac{S_n^2 - 5}{2} + \sqrt{S_n(S_n^2 - 2S_n + 5)}\right]$

ITERATIVE ALGORITHMS that yield extremely accurate values of pi were developed by the authors. (An iterative algorithm is a sequence of operations repeated in such a way that the ouput of one cycle is taken as the input for the next.) Algorithm a converges to $1/\pi$ quadratically: the number of correct digits given by α_n more than doubles each time n is increased by 1. Algorithm b converges quartically and algorithm c converges quintically, so that the number of coinciding digits given by each iteration increases respectively by more than a factor of four and by more than a factor of five. Algorithm b is possibly the most efficient known algorithm for calculating pi; it was run on supercomputers in the last three record-setting calculations. As the authors worked on the algorithms it became clear to them that Ramanujan had pursued similar methods in coming up with his approximations for pi. In fact, the computation of s_n in algorithm c rests on a remarkable fifth-order modular equation discovered by Ramanujan.

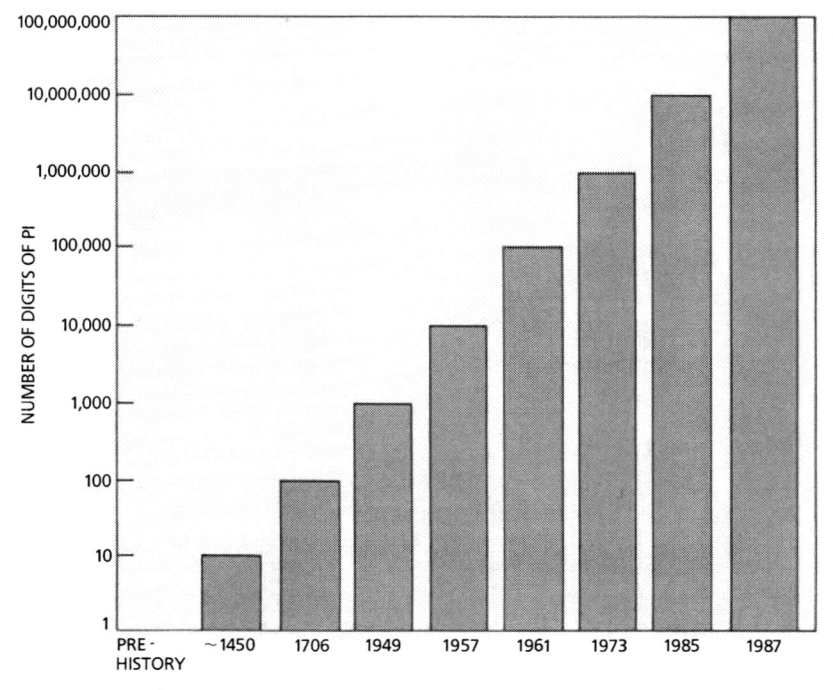

NUMBER OF KNOWN DIGITS of pi has increased by two orders of magnitude (factors of 10) in the past decade as a result of the development of iterative algorithms that can be run on supercomputers equipped with new, efficient methods of multiplication.

its of pi to be calculated practically a beautiful formula known a century and a half earlier to Carl Friedrich Gauss had to be rediscovered. In the mid-1970's Richard P. Brent and Eugene Salamin independently noted that the formula produced an algorithm for pi that converged quadratically, that is, the number of digits doubled with each iteration. Between 1983 and the present Yasumasa Kanada and his colleagues at the University of Tokyo have employed this algorithm to set several world records for the number of digits of pi.

We wondered what underlies the remarkably fast convergence to pi of the Gauss-Brent-Salamin algorithm, and in studying it we developed general techniques for the construction of similar algorithms that rapidly converge to pi as well as to other quantities. Building on a theory outlined by the German mathematician Karl Gustav Jacob Jacobi in 1829, we realized we could in principle arrive at a value for pi by evaluating integrals of a class called elliptic integrals, which can serve to calculate the perimeter of an ellipse. (A circle, the geometric setting of previous efforts to approximate pi, is simply an ellipse with axes of equal length.)

Elliptic integrals cannot generally be evaluated as integrals, but they can be easily approximated through iterative procedures that rely on modular equations. We found that the Gauss-Brent-Salamin algorithm is actually a specific case of our more general technique relying on a second-order modular equation. Quicker convergence to the value of the integral, and thus a faster algorithm for pi, is possible if higher-order modular equations are used, and so we have also constructed various algorithms based on modular equations of third, fourth and higher orders.

In January, 1986, David H. Bailey of the National Aeronautics and Space Administration's Ames Research Center produced 29,360,000 decimal places of pi by iterating one of our algorithms 12 times on a Cray-2 supercomputer. Because the algorithm is based on a fourth-order modular equation, it converges on pi quartically, more than quadrupling the number of digits with each iteration. A year later Kanada and his colleagues carried out one more iteration to attain 134,217,000 places on an NEC SX-2 supercomputer and thereby verified a similar computation they had done earlier using the Gauss-Brent-Salamin algorithm. (Iterating our algorithm twice more—a feat entirely feasible if one could somehow monopolize a supercomputer for a few weeks—would yield more than two billion digits of pi.)

Iterative methods are best suited for calculating pi on a computer, and so it is not surprising that Ramanujan never bothered to pursue them. Yet the basic ingredients of the iterative algorithms for pi—modular equations in particular—are to be found in Ramanujan's work. Parts of his original derivation of infinite series and approximations for pi more than three-quarters of a century ago must have paralleled our own efforts to come up with algorithms for pi. Indeed, the formulas he lists in his paper on pi and in the "Notebooks" helped us greatly in the construction of some of our algorithms. For example, although we were able to prove that an 11th-order algorithm exists and knew its general formulation, it was not until we stumbled on Ramanujan's modular equations of the same order that we discovered its unexpectedly simple form.

Conversely, we were also able to derive all Ramanujan's series from the general formulas we had developed. The derivation of one, which converged to pi faster than any other series we knew at the time, came about with a little help from an unexpected source. We had justified all the quantities in the expression for the series except one: the coefficient 1,103, which appears in the numerator of the expression [see illustration on page 113]. We were convinced—as Ramanujan must have been—that 1,103 had to be correct. To prove it we had either to simplify a daunting equation containing variables raised to powers of several thousand or

---

HOW TO GET TWO BILLION DIGITS OF PI WITH A CALCULATOR*

Let

$y_0 = \sqrt{2} - 1$

$y_1 = [1 - \sqrt[4]{1 - y_0^4}]/[1 + \sqrt[4]{1 - y_0^4}]$

$y_2 = [1 - \sqrt[4]{1 - y_1^4}]/[1 + \sqrt[4]{1 - y_1^4}]$

$y_3 = [1 - \sqrt[4]{1 - y_2^4}]/[1 + \sqrt[4]{1 - y_2^4}]$

$y_4 = [1 - \sqrt[4]{1 - y_3^4}]/[1 + \sqrt[4]{1 - y_3^4}]$

$y_5 = [1 - \sqrt[4]{1 - y_4^4}]/[1 + \sqrt[4]{1 - y_4^4}]$

$y_6 = [1 - \sqrt[4]{1 - y_5^4}]/[1 + \sqrt[4]{1 - y_5^4}]$

$y_7 = [1 - \sqrt[4]{1 - y_6^4}]/[1 + \sqrt[4]{1 - y_6^4}]$

$y_8 = [1 - \sqrt[4]{1 - y_7^4}]/[1 + \sqrt[4]{1 - y_7^4}]$

$y_9 = [1 - \sqrt[4]{1 - y_8^4}]/[1 + \sqrt[4]{1 - y_8^4}]$

$y_{10} = [1 - \sqrt[4]{1 - y_9^4}]/[1 + \sqrt[4]{1 - y_9^4}]$

$y_{11} = [1 - \sqrt[4]{1 - y_{10}^4}]/[1 + \sqrt[4]{1 - y_{10}^4}]$

$y_{12} = [1 - \sqrt[4]{1 - y_{11}^4}]/[1 + \sqrt[4]{1 - y_{11}^4}]$

$y_{13} = [1 - \sqrt[4]{1 - y_{12}^4}]/[1 + \sqrt[4]{1 - y_{12}^4}]$

$y_{14} = [1 - \sqrt[4]{1 - y_{13}^4}]/[1 + \sqrt[4]{1 - y_{13}^4}]$

$y_{15} = [1 - \sqrt[4]{1 - y_{14}^4}]/[1 + \sqrt[4]{1 - y_{14}^4}]$

$\alpha_0 = 6 - 4\sqrt{2}$

$\alpha_1 = (1 + y_1)^4 \alpha_0 - 2^3 y_1 (1 + y_1 + y_1^2)$

$\alpha_2 = (1 + y_2)^4 \alpha_1 - 2^5 y_2 (1 + y_2 + y_2^2)$

$\alpha_3 = (1 + y_3)^4 \alpha_2 - 2^7 y_3 (1 + y_3 + y_3^2)$

$\alpha_4 = (1 + y_4)^4 \alpha_3 - 2^9 y_4 (1 + y_4 + y_4^2)$

$\alpha_5 = (1 + y_5)^4 \alpha_4 - 2^{11} y_5 (1 + y_5 + y_5^2)$

$\alpha_6 = (1 + y_6)^4 \alpha_5 - 2^{13} y_6 (1 + y_6 + y_6^2)$

$\alpha_7 = (1 + y_7)^4 \alpha_6 - 2^{15} y_7 (1 + y_7 + y_7^2)$

$\alpha_8 = (1 + y_8)^4 \alpha_7 - 2^{17} y_8 (1 + y_8 + y_8^2)$

$\alpha_9 = (1 + y_9)^4 \alpha_8 - 2^{19} y_9 (1 + y_9 + y_9^2)$

$\alpha_{10} = (1 + y_{10})^4 \alpha_9 - 2^{21} y_{10} (1 + y_{10} + y_{10}^2)$

$\alpha_{11} = (1 + y_{11})^4 \alpha_{10} - 2^{23} y_{11} (1 + y_{11} + y_{11}^2)$

$\alpha_{12} = (1 + y_{12})^4 \alpha_{11} - 2^{25} y_{12} (1 + y_{12} + y_{12}^2)$

$\alpha_{13} = (1 + y_{13})^4 \alpha_{12} - 2^{27} y_{13} (1 + y_{13} + y_{13}^2)$

$\alpha_{14} = (1 + y_{14})^4 \alpha_{13} - 2^{29} y_{14} (1 + y_{14} + y_{14}^2)$

$\alpha_{15} = (1 + y_{15})^4 \alpha_{14} - 2^{31} y_{15} (1 + y_{15} + y_{15}^2)$

$1/\alpha_{15}$ agress with π for more than two billion decimal digits

*Of course, the calculator needs to have a two-billion-digit display; on a pocket calculator the computation would not be very interesting after the second iteration.

EXPLICIT INSTRUCTIONS for executing algorithm *b* in the top illustration on the preceding page makes it possible in principle to compute the first two billion digits of pi in a matter of minutes. All one needs is a calculator that has two memory registers and the usual capacity to add, subtract, multiply, divide and extract roots. Unfortunately most calculators come with only an eight-digit display, which makes the computation moot.

to delve considerably further into somewhat arcane number theory.

By coincidence R. William Gosper, Jr., of Symbolics, Inc., had decided in 1985 to exploit the same series of Ramanujan's for an extended-accuracy value for pi. When he carried out the calculation to more than 17 million digits (a record at the time), there was to his knowledge no proof that the sum of the series actually converged to pi. Of course, he knew that millions of digits of his value coincided with an earlier Gauss-Brent-Salamin calculation done by Kanada. Hence the possibility of error was vanishingly small.

As soon as Gosper had finished his calculation and verified it against Kanada's, however, we had what we needed to prove that 1,103 was the number needed to make the series true to within one part in $10^{10,000,000}$. In much the same way that a pair of integers differing by less than 1 must be equal, his result sufficed to specify the number: it is precisely 1,103. In effect, Gosper's computation became part of our proof. We knew that the series (and its associated algorithm) is so sensitive to slight inaccuracies that if Gosper had used any other value for the coefficient or, for that matter, if the computer had introduced a single-digit error during the calculation, he would have ended up with numerical nonsense instead of a value for pi.

Ramanujan-type algorithms for approximating pi can be shown to be very close to the best possible. If all the operations involved in the execution of the algorithms are totaled (assuming that the best techniques known for addition, multiplication and root extraction are applied), the bit complexity of computing n digits of pi is only marginally greater than that of multiplying two n-digit numbers. But multiplying two n-digit numbers by means of an FFT-based technique is only marginally more complicated than summing two n-digit numbers, which is the simplest of the arithmetic operations possible on a computer.

Mathematics has probably not yet felt the full impact of Ramanujan's genius. There are many other wonderful formulas contained in the "Notebooks" that revolve around integrals, infinite series and continued fractions (a number plus a fraction, whose denominator can be expressed as a number plus a fraction, whose denominator can be expressed as a number plus a fraction, and so on). Unfortunately they are listed with little—if any—indication of the method by which Ramanujan proved them. Littlewood wrote: "If a significant piece of reasoning occurred somewhere, and the total mixture of evidence and intuition gave him certainty, he looked no further."

The herculean task of editing the "Notebooks," initiated 60 years ago by the British analysts G. N. Watson and B. N. Wilson and now being completed by Bruce Berndt, requires providing a proof, a source or an occasional correction for each of many thousands of asserted theorems and identities. A single line in the "Notebooks" can easily elicit many pages of commentary. The task is made all the more difficult by the nonstandard mathematical notation in which the formulas are written. Hence a great deal of Ramanujan's work will not become accessible to the mathematical community until Berndt's project is finished.

Ramanujan's unique capacity for working intuitively with complicated formulas enabled him to plant seeds in a mathematical garden (to borrow a metaphor from Freeman Dyson) that is only now coming into bloom. Along with many other mathematicians, we look forward to seeing which of the seeds will germinate in future years and further beautify the garden.

RAMANUJAN'S "NOTEBOOKS" were personal records in which he jotted down many of his formulas. The page shown contains various third-order modular equations—all in Ramanujan's nonstandard notation. Unfortunately Ramanujan did not bother to include formal proofs for the equations; others have had to compile, edit and prove them. The formulas in the "Notebooks" embody subtle relations among numbers and functions that can be applied in other fields of mathematics or even in theoretical physics.

4. Ramanujan, modular equations, and approximations to pi or how to compute a billion digits of pi

Discussion

This article, much like the previous article, describes Ramanujan's work and much more. Described as a fine example of "tip of the iceberg writing," the article won the 1993 Chauvenet prize of the American Mathematical Society given for the "outstanding survey or expository mathematics paper" published in a North American journal. It was also awarded the 1993 Hasse Prize of the Mathematical Association of America given for a "noteworthy expository paper in an Association Journal one of whose authors is younger than forty." The Chauvenet prize has been awarded regularly since 1925[1] and the winners are a group that we are honored to be counted amongst. Amusingly, despite having worked together for a decade, we first met David Bailey in person when we accepted the award together in San Antonio.

In his 1992 essay "The Definition of Numerical Analysis," [2] Lloyd N. Trefethen engagingly demolishes the conventional definition of Numerical Analysis as "the science of rounding errors." He explores how this hyperbolic view emerged and finishes by writing:

> I believe that the existence of finite algorithms for certain problems, together with other historical forces, has distracted us for decades from a balanced view of numerical analysis. Rounding errors and instability are important, and numerical analysts will always be the experts in these subjects and at pains to ensure that the unwary are not tripped up by them. But our central mission is to compute quantities that are typically uncomputable, from an analytical point of view, and to do it with lightning speed. For guidance to the future we should study not Gaussian elimination and its beguiling stability properties, but the diabolically fast conjugate gradient iteration, or Greengard and Rokhlin's $O(N)$ multipole algorithm for particle simulations, or the exponential convergence of spectral methods for solving certain PDEs, or the convergence in $O(N)$ iteration achieved by multigrid methods for many kinds of problems, or even Borwein and Borwein's[3] magical AGM iteration for determining 1,000,000 digits of π in the blink of an eye. That is the *heart* of numerical analysis.

Source

J.M. Borwein, P.B. Borwein, and D.H. Bailey, "Ramanujan, modular equations, and approximations to pi or how to compute a billion digits of pi," *MAA Monthly*, **96** (1989), 201–219. Also available online at http://www.cecm.sfu.ca/organics/papers/borwein/index.html.

---

[1] See http://www.maa.org/Awards/chauvent.html.
[2] *SIAM News* November 1992.
[3] As in many cases this eponymy is inaccurate, if flattering, and really should be to Gauss-Brent-Salamin.

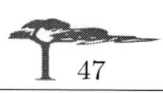

Ramanujan, Modular Equations, and Approximations to Pi or How to Compute One Billion Digits of Pi

J. M. BORWEIN AND P. B. BORWEIN
Mathematics Department, Dalhousie University, Halifax, N.S. B3H 3J5 Canada

and

D. H. BAILEY
NASA Ames Research Center, Moffett Field, CA 94035

Preface. The year 1987 was the centenary of Ramanujan's birth. He died in 1920 Had he not died so young, his presence in modern mathematics might be more immediately felt. Had he lived to have access to powerful algebraic manipulation software, such as MACSYMA, who knows how much more spectacular his already astonishing career might have been.

This article will follow up one small thread of Ramanujan's work which has found a modern computational context, namely, one of his approaches to approximating pi. Our experience has been that as we have come to understand these pieces of Ramanujan's work, as they have become mathematically demystified, and as we have come to realize the intrinsic complexity of these results, we have come to realize how truly singular his abilities were. This article attempts to present a considerable amount of material and, of necessity, little is presented in detail. We have, however, given much more detail than Ramanujan provided. Our intention is that the circle of ideas will become apparent and that the finer points may be pursued through the indicated references.

1. Introduction. There is a close and beautiful connection between the transformation theory for elliptic integrals and the very rapid approximation of pi. This connection was first made explicit by Ramanujan in his 1914 paper "Modular Equations and Approximations to π" [26]. We might emphasize that Algorithms 1 and 2 are not to be found in Ramanujan's work, indeed no recursive approximation of π is considered, but as we shall see they are intimately related to his analysis. Three central examples are:

Sum 1. (Ramanujan)

$$\frac{1}{\pi} = \frac{\sqrt{8}}{9801} \sum_{n=0}^{\infty} \frac{(4n)!}{(n!)^4} \frac{[1103 + 26390n]}{396^{4n}}.$$

Algorithm 1. Let $\alpha_0 := 6 - 4\sqrt{2}$ and $y_0 := \sqrt{2} - 1$.
Let

$$y_{n+1} := \frac{1 - (1 - y_n^4)^{1/4}}{1 + (1 - y_n^4)^{1/4}}$$

and

$$\alpha_{n+1} := (1 + y_{n+1})^4 \alpha_n - 2^{2n+3} y_{n+1}(1 + y_{n+1} + y_{n+1}^2).$$

Then
$$0 < \alpha_n - 1/\pi < 16 \cdot 4^n e^{-2 \cdot 4^n \pi}$$
and α_n converges to $1/\pi$ *quartically* (that is, with order four).

Algorithm 2. Let $s_0 := 5(\sqrt{5} - 2)$ and $\alpha_0 := 1/2$.
Let
$$s_{n+1} := \frac{25}{(z + x/z + 1)^2 s_n},$$
where
$$x := 5/s_n - 1 \qquad y := (x - 1)^2 + 7$$
and
$$z := \left[\frac{1}{2}x\left(y + \sqrt{y^2 - 4x^3}\right)\right]^{1/5}.$$
Let
$$\alpha_{n+1} := s_n^2 \alpha_n - 5^n \left\{ \frac{s_n^2 - 5}{2} + \sqrt{s_n(s_n^2 - 2s_n + 5)} \right\}.$$
Then
$$0 < \alpha_n - \frac{1}{\pi} < 16 \cdot 5^n e^{-5^n \pi}$$
and α_n converges to $1/\pi$ *quintically* (that is, with order five).

Each additional term in Sum 1 adds roughly eight digits, each additional iteration of Algorithm 1 quadruples the number of correct digits, while each additional iteration of Algorithm 2 quintuples the number of correct digits. Thus a mere thirteen iterations of Algorithm 2 provide in excess of one billion decimal digits of pi. In general, for us, pth-order convergence of a sequence $\{\alpha_n\}$ to α means that α_n tends to α and that
$$|\alpha_{n+1} - \alpha| \leq C|\alpha_n - \alpha|^p$$
for some constant $C > 0$. Algorithm 1 is arguably the most efficient algorithm currently known for the extended precision calculation of pi. While the rates of convergence are impressive, it is the subtle and thoroughly nontransparent nature of these results and the beauty of the underlying mathematics that intrigue us most.

Watson [37], commenting on certain formulae of Ramanujan, talks of

a thrill which is indistinguishable from the thrill which I feel when I enter the Sagrestia Nuovo of the Capella Medici and see before me the austere beauty of the four statues representing "Day," "Night," "Evening," and "Dawn" which Michelangelo has set over the tomb of Giuliano de'Medici and Lorenzo de'Medici.

Sum 1 is directly due to Ramanujan and appears in [26]. It rests on a modular identity of order 58 and, like much of Ramanujan's work, appears without proof and with only scanty motivation. The first complete derivation we know of appears in [11]. Algorithms 1 and 2 are based on modular identities of orders 4 and 5, respectively. The underlying quintic modular identity in Algorithm 2 (the relation for s_n) is also due to Ramanujan, though the first proof is due to Berndt and will appear in [7].

One intention in writing this article is to explain the genesis of Sum 1 and of Algorithms 1 and 2. It is not possible to give a short self-contained account without assuming an unusual degree of familiarity with modular function theory. Also, parts of the derivation involve considerable algebraic calculation and may most easily be done with the aid of a symbol manipulation package (MACSYMA, MAPLE, REDUCE, etc.). We hope however to give a taste of methods involved. The full details are available in [11].

A second intention is very briefly to describe the role of these and related approximations in the recent extended precision calculations of pi. In part this entails a short discussion of the complexity and implementation of such calculations. This centers on a discussion of multiplication by fast Fourier transform methods. Of considerable related interest is the fact that these algorithms for π are provably close to the theoretical optimum.

2. The State of Our Current Ignorance. Pi is almost certainly the most natural of the transcendental numbers, arising as the circumference of a circle of unit diameter. Thus, it is not surprising that its properties have been studied for some twenty-five hundred years. What is surprising is how little we actually know.

We know that π is irrational, and have known this since Lambert's proof of 1771 (see [5]). We have known that π is transcendental since Lindemann's proof of 1882 [23]. We also know that π is not a Liouville number. Mahler proved this in 1953. An irrational number β is *Liouville* if, for any n, there exist integers p and q so that

$$0 < \left| \beta - \frac{p}{q} \right| < \frac{1}{q^n}.$$

Liouville showed these numbers are all transcendental. In fact we know that

$$\left| \pi - \frac{p}{q} \right| > \frac{1}{q^{14.65}} \tag{2.1}$$

for p, q integral with q sufficiently large. This *irrationality estimate*, due to Chudnovsky and Chudnovsky [16] is certainly not best possible. It is likely that 14.65 should be replaced by $2 + \varepsilon$ for any $\varepsilon > 0$. Almost all transcendental numbers satisfy such an inequality. We know a few related results for the rate of algebraic approximation. The results may be pursued in [4] and [11].

We know that e^{π} is transcendental. This follows by noting that $e^{\pi} = (-1)^{-i}$ and applying the Gelfond-Schneider theorem [4]. We know that $\pi + \log 2 + \sqrt{2} \log 3$ is transcendental. This result is a consequence of the work that won Baker a Fields Medal in 1970. And we know a few more than the first two hundred million digits of the decimal expansion for π (Kanada, see Section 3).

The state of our ignorance is more profound. We do not know whether such basic constants as $\pi + e$, π/e, or $\log \pi$ are irrational, let alone transcendental. The best we can say about these three particular constants is that they cannot satisfy any polynomial of degree eight or less with integer coefficients of average size less than 10^9 [3]. This is a consequence of some recent computations employing the

Ferguson-Forcade algorithm [17]. We don't know anything of consequence about the single continued fraction of pi, except (numerically) the first 17 million terms, which Gosper computed in 1985 using Sum 1. Likewise, apart from listing the first many millions of digits of π, we know virtually nothing about the decimal expansion of π. It is possible, albeit not a good bet, that all but finitely many of the decimal digits of pi are in fact 0's and 1's. Carl Sagan's recent novel *Contact* rests on a similar possibility. Questions concerning the normality of or the distribution of digits of particular transcendentals such as π appear completely beyond the scope of current mathematical techniques. The evidence from analysis of the first thirty million digits is that they are very uniformly distributed [2]. The next one hundred and seventy million digits apparently contain no surprises.

In part we perhaps settle for computing digits of π because there is little else we can currently do. We would be amiss, however, if we did not emphasize that the extended precision calculation of pi has substantial application as a test of the "global integrity" of a supercomputer. The extended precision calculations described in Section 3 uncovered hardware errors which had to be corrected before those calculations could be successfully run. Such calculations, implemented as in Section 4, are apparently now used routinely to check supercomputers before they leave the factory. A large-scale calculation of pi is entirely unforgiving; it soaks into all parts of the machine and a single bit awry leaves detectable consequences.

3. Matters Computational

I am ashamed to tell you to how many figures I carried these calculations, having no other business at the time.

Isaac Newton

Newton's embarrassment at having computed 15 digits, which he did using the arcsinlike formula

$$\pi = \frac{3\sqrt{3}}{4} + 24\left(\frac{1}{12} - \frac{1}{5 \cdot 2^5} - \frac{1}{28 \cdot 2^7} - \frac{1}{72 \cdot 2^9} - \cdots\right)$$

$$= \frac{3\sqrt{3}}{4} + 24\int_0^{\frac{1}{4}} \sqrt{x - x^2}\, dx,$$

is indicative both of the spirit in which people calculate digits and the fact that a surprising number of people have succumbed to the temptation [5].

The history of efforts to determine an accurate value for the constant we now know as π is almost as long as the history of civilization itself. By 2000 B.C. both the Babylonians and the Egyptians knew π to nearly two decimal places. The Babylonians used, among others, the value 3 1/8 and the Egyptians used 3 13/81. Not all ancient societies were as accurate, however—nearly 1500 years later the Hebrews were perhaps still content to use the value 3, as the following quote suggests.

Also, he made a molten sea of ten cubits from brim to brim, round in compass, and five cubits the height thereof; and a line of thirty cubits did compass it round about.

Old Testament, 1 Kings 7:23

Despite the long pedigree of the problem, all nonempirical calculations have employed, up to minor variations, only three techniques.

i) The first technique due to Archimedes of Syracuse (287–212 B.C.) is, recursively, to calculate the length of circumscribed and inscribed regular $6 \cdot 2^n$-gons about a circle of diameter 1. Call these quantities a_n and b_n, respectively. Then $a_0 := 2\sqrt{3}$, $b_0 := 3$ and, as Gauss's teacher Pfaff discovered in 1800,

$$a_{n+1} := \frac{2a_n b_n}{a_n + b_n} \quad \text{and} \quad b_{n+1} := \sqrt{a_{n+1} b_n}.$$

Archimedes, with $n = 4$, obtained

$$3\tfrac{10}{71} < \pi < 3\tfrac{1}{7}.$$

While hardly better than estimates one could get with a ruler, this is the first method that can be used to generate an arbitrary number of digits, and to a nonnumerical mathematician perhaps the problem ends here. Variations on this theme provided the basis for virtually all calculations of π for the next 1800 years, culminating with a 34 digit calculation due to Ludolph van Ceulen (1540–1610). This demands polygons with about 2^{60} sides and so is extraordinarily time consuming.

ii) Calculus provides the basis for the second technique. The underlying method relies on Gregory's series of 1671

$$\arctan x = \int_0^x \frac{dt}{1+t^2} = x - \frac{x^3}{3} + \frac{x^5}{5} - \cdots \qquad |x| \leq 1$$

coupled with a formula which allows small x to be used, like

$$\frac{\pi}{4} = 4 \arctan\left(\frac{1}{5}\right) - \arctan\left(\frac{1}{239}\right).$$

This particular formula is due to Machin and was employed by him to compute 100 digits of π in 1706. Variations on this second theme are the basis of all the calculations done until the 1970's including William Shanks' monumental hand-calculation of 527 digits. In the introduction to his book [32], which presents this calculation, Shanks writes:

> *Towards the close of the year 1850 the Author first formed the design of rectifying the circle to upwards of 300 places of decimals. He was fully aware at that time, that the accomplishment of his purpose would add little or nothing to his fame as a Mathematician though it might as a Computer; nor would it be productive of anything in the shape of pecuniary recompense.*

Shanks actually attempted to hand-calculate 707 digits but a mistake crept in at the 527th digit. This went unnoticed until 1945, when D. Ferguson, in one of the last "nondigital" calculations, computed 530 digits. Even with machine calculations mistakes occur, so most record-setting calculations are done twice—by sufficiently different methods.

The advent of computers has greatly increased the scope and decreased the toil of such calculations. Metropolis, Reitwieser, and von Neumann computed and analyzed 2037 digits using Machin's formula on ENIAC in 1949. In 1961, Dan Shanks and Wrench calculated 100,000 digits on an IBM 7090 [31]. By 1973, still using Machin-like arctan expansions, the million digit mark was passed by Guillard and Bouyer on a CDC 7600.

iii) The third technique, based on the transformation theory of elliptic integrals, provides the algorithms for the most recent set of computations. The most recent records are due separately to Gosper, Bailey, and Kanada. Gosper in 1985 calculated over 17 million digits (in fact over 17 million terms of the continued fraction) using a carefully orchestrated evaluation of Sum 1.

Bailey in January 1986 computed over 29 million digits using Algorithm 1 on a Cray 2 [2]. Kanada, using a related quadratic algorithm (due in basis to Gauss and made explicit by Brent [12] and Salamin [27]) and using Algorithm 1 for a check, verified 33,554,000 digits. This employed a HITACHI S-810/20, took roughly eight hours, and was completed in September of 1986. In January 1987 Kanada extended his computation to 2^{27} decimal places of π and the hundred million digit mark had been passed. The calculation took roughly a day and a half on a NEC SX2 machine. Kanada's most recent feat (Jan. 1988) was to compute 201,326,000 digits, which required only six hours on a new Hitachi S-820 supercomputer. Within the next few years many hundreds of millions of digits will no doubt have been similarly computed. Further discussion of the history of the computation of pi may be found in [5] and [9].

4. Complexity Concerns. One of the interesting morals from theoretical computer science is that many familiar algorithms are far from optimal. In order to be more precise we introduce the notion of *bit complexity*. Bit complexity counts the number of single operations required to complete an algorithm. The single-digit operations we count are $+, -, \times$. (We could, if we wished, introduce storage and logical comparison into the count. This, however, doesn't affect the order of growth of the algorithms in which we are interested.) This is a good measure of time on a serial machine. Thus, addition of two n-digit integers by the usual method has bit complexity $O(n)$, and straightforward uniqueness considerations show this to be asymptotically best possible.

Multiplication is a different story. Usual multiplication of two n-digit integers has bit complexity $O(n^2)$ and no better. However, it is possible to multiply two n-digit integers with complexity $O(n(\log n)(\log \log n))$. This result is due to Schönhage and Strassen and dates from 1971 [29]. It provides the best bound known for multiplication. No multiplication can have speed better than $O(n)$. Unhappily, more exact results aren't available.

The original observation that a faster than $O(n^2)$ multiplication is possible was due to Karatsuba in 1962. Observe that

$$(a + b10^n)(c + d10^n) = ac + [(a - b)(c - d) - ac - bd]10^n + bd10^{2n},$$

and thus multiplication of two $2n$-digit integers can be reduced to three multiplications of n-digit integers and a few extra additions. (Of course multiplication by 10^n is just a shift of the decimal point.) If one now proceeds recursively one produces a multiplication with bit complexity

$$O(n^{\log_2 3}).$$

Note that $\log_2 3 = 1.58\ldots < 2$.

We denote by $M(n)$ the bit complexity of multiplying two n-digit integers together by any method that is at least as fast as usual multiplication.

The trick to implementing high precision arithmetic is to get the multiplication right. Division and root extraction piggyback off multiplication using Newton's method. One may use the iteration

$$x_{k+1} = 2x_k - x_k^2 y$$

to compute $1/y$ and the iteration

$$x_{k+1} = \frac{1}{2}\left(x_k + \frac{y}{x_k}\right)$$

to compute $\sqrt{y}$. One may also compute $1/\sqrt{y}$ from

$$x_{k+1} = \frac{x_k(3 - yx_k^2)}{2}$$

and so avoid divisions in the computation of $\sqrt{y}$. Not only do these iterations converge quadratically but, because Newton's method is self-correcting (a slight perturbation in x_k does not change the limit), it is possible at the kth stage to work only to precision 2^k. If division and root extraction are so implemented, they both have bit complexity $O(M(n))$, in the sense that n-digit input produces n-digit accuracy in a time bounded by a constant times the speed of multiplication. This extends in the obvious way to the solution of any algebraic equation, with the startling conclusion that every algebraic number can be computed (to n-digit accuracy) with bit complexity $O(M(n))$. Writing down n-digits of $\sqrt{2}$ or $3\sqrt{7}$ is (up to a constant) no more complicated than multiplication.

The Schönhage-Strassen multiplication is hard to implement. However, a multiplication with complexity $O((\log n)^{2+\varepsilon} n)$ based on an ordinary complex (floating point) fast Fourier transform is reasonably straightforward. This is Kanada's approach, and the recent records all rely critically on some variations of this technique.

To see how the fast Fourier transform may be used to accelerate multiplication, let $x := (x_0, x_1, x_2, \ldots, x_{n-1})$ and $y := (y_0, y_1, y_2, \ldots, y_{n-1})$ be the representations of two high-precision numbers in some radix b. The radix b is usually selected to be some power of 2 or 10 whose square is less than the largest integer exactly representable as an ordinary floating-point number on the computer being used. Then, except for releasing each "carry," the product $z := (z_0, z_1, z_2, \ldots, z_{2n-1})$ of x and y may be written as

$$\begin{aligned}
z_0 &= x_0 y_0 \\
z_1 &= x_0 y_1 + x_1 y_0 \\
z_2 &= x_0 y_2 + x_1 y_1 + x_2 y_0 \\
&\vdots \\
z_{n-1} &= x_0 y_{n-1} + x_1 y_{n-2} + \cdots + x_{n-1} y_0 \\
&\vdots \\
z_{2n-3} &= x_{n-1} y_{n-2} + x_{n-2} y_{n-1} \\
z_{2n-2} &= x_{n-1} y_{n-1} \\
z_{2n-1} &= 0.
\end{aligned}$$

Now consider x and y to have n zeros appended, so that x, y, and z all have length $N = 2n$. Then a key observation may be made: the product sequence z is precisely the discrete convolution $C(x, y)$:

$$z_k = C_k(x, y) = \sum_{j=0}^{N-1} x_j y_{k-j},$$

where the subscript $k - j$ is to be interpreted as $k - j + N$ if $k - j$ is negative.

Now a well-known result of Fourier analysis may be applied. Let $F(x)$ denote the *discrete Fourier transform* of the sequence x, and let $F^{-1}(x)$ denote the inverse discrete Fourier transform of x:

$$F_k(x) := \sum_{j=0}^{N-1} x_j e^{-2\pi i j k / N}$$

$$F_k^{-1}(x) := \frac{1}{N} \sum_{j=0}^{N-1} x_j e^{2\pi i j k / N}.$$

Then the "convolution theorem," whose proof is a straightforward exercise, states that

$$F[C(x, y)] = F(x)F(y)$$

or, expressed another way,

$$C(x, y) = F^{-1}[F(x)F(y)].$$

Thus the entire multiplication pyramid z can be obtained by performing two forward discrete Fourier transforms, one vector complex multiplication and one inverse transform, each of length $N = 2n$. Once the real parts of the resulting complex numbers have been rounded to the nearest integer, the final multiprecision product may be obtained by merely releasing the carries modulo b. This may be done by starting at the end of the z vector and working backward, as in elementary school arithmetic, or by applying other schemes suitable for vector processing on more sophisticated computers.

A straightforward implementation of the above procedure would not result in any computational savings—in fact, it would be several times more costly than the usual "schoolboy" scheme. The reason this scheme is used is that the discrete Fourier transform may be computed much more rapidly using some variation of the well-known "fast Fourier transform" (FFT) algorithm [13]. In particular, if $N = 2^m$, then the discrete Fourier transform may be evaluated in only $5m 2^m$ arithmetic operations using an FFT. Direct application of the definition of the discrete Fourier transform would require 2^{2m+3} floating-point arithmetic operations, even if it is assumed that all powers of $e^{-2\pi i/N}$ have been precalculated.

This is the basic scheme for high-speed multiprecision multiplication. Many details of efficient implementations have been omitted. For example, it is possible to take advantage of the fact that the input sequences x and y and the output sequence z are all purely real numbers, and thereby sharply reduce the operation count. Also, it is possible to dispense with complex numbers altogether in favor of performing computations in fields of integers modulo large prime numbers. Interested readers are referred to [2], [8], [13], and [22].

When the costs of all the constituent operations, using the best known techniques, are totalled both Algorithms 1 and 2 compute n digits of π with bit complexity $O(M(n)\log n)$, and use $O(\log n)$ full precision operations.

The bit complexity for Sum 1, or for π using any of the arctan expansions, is between $O((\log n)^2 M(n))$ and $O(nM(n))$ depending on implementation. In each case, one is required to sum $O(n)$ terms of the appropriate series. Done naively, one obtains the latter bound. If the calculation is carefully orchestrated so that the terms are grouped to grow evenly in size (as rational numbers) then one can achieve the former bound, but with no corresponding reduction in the number of operations.

The Archimedean iteration of section 2 converges like $1/4^n$ so in excess of n iterations are needed for n-digit accuracy, and the bit complexity is $O(nM(n))$.

Almost any familiar transcendental number such as e, γ, $\zeta(3)$, or Catalan's constant (presuming the last three to be nonalgebraic) can be computed with bit complexity $O((\log n)M(n))$ or $O((\log n)^2 M(n))$. None of these numbers is known to be computable essentially any faster than this. In light of the previous observation that algebraic numbers are all computable with bit complexity $O(M(n))$, a proof that π cannot be computed with this speed would imply the transcendence of π. It would, in fact, imply more, as there are transcendental numbers which have complexity $O(M(n))$. An example is $0.10100100001\ldots$.

It is also reasonable to speculate that computing the nth digit of π is not very much easier than computing all the first n digits. We think it very probable that computing the nth digit of π cannot be $O(n)$.

5. The Miracle of Theta Functions

When I was a student, abelian functions were, as an effect of the Jacobian tradition, considered the uncontested summit of mathematics, and each of us was ambitious to make progress in this field. And now? The younger generation hardly knows abelian functions.

Felix Klein [21]

Felix Klein's lament from a hundred years ago has an uncomfortable timelessness to it. Sadly, it is now possible never to see what Bochner referred to as "the miracle of the theta functions" in an entire university mathematics program. A small piece of this miracle is required here [6], [11], [28]. First some standard notations. The *complete elliptic integrals of the first and second kind*, respectively,

$$K(k) := \int_0^{\pi/2} \frac{dt}{\sqrt{1 - k^2 \sin^2 t}} \qquad (5.1)$$

and

$$E(k) := \int_0^{\pi/2} \sqrt{1 - k^2 \sin^2 t}\, dt. \qquad (5.2)$$

The second integral arises in the rectification of the ellipse, hence the name elliptic integrals. The *complementary modulus* is

$$k' := \sqrt{1 - k^2}$$

and the *complementary integrals* K' and E' are defined by

$$K'(k) := K(k') \quad \text{and} \quad E'(k) := E(k').$$

The first remarkable identity is *Legendre's relation* namely

$$E(k)K'(k) + E'(k)K(k) - K(k)K'(k) = \frac{\pi}{2} \qquad (5.3)$$

(for $0 < k < 1$), which is pivotal in relating these quantities to pi. We also need to define two *Jacobian theta functions*

$$\Theta_2(q) := \sum_{n=-\infty}^{\infty} q^{(n+1/2)^2} \tag{5.4}$$

and

$$\Theta_3(q) := \sum_{n=-\infty}^{\infty} q^{n^2}. \tag{5.5}$$

These are in fact specializations with $(t = 0)$ of the general theta functions. More generally

$$\Theta_3(t, q) := \sum_{n=-\infty}^{\infty} q^{n^2} e^{\mathrm{i} m t} \qquad (\mathrm{im}\, t > 0)$$

with similar extensions of Θ_2. In Jacobi's approach these general theta functions provide the basic building blocks for elliptic functions, as functions of t (see [11], [39]).

The complete elliptic integrals and the special theta functions are related as follows. For $|q| < 1$

$$K(k) = \frac{\pi}{2} \Theta_3^2(q) \tag{5.6}$$

and

$$E(k) = (k')^2 \left[K(k) + k \frac{dK(k)}{dk} \right], \tag{5.7}$$

where

$$k := k(q) = \frac{\Theta_2^2(q)}{\Theta_3^2(q)}, \qquad k' := k'(q) = \frac{\Theta_3^2(-q)}{\Theta_3^2(q)} \tag{5.8}$$

and

$$q = e^{-\pi K'(k)/K(k)}. \tag{5.9}$$

The *modular function* λ is defined by

$$\lambda(t) := \lambda(q) := k^2(q) := \left[\frac{\Theta_2(q)}{\Theta_3(q)} \right]^4, \tag{5.10}$$

where

$$q := e^{i\pi t}.$$

We wish to make a few comments about modular functions in general before restricting our attention to the particular modular function λ. *Modular functions* are functions which are meromorphic in H, the upper half of the complex plane, and which are invariant under a group of linear fractional transformations, G, in the sense that

$$f(g(z)) = f(z) \qquad \forall g \in G.$$

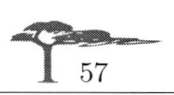

[Additional growth conditions on f at certain points of the associated fundamental region (see below) are also demanded.] We restrict G to be a subgroup of the *modular group* Γ where Γ is the set of all transformations w of the form

$$w(t) = \frac{at+b}{ct+d},$$

with a, b, c, d integers and $ad - bc = 1$. Observe that Γ is a group under composition. A *fundamental region* F_G is a set in H with the property that any element in H is uniquely the image of some element in F_G under the action of G. Thus the behaviour of a modular function is uniquely determined by its behaviour on a fundamental region.

Modular functions are, in a sense, an extension of elliptic (or doubly periodic) functions—functions such as sn which are invariant under linear transformations and which arise naturally in the inversion of elliptic integrals.

The definitions we have given above are not complete. We will be more precise in our discussion of λ. One might bear in mind that much of the theory for λ holds in considerably greater generality.

The *fundamental region* F we associate with λ is the set of complex numbers

$$F := \{\operatorname{im} t \geq 0\} \cap [\{|\operatorname{re} t| < 1 \text{ and }$$
$$|2t \pm 1| > 1\} \cup \{\operatorname{re} t = -1\} \cup \{|2t + 1| = 1\}].$$

The λ-*group* (or theta-subgroup) is the set of linear fractional transformations w satisfying

$$w(t) := \frac{at+b}{ct+d},$$

where a, b, c, d are integers and $ad - bc = 1$, while in addition a and d are odd and b and c are even. Thus the corresponding matrices are unimodular. What makes λ a λ-modular function is the fact that λ is meromorphic in $\{\operatorname{im} t > 0\}$ and that

$$\lambda(w(t)) := \lambda(t)$$

for all w in the λ-group, plus the fact that λ tends to a definite limit (possibly infinite) as t tends to a vertex of the fundamental region (one of the three points $(0, -1), (0, 0), (i, \infty)$). Here we only allow convergence from within the fundamental region.

Now some of the miracle of modular functions can be described. Largely because every point in the upper half plane is the image of a point in F under an element of the λ-group, one can deduce that any λ-modular function that is bounded on F is constant. Slightly further into the theory, but relying on the above, is the result that any two modular functions are algebraically related, and resting on this, but further again into the field, is the following remarkable result. Recall that q is given by (5.9).

THEOREM 1. *Let z be a primitive pth root of unity for p an odd prime. Consider the pth order modular equation for λ as defined by*

$$W_p(x, \lambda) := (x - \lambda_0)(x - \lambda_1) \cdots (x - \lambda_p), \qquad (5.11)$$

where
$$\lambda_i := \lambda(z^i q^{1/p}) \qquad i < p$$
and
$$\lambda_p := \lambda(q^p).$$

Then the function W_p is a polynomial in x and λ (**independent** of q), which has integer coefficients and is of degree $p + 1$ in both x and λ.

The modular equation for λ usually has a simpler form in the associated variables $u := x^{1/8}$ and $v := \lambda^{1/8}$. In this form the 5th-order modular equation is given by

$$\Omega_5(u, v) := u^6 - v^6 + 5u^2v^2(u^2 - v^2) + 4uv(1 - u^4v^4). \qquad (5.12)$$

In particular

$$\frac{\Theta_2(q^p)}{\Theta_3(q^p)} = v^2 \quad \text{and} \quad \frac{\Theta_2(q)}{\Theta_3(q)} = u^2$$

are related by an algebraic equation of degree $p + 1$.

The miracle is not over. The *pth-order multiplier* (for λ) is defined by

$$M_p(k(q), k(q^p)) := \frac{K(k(q^p))}{K(k(q))} = \left[\frac{\Theta_3(q^p)}{\Theta_3(q)}\right]^2 \qquad (5.13)$$

and turns out to be a rational function of $k(q^p)$ and $k(q)$.

One is now in possession of a pth-order algorithm for K/π, namely: Let $k_i := k(q^{p^i})$. Then

$$\frac{2K(k_0)}{\pi} = M_p^{-1}(k_0, k_1) M_p^{-1}(k_1, k_2) M_p^{-1}(k_2, k_3) \cdots.$$

This is an entirely algebraic algorithm. One needs to know the pth-order modular equation for λ to compute k_{i+1} from k_i and one needs to know the rational multiplier M_p. The speed of convergence ($O(c^{p^i})$, for some $c < 1$) is easily deduced from (5.13) and (5.9).

The function $\lambda(t)$ is 1-1 on F and has a well-defined inverse, λ^{-1}, with branch points only at 0, 1 and ∞. This can be used to provide a one line proof of the "big" Picard theorem that a nonconstant entire function misses at most one value (as does exp). Indeed, suppose g is an entire function and that it is never zero or one; then $\exp(\lambda^{-1}(g(z)))$ is a bounded entire function and is hence constant.

Littlewood suggested that, at the right point in history, the above would have been a strong candidate for a 'one line doctoral thesis'.

6. Ramanujan's Solvable Modular Equations.

Hardy [19] commenting on Ramanujan's work on elliptic and modular functions says

It is here that both the profundity and limitations of Ramanujan's knowledge stand out most sharply.

We present only one of Ramanujan's modular equations.

THEOREM 2.
$$\frac{5\Theta_3(q^{25})}{\Theta_3(q)} = 1 + r_1^{1/5} + r_2^{1/5}, \tag{6.1}$$

where for $i = 1$ and 2
$$r_i := \tfrac{1}{2}x\left(y \pm \sqrt{y^2 - 4x^3}\right)$$

with
$$x := \frac{5\Theta_3(q^5)}{\Theta_3(q)} - 1 \quad \text{and} \quad y := (x-1)^2 + 7.$$

This is a slightly rewritten form of entry 12(iii) of Chapter 19 of Ramanujan's *Second Notebook* (see [7], where Berndt's proofs may be studied). One can think of Ramanujan's quintic modular equation as an equation in the multiplier M_p of (5.13). The initial surprise is that it is solvable. The quintic modular relation for λ, W_5, and the related equation for $\lambda^{1/8}$, Ω_5, are both nonsolvable. The Galois group of the sixth-degree equation Ω_5 (see (5.12)) over $\mathbb{Q}(v)$ is A_5 and is nonsolvable. Indeed both Hermite and Kronecker showed, in the middle of the last century, that the solution of a general quintic may be effected in terms of the solution of the 5th-order modular equation (5.12) and the roots may thus be given in terms of the theta functions.

In fact, in general, the Galois group for W_p of (5.11) has order $p(p+1)(p-1)$ and is never solvable for $p \geq 5$. The group is quite easy to compute, it is generated by two permutations. If

$$q := e^{i\pi t}, \quad \text{then} \quad \tau \to \tau + 2 \quad \text{and} \quad \tau \to \frac{\tau}{(2\tau + 1)}$$

are both elements of the λ-group and induce permutations on the λ_i of Theorem 1. For any fixed p, one can use the q-expansion of (5.10) to compute the effect of these transformations on the λ_i, and can thus easily write down the Galois group.

While W_p is not solvable over $\mathbb{Q}(\lambda)$, it is solvable over $\mathbb{Q}(\lambda, \lambda_0)$. Note that λ_0 is a root of W_p. It is of degree $p + 1$ because W_p is irreducible. Thus the Galois group for W_p over $\mathbb{Q}(\lambda, \lambda_0)$ has order $p(p-1)$. For $p = 5, 7,$ and 11 this gives groups of order 20, 42, and 110, respectively, which are obviously solvable and, in fact, for general primes, the construction always produces a solvable group.

From (5.8) and (5.10) one sees that Ramanujan's modular equation can be rewritten to give λ_5 solvable in terms of λ_0 and λ. Thus, we can hope to find an explicit solvable relation for λ_p in terms of λ and λ_0. For $p = 3$, W_p is of degree 4 and is, of course, solvable. For $p = 7$, Ramanujan again helps us out, by providing a solvable seventh-order modular identity for the closely related *eta function* defined by

$$\eta(q) := q^{\frac{1}{12}} \prod_{n=1}^{\infty} (1 - q^{2n}).$$

The first interesting prime for which an explicit solvable form is not known is the "endecadic" ($p = 11$) case. We consider only prime values because for nonprime values the modular equation factors.

This leads to the interesting problem of mechanically constructing these equations. In principle, and to some extent in practice, this is a purely computational problem. Modular equations can be computed fairly easily from (5.11) and even more easily in the associated variables u and v. Because one knows a priori bounds on the size of the (integer) coefficients of the equations one can perform these calculations exactly. The coefficients of the equation, in the variables u and v, grow at most like 2^n. (See [11].) Computing the solvable forms and the associated computational problems are a little more intricate—though still in principle entirely mechanical. A word of caution however: in the variables u and v the endecadic modular equation has largest coefficient 165, a three digit integer. The endecadic modular equation for the intimately related function J (Klein's *absolute invariant*) has coefficients as large as

$$270909647855313899315632002810352263119290522227303 \times 2^{92}3^{19}5^{20}11^{2}53.$$

It is, therefore, one thing to solve these equations, it is entirely another matter to present them with the economy of Ramanujan.

The paucity of Ramanujan's background in complex analysis and group theory leaves open to speculation Ramanujan's methods. The proofs given by Berndt are difficult. In the seventh-order case, Berndt was aided by MACSYMA—a sophisticated algebraic manipulation package. Berndt comments after giving the proof of various seventh-order modular identities:

Of course, the proof that we have given is quite unsatisfactory because it is a verification that could not have been achieved without knowledge of the result. Ramanujan obviously possessed a more natural, transparent, and ingenious proof.

7. Modular Equations and Pi. We wish to connect the modular equations of Theorem 1 to pi. This we contrive via the function *alpha* defined by:

$$\alpha(r) := \frac{E'(k)}{K(k)} - \frac{\pi}{(2K(k))^2}, \tag{7.1}$$

where

$$k := k(q) \quad \text{and} \quad q := e^{-\pi\sqrt{r}}.$$

This allows one to rewrite Legendre's equation (5.3) in a one-sided form without the conjugate variable as

$$\frac{\pi}{4} = K\left[\sqrt{r}\,E - \left(\sqrt{r} - \alpha(r)\right)K\right]. \tag{7.2}$$

We have suppressed, and will continue to suppress, the k variable. With (5.6) and (5.7) at hand we can write a q-expansion for α, namely,

$$\alpha(r) = \frac{\dfrac{1}{\pi} - \sqrt{r}\,4\dfrac{\displaystyle\sum_{n=-\infty}^{\infty} n^2(-q)^{n^2}}{\displaystyle\sum_{n=-\infty}^{\infty}(-q)^{n^2}}}{\left[\displaystyle\sum_{n=-\infty}^{\infty} q^{n^2}\right]^4}, \tag{7.3}$$

and we can see that as r tends to infinity $q = e^{-\pi\sqrt{r}}$ tends to zero and $\alpha(r)$ tends to $1/\pi$. In fact

$$\alpha(r) - \frac{1}{\pi} \approx 8\left(\sqrt{r} - \frac{1}{\pi}\right)e^{-\pi\sqrt{r}}. \tag{7.4}$$

The key now is iteratively to calculate α. This is the content of the next theorem.

THEOREM 3. *Let* $k_0 := k(q)$, $k_1 := k(q^p)$ *and* $M_p := M_p(k_0, k_1)$ *as in* (5.13). *Then*

$$\alpha(p^2 r) = \frac{\alpha(r)}{M_p^2} - \sqrt{r}\left[\frac{k_0^2}{M_p^2} - pk_1^2 + \frac{pk_1'^2 k_1 \dot{M}_p}{M_p}\right],$$

where represents the full derivative of M_p *with respect to* k_0. *In particular*, α *is algebraic for rational arguments.*

We know that $K(k_1)$ is related via M_p to $K(k)$ and we know that $E(k)$ is related via differentiation to K. (See (5.7) and (5.13).) Note that $q \to q^p$ corresponds to $r \to p^2 r$. Thus from (7.2) some relation like that of the above theorem must exist. The actual derivation requires some careful algebraic manipulation. (See [11], where it has also been made entirely explicit for $p := 2, 3, 4, 5$, and 7, and where numerous algebraic values are determined for $\alpha(r)$.) In the case $p := 5$ we can specialize with some considerable knowledge of quintic modular equations to get:

THEOREM 4. *Let* $s := 1/M_5(k_0, k_1)$. *Then*

$$\alpha(25r) = s^2 \alpha(r) - \sqrt{r}\left[\frac{(s^2 - 5)}{2} + \sqrt{s(s^2 - 2s + 5)}\right].$$

This couples with Ramanujan's quintic modular equation to provide a derivation of Algorithm 2.

Algorithm 1 results from specializing Theorem 3 with $p := 4$ and coupling it with a quartic modular equation. The quartic equation in question is just two steps of the corresponding quadratic equation which is Legendre's form of the *arithmetic geometric mean iteration*, namely:

$$k_1 = \frac{2\sqrt{k}}{1 + k}.$$

An algebraic pth-order algorithm for π is derived from coupling Theorem 3 with a pth-order modular equation. The substantial details which are skirted here are available in [11].

8. Ramanujan's sum.
This amazing sum,

$$\frac{1}{\pi} = \frac{\sqrt{8}}{9801} \sum_{n=0}^{\infty} \frac{(4n)!}{(n!)^4} \frac{[1103 + 26390n]}{396^{4n}}$$

is a specialization ($N = 58$) of the following result, which gives reciprocal series for π in terms of our function alpha and related modular quantities.

THEOREM 5.

$$\frac{1}{\pi} = \sum_{n=0}^{\infty} \frac{\left(\frac{1}{4}\right)_n \left(\frac{1}{2}\right)_n \left(\frac{3}{4}\right)_n d_n(N)}{(n!)^3} x_N^{2n+1}, \qquad (8.1)$$

where,

$$x_N := \frac{4k_N(k_N')^2}{(1+k_N^2)^2} := \left(\frac{g_N^{12} + g_N^{-12}}{2}\right)^{-1},$$

with

$$d_n(N) = \left[\frac{\alpha(N) x_N^{-1}}{1 + k_N^2} - \frac{\sqrt{N}}{4} g_N^{-12}\right] + n\sqrt{N}\left(\frac{g_N^{12} - g_N^{-12}}{2}\right)$$

and

$$k_N := k(e^{-\pi\sqrt{N}}), \qquad g_N^{12} = (k_N')^2/(2k_N).$$

Here $(c)_n$ is the rising factorial: $(c)_n := c(c+1)(c+2)\cdots(c+n-1)$.

Some of the ingredients for the proof of Theorem 5, which are detailed in [11], are the following. Our first step is to write (7.2) as a sum after replacing the E by K and dK/dk using (5.7). One then uses an identity of Clausen's which allows one to write the square of a hypergeometric function $_2F_1$ in terms of a generalized hypergeometric $_3F_2$, namely, for all k one has

$$(1+k^2)\left[\frac{2K(k)}{\pi}\right]^2 = {}_3F_2\left(\frac{1}{4}, \frac{3}{4}, \frac{1}{2}, 1, 1 : \left(\frac{2}{g^{12} + g^{-12}}\right)^2\right)$$

$$= \sum_{n=0}^{\infty} \frac{\left(\frac{1}{4}\right)_n \left(\frac{3}{4}\right)_n \left(\frac{1}{2}\right)_n}{(1)_n (1)_n} \frac{\left(\frac{2}{g^{12} + g^{-12}}\right)^{2n}}{n!}.$$

Here g is related to k by

$$\frac{4k(k')^2}{(1+k^2)^2} = \left(\frac{g^{12} + g^{-12}}{2}\right)^{-1}$$

as required in Theorem 5. We have actually done more than just use Clausen's identity, we have also transformed it once using a standard hypergeometric substitution due to Kummer. Incidentally, Clausen was a nineteenth-century mathematician who, among other things, computed 250 digits of π in 1847 using Machin's formula. The desired formula (8.1) is obtained on combining these pieces.

Even with Theorem 5, our work is not complete. We still have to compute

$$k_{58} := k(e^{-\pi\sqrt{58}}) \quad \text{and} \quad \alpha_{58} := \alpha(58).$$

In fact

$$g_{58}^2 = \left(\frac{\sqrt{29} + 5}{2}\right)$$

is a well-known *invariant* related to the fundamental solution to Pell's equation for 29 and it turns out that

$$\alpha_{58} = \left(\frac{\sqrt{29} + 5}{2} \right)^6 (99\sqrt{29} - 444)(99\sqrt{2} - 70 - 13\sqrt{29}).$$

One can, in principle, and for $N := 58$, probably in practice, solve for k_N by directly solving the Nth-order equation

$$W_N(k_N^2, 1 - k_N^2) = 0.$$

For $N = 58$, given that Ramanujan [26] and Weber [38] have calculated g_{58} for us, verification by this method is somewhat easier though it still requires a tractable form of W_{58}. Actually, more sophisticated number-theoretic techniques exist for computing k_N (these numbers are called *singular moduli*). A description of such techniques, including a reconstruction of how Ramanujan might have computed the various singular moduli he presents in [26], is presented by Watson in a long series of papers commencing with [36]; and some more recent derivations are given in [11] and [30]. An inspection of Theorem 5 shows that all the constants in Series 1 are determined from g_{58}. Knowing α is equivalent to determining that the number 1103 is correct.

It is less clear how one explicitly calculates α_{58} in algebraic form, except by brute force, and a considerable amount of brute force is required; but a numerical calculation to any reasonable accuracy is easily obtained from (7.3) and 1103 appears! The reader is encouraged to try this to, say, 16 digits. This presumably is what Ramanujan observed. Ironically, when Gosper computed 17 million digits of π using Sum 1, he had no mathematical proof that Sum 1 actually converged to $1/\pi$. He compared ten million digits of the calculation to a previous calculation of Kanada et al. This verification that Sum 1 is correct to ten million places also provided the first complete proof that α_{58} is as advertised above. A nice touch—that the calculation of the sum should prove itself as it goes.

Roughly this works as follows. One knows enough about the exact algebraic nature of the components of $d_n(N)$ and x_N to know that if the purported sum (of positive terms) were incorrect, that before one reached 3 million digits, this sum must have ceased to agree with $1/\pi$. Notice that the components of Sum 1 are related to the solution of an equation of degree 58, but virtually no irrationality remains in the final packaging. Once again, there are very good number-theoretic reasons, presumably unknown to Ramanujan, why this must be so (58 is at least a good candidate number for such a reduction). Ramanujan's insight into this marvellous simplification remains obscure.

Ramanujan [26] gives 14 other series for $1/\pi$, some others almost as spectacular as Sum 1—and one can indeed derive some even more spectacular related series.* He gives almost no explanation as to their genesis, saying only that there are "corresponding theories" to the standard theory (as sketched in section 5) from which they follow. Hardy, quoting Mordell, observed that "it is unfortunate that Ramanujan has not developed the corresponding theories." By methods analogous

---

*(Added in proof) Many related series due to Borwein and Borwein and to Chudnovsky and Chudnovsky appear in papers in *Ramanujan Revisited*, Academic Press, 1988.

to those used above, all his series can be derived from the classical theory [11]. Again it is unclear what passage Ramanujan took to them, but it must in some part have diverged from ours.

We conclude by writing down another extraordinary series of Ramanujan's, which also derives from the same general body of theory,

$$\frac{1}{\pi} = \sum_{n=0}^{\infty} \binom{2n}{n}^3 \frac{42n+5}{2^{12n+4}}.$$

This series is composed of fractions whose numerators grow like 2^{6n} and whose denominators are exactly $16 \cdot 2^{12n}$. In particular this can be used to calculate the second block of n binary digits of π without calculating the first n binary digits. This beautiful observation, due to Holloway, results, disappointingly, in no intrinsic reduction in complexity.

9. Sources. References [7], [11], [19], [26], [36], and [37] relate directly to Ramanujan's work. References [2], [8], [9], [10], [12], [22], [24], [27], [29], and [31] discuss the computational concerns of the paper.

Material on modular functions and special functions may be pursued in [1], [6], [9], [14], [15], [18], [20], [28], [34], [38], and [39]. Some of the number-theoretic concerns are touched on in [3], [6], [9], [11], [16], [23], and [35].

Finally, details of all derivations are given in [11].

REFERENCES

1. M. Abramowitz and I. Stegun, Handbook of Mathematical Functions, Dover, New York, 1964.
2. D. H. Bailey, The Computation of π to 29,360,000 decimal digits using Borweins' quartically convergent algorithm, Math. Comput., 50 (1988) 283-96.
3. _____, Numerical results on the transcendence of constants involving π, e, and Euler's constant, Math. Comput., 50 (1988) 275-81.
4. A. Baker, Transcendental Number Theory, Cambridge Univ. Press, London, 1975.
5. P. Beckmann, A History of Pi, 4th ed., Golem Press, Boulder, CO, 1977.
6. R. Bellman, A Brief Introduction to Theta Functions, Holt, Reinhart and Winston, New York, 1961.
7. B. C. Berndt, Modular Equations of Degrees 3, 5, and 7 and Associated Theta Functions Identities, chapter 19 of Ramanujan's Second Notebook, Springer—to be published.
8. A. Borodin and I. Munro, The Computational Complexity of Algebraic and Numeric Problems, American Elsevier, New York, 1975.
9. J. M. Borwein and P. B. Borwein, The arithmetic-geometric mean and fast computation of elementary functions, SIAM Rev., 26 (1984), 351-365.
10. _____, An explicit cubic iteration for pi, BIT, 26 (1986) 123-126.
11. _____, Pi and the AGM—A Study in Analytic Number Theory and Computational Complexity, Wiley, N.Y., 1987.
12. R. P. Brent, Fast multiple-precision evaluation of elementary functions, J. ACM, 23 (1976) 242-251.
13. E. O. Brigham, The Fast Fourier Transform, Prentice-Hall, Englewood Cliffs, N.J., 1974.
14. A. Cayley, An Elementary Treatise on Elliptic Functions, Bell and Sons, 1885; reprint Dover, 1961.
15. A. Cayley, A memoir on the transformation of elliptic functions, Phil. Trans. T., 164 (1874) 397-456.
16. D. V. Chudnovsky and G. V. Chudnovsky, Padé and Rational Approximation to Systems of Functions and Their Arithmetic Applications, Lecture Notes in Mathematics 1052, Springer, Berlin, 1984.
17. H. R. P. Ferguson and R. W. Forcade, Generalization of the Euclidean algorithm for real numbers to all dimensions higher than two, Bull. AMS, 1 (1979) 912-914.
18. C. F. Gauss, Werke, Göttingen 1866-1933, Bd 3, pp. 361-403.

19. G. H. Hardy, Ramanujan, Cambridge Univ. Press, London, 1940.
20. L. V. King, On The Direct Numerical Calculation of Elliptic Functions and Integrals, Cambridge Univ. Press, 1924.
21. F. Klein, Development of Mathematics in the 19th Century, 1928, Trans Math Sci. Press, R. Hermann ed., Brookline, MA, 1979.
22. D. Knuth, The Art of Computer Programming, vol. 2: Seminumerical Algorithms, Addison-Wesley, Reading, MA, 1981.
23. F. Lindemann, Über die Zahl π, Math. Ann., 20 (1882) 213–225.
24. G. Miel, On calculations past and present: the Archimedean algorithm, Amer. Math. Monthly, 90 (1983) 17–35.
25. D. J. Newman, Rational Approximation Versus Fast Computer Methods, in Lectures on Approximation and Value Distribution, Presses de l'Université de Montreal, 1982, pp. 149–174.
26. S. Ramanujan, Modular equations and approximations to π, Quart. J. Math, 45 (1914) 350–72.
27. E. Salamin, Computation of π using arithmetic-geometric mean, Math. Comput., 30 (1976) 565–570.
28. B. Schoenberg, Elliptic Modular Functions, Springer, Berlin, 1976.
29. A. Schönhage and V. Strassen, Schnelle Multiplikation Grosser Zahlen, Computing, 7 (1971) 281–292.
30. D. Shanks, Dihedral quartic approximations and series for π, J. Number Theory, 14 (1982) 397–423.
31. D. Shanks and J. W. Wrench, Calculation of π to 100,000 decimals, Math Comput., 16 (1962) 76–79.
32. W. Shanks, Contributions to Mathematics Comprising Chiefly of the Rectification of the Circle to 607 Places of Decimals, G. Bell, London, 1853.
33. Y. Tamura and Y. Kanada, Calculation of π to 4,196,393 decimals based on Gauss-Legendre algorithm, preprint (1983).
34. J. Tannery and J. Molk, Fonctions Elliptiques, vols. 1 and 2, 1893; reprint Chelsea, New York, 1972.
35. S. Wagon, Is π normal?, The Math Intelligencer, 7 (1985) 65–67.
36. G. N. Watson, Some singular moduli (1), Quart. J. Math., 3 (1932) 81–98.
37. _____, The final problem: an account of the mock theta functions, J. London Math. Soc., 11 (1936) 55–80.
38. H. Weber, Lehrbuch der Algebra, Vol. 3, 1908; reprint Chelsea, New York, 1980.
39. E. T. Whittaker and G. N. Watson, A Course of Modern Analysis, 4th ed, Cambridge Univ. Press, London, 1927.

5. Strange series and high precision fraud

Discussion

This article is one of the authors' favorites. Mathematics has frequently seen alternating periods of theory building and periods of pathology hunting. The first without the second leads to sterile structures save for a few Grothendiecks. The second without the first runs out of steam, and one is left only with something akin to a pre-Linnaean taxonomy, in which no structures are to be discerned. In his wonderful article "Birds and Frogs,"[1] Freeman Dyson makes the same point forcibly and elegantly. In Dyson's terms, we are unabashed frogs who consume the droppings of friendly birds thereby enriching the pond's nutrients for future visiting birds:

> Some mathematicians are birds, others are frogs. Birds fly high in the air and survey broad vistas of mathematics out to the far horizon. They delight in concepts that unify our thinking and bring together diverse problems from different parts of the landscape. Frogs live in the mud below and see only the flowers that grow nearby. They delight in the details of particular objects, and they solve problems one at a time. I happen to be a frog, but many of my best friends are birds. The main theme of my talk tonight is this. Mathematics needs both birds and frogs. Mathematics is rich and beautiful because birds give it broad visions and frogs give it intricate details. Mathematics is both great art and important science, because it combines generality of concepts with depth of structures. It is stupid to claim that birds are better than frogs because they see farther, or that frogs are better than birds because they see deeper. The world of mathematics is both broad and deep, and we need birds and frogs working together to explore it.

Source

J.M. Borwein and P.B. Borwein, "Strange series and high precision fraud," *MAA Monthly*, **99** (1992), 622–640.

---

[1] Published in the *Notices of the AMS*, February 2009, 212–223, the paper is available at http://www.ams.org/notices/200902/rtx090200212p.pdf. Great frogs include von Neumann and Besicovitch, birds include Yang and Witten.

Strange Series and High Precision Fraud

J. M. Borwein and P. B. Borwein

INTRODUCTION. Five of the following twelve series approximations are exact. The remaining seven are not identities but are approximations that are correct to at least 30 digits. One in fact is correct to over 18,000 digits and another to in excess of a billion digits. The reader is invited to separate the true from the bogus. (For answers see the end of the introduction.) Most of these series are easily amenable to high precision calculation in one's favorite high precision environment, such as Maple or MACSYMA, and provide examples of "caveat computat." Things are not always as they appear.

Sum 1

$$\sum_{n=1}^{\infty} \frac{a(2^n)}{2^n} \doteq \frac{1}{99}$$

where $a(n)$ counts the number of odd digits in odd places in the decimal expansion of n. ($a(901) = 2$, $a(210) = 0$, $a(811) = 1$, here the 1st digit is the 1st to the left of the decimal point.)

Sum 2

$$\sum_{n=1}^{\infty} \frac{a(n)}{10^n} \doteq \frac{10}{99}$$

where $a(n)$ is as above.

Sum 3

$$\sum_{n=1}^{\infty} b(n)\left(\frac{1}{n^2} - \frac{1}{(n+1)^2}\right) \doteq \frac{25\pi^2}{297}$$

where $b(n)$ counts the number of odd digits in n ($b(901) = 2$, $b(811) = 2$, ($b(406) = 0$).

Sum 4

$$\sum_{n=1}^{\infty} \frac{c(n)}{2^n} \doteq \frac{511}{8184}$$

where $c(n) := 32c_1(n) - c_2(n)/32$, and $c_1(n)$ counts the number of nines in n, while $c_2(n)$ counts the number of eights in n ($c(8199) = 32 \cdot 2 - 1/32$).

Chapter 5

Sum 5

$$1 + \sum_{n=1}^{\infty} \frac{(-1)^{\delta(n)}}{16^{\delta(n)}(D(n))^4} \doteq 2\frac{(e^{\pi/2} - e^{-\pi/2})}{\pi^2}$$

where $\delta(n)$ is the number of ones in the binary expansion of n and $D(n)$ is the product $\prod_i \max\{i\delta_i(n), 1\}$ where $\delta_i(n)$ is the ith binary digit of n ($\delta(1011_2) = 3$, $D(1011_2) = 4 \cdot 2 \cdot 1 = 8$).

Sum 6

$$\sum_{n=1}^{\infty} \frac{e(n)}{n(n+1)} \doteq \frac{10}{99} \log 10$$

where $e(n)$ "reflects" n through the decimal point ($e(123) = .321$, $e(90140) = .04109$).

Sum 7

$$\sum_{n=1}^{\infty} \frac{b(2^n)}{2^n} \doteq \frac{1}{9}$$

where $b(n)$ counts the number of odd digits in n (as in Sum 3).

Sum 8

$$\sum_{n=1}^{\infty} \frac{e(2^n)}{2^n} \doteq \frac{3166}{3069}$$

where $e(n)$ counts the number of even digits in n.

Sum 9

$$\sum_{n=1}^{\infty} \frac{\lfloor n \tanh \pi \rfloor}{10^n} \doteq \frac{1}{81}$$

where $\lfloor \ \rfloor$ is the greatest integer function ($\lfloor 3.7 \rfloor = 3$).

Sum 10

$$\sum_{n=1}^{\infty} \frac{\lfloor ne^{\pi\sqrt{163/9}} \rfloor}{2^n} \doteq 1280640$$

Sum 11

$$\sum_{-\infty}^{\infty} \frac{1}{10^{(n/100)^2}} \doteq 100\sqrt{\frac{\pi}{\log 10}}$$

Sum 12

$$\left(\frac{1}{10^5} \sum_{n=-\infty}^{\infty} e^{-(n^2/10^{10})}\right)^2 \doteq \pi$$

These sums break into four types. Sums 2, 3, 4, 5, and 6 are all specializations of generating functions for digit sums, more-or-less of the type:

$$\prod_{n=0}^{\infty} (1 + q^{2^n}) = \sum_{n=0}^{\infty} x^{\delta(n)} q^n \qquad (1.1)$$

where $\delta(n)$ counts the number of ones in the binary expansion of n. These are treated in section 2. See also [14].

Sums 1 and 7 are related to a problem independently due to E. Levine (*College Math Journal*, Vol. 19, number 5, 1989) and to D. Bowman and T. White (*Amer. Math. Monthly*, Vol. 96 1989, p. 743), which asks if

$$\sum_{n=0}^{\infty} \frac{g(2^n)}{2^n} = \frac{2}{9}$$

where $g(n)$ counts the number of digits ≥ 5 in n. The key to the solution we provide is due to our colleague A. C. Thompson. See section 3.

The sums 8, 9 and 10 revolve around the fact that

$$\sum_{n=0}^{\infty} w^{\lfloor n\alpha \rfloor} q^n$$

has a particularly attractive and rapidly convergent generating function that is related to the continued fraction expansion of α. This is essentially an observation of Mahler's [11], though the development we offer in section 4 is quite distinct. See also [10], [3]. This is closely related to problem #E3353 in the *MAA Monthly* due to H. Diamond [6].

The last section deals with series like Sums 11 and 12. There are consequences of the fact that $f(t) := \sum_{n=-\infty}^{\infty} e^{-n^2 t \pi}$ is a modular form and satisfies a simple functional equation linking $f(t)$ and $f(1/t)$.

The fradulent series are: Sum 2 (correct to 99 digits), Sum 4 (correct to 240 digits), Sum 8 (correct to 30 digits), Sum 9 (correct to 267 digits), Sum 10 (correct to at least half a billion digits), Sum 11 (correct to at least 18,000 digits), and Sum 12 (correct to at least 42 billion digits).

GENERATING FUNCTIONS—PART ONE. Many digit sums are generated by the following type of argument.

Example 2.1. Let $b(n)$ count the number of odd digits in n base 10 (as in Sums 3 and 7). Then for $|q| < 1$,

$$\sum_{n=0}^{\infty} x^{b(n)} q^n = \prod_{n=0}^{\infty} (1 + xq^{10^n} + q^{2 \cdot 10^n} + xq^{3 \cdot 10^n} + q^{4 \cdot 10^n} + xq^{5 \cdot 10^n} + q^{6 \cdot 10^n}$$

$$+ xq^{7 \cdot 10^n} + q^{8 \cdot 10^n} + xq^{9 \cdot 10^n})$$

$$=: \prod_{n=0}^{\infty} r(x, q^{10^n}). \qquad (2.1)$$

To see this, observe that in the expansion of the product each power of q^m arises in exactly one way. This is just the unique expansion of m base 10. The coefficient of q^m is just a product of x's, one for each odd digit in m. If we differentiate (2.1) with respect to x as is legitimate since $b(n) = O(n)$ and the derivatives converge uniformly, we get

$$\frac{\sum_{n=0}^{\infty} b(n) x^{b(n)-1} q^n}{\sum_{n=0}^{\infty} x^{b(n)} q^n} = \sum_{n=0}^{\infty} \frac{q^{10^n} + q^{3\cdot 10^n} + q^{5\cdot 10^n} + q^{7\cdot 10^n} + q^{9\cdot 10^n}}{1 + xq^{10^n} + q^{2\cdot 10^n} + \cdots + q^{8\cdot 10^n} + xq^{9\cdot 10^n}} \quad (2.2)$$

and at $x := 1$

$$\frac{\sum_{n=0}^{\infty} b(n) q^n}{(1-q)^{-1}} = \sum_{n=0}^{\infty} \frac{q^{10^n} + q^{3\cdot 10^n} + \cdots + q^{9\cdot 10^n}}{1 + q^{10^n} + q^{2\cdot 10^n} + \cdots + q^{9\cdot 10^n}} \quad (2.3)$$

$$= \sum_{n=0}^{\infty} \frac{q^{10^n}}{1 + q^{10^n}}$$

$$=: \sum_{n=0}^{\infty} R(q^{10^n})$$

where the second last equality follows on factoring each term. It is apparent from this representation for example that

$$\sum_{n=0}^{\infty} b(n) q^n = \frac{1}{1-q} \left(\frac{q^1}{1+q^1} + \frac{q^{10}}{1+q^{10}} \right) + O(q^{100}). \quad (2.4)$$

We need the following observation which we encapsulate as Lemma 2.1.

Lemma 2.1. *Suppose $R(q)$ is a non-negative, measurable function on $[0,1]$. If $b > 1$ and*

$$f(q) := \sum_{n=0}^{\infty} R(q^{b^n}) \qquad |q| < 1$$

then

$$\int_0^1 \frac{f(q)}{q} \, dq = \frac{b}{b-1} \int_0^1 \frac{R(q)}{q} \, dq.$$

Proof:

$$\int_0^1 \frac{f(q)}{q} \, dq = \int_0^1 \sum_{n=0}^{\infty} \frac{R(q^{b^n})}{q} \, dq$$

$$= \sum_{n=0}^{\infty} \int_0^1 \frac{R(q^{b^n})}{q} \, dq$$

$$= \sum_{n=0}^{\infty} \int_0^1 \frac{S(q^{b^n}) q^{b^n}}{q} \, dq$$

where $S(q) := R(q)/q$.

Now set $u = q^{b^n}$ and observe that

$$\int_0^1 \frac{f(q)}{q} \, dq = \sum_{n=0}^{\infty} \int_0^1 \frac{S(u)}{b^n} \, du$$

and the lemma is proved. (The interchange of sum and integral is just the monotone convergence theorem.) ∎

From (2.3) we have

$$\sum_{n=0}^{\infty} b(n)q^{n-1}(1-q) = \sum_{n=0}^{\infty} \frac{R(q^{10^n})}{q} \qquad (2.5)$$

and with Lemma 2.1,

$$\sum_{n=1}^{\infty} b(n)\left(\frac{1}{n} - \frac{1}{n+1}\right) = \frac{10}{9}\int_0^1 \frac{1}{1+q}\,dq$$

or

$$\sum_{n=1}^{\infty} \frac{b(n)}{n(n+1)} = \frac{10}{9}\log 2. \qquad (2.6)$$

Indeed this process iterates, in the sense that we can keep dividing by q and integrating in (2.5). This yields with some effort the following

Sum 13. For k a positive integer

$$\sum_{n=1}^{\infty} b(n)\left(\frac{1}{n^k} - \frac{1}{(n+1)^k}\right) = \frac{10^k}{10^k - 1}\alpha(k)$$

where, α is the alternating zeta function,

$$\alpha(s) := (1 - 2^{1-s})\zeta(s) := \sum_{n=1}^{\infty} \frac{(-1)^{n+1}}{n^s}.$$

Note that Sum 3 is just the $k := 2$ case of the above, while $k := 1$ gives (2.6).

A direct derivation of Sum 13 valid for non-integer k can be based on the fact that:

$$\alpha(s)\sum_{n=1}^{\infty} a_n n^{-s} = \sum_{n=1}^{\infty} b_n n^{-s}$$

if and only if

$$\sum_{n=1}^{\infty} a_n \frac{x^n}{1+x^n} = \sum_{n=1}^{\infty} b_n x^n.$$

This identity is now coupled with (2.3). See [18].

Example 2.2. The generating function for q, the number of odd digits in odd places (as in Sums 1 and 2), is given by

$$\sum_{n=0}^{\infty} x^{a(n)} q^n = \prod_{n=0}^{\infty} r(x, q^{10^{2n}})$$

where

$$r(x,q) := (1 + xq + q^2 + xq^3 + q^4 + \cdots + xq^9)$$
$$\cdot (1 + q^{10} + q^{2\cdot 10} + q^{3\cdot 10} + \cdots + q^{9\cdot 10})$$

and leads, as in (2.3), to the series

$$\sum_{n=0}^{\infty} a(n)q^n = \frac{1}{1-q}\sum_{n=0}^{\infty} \frac{q^{100^n}}{1+q^{100^n}}. \qquad (2.7)$$

Sum 2 now appears on taking $q := \frac{1}{10}$ and using the first term of the above expansion. It is apparent that the remainder is positive of size very close to $\frac{1}{9} \cdot 10^{-99}$. This gives the nature of the estimate in Sum 2.

In similar fashion

$$\sum_{n=0}^{\infty} A_k(n) q^n = \frac{1}{1-q} \sum_{n=0}^{\infty} \frac{q^{(10^k)^n}}{1 + q^{(10^k)^n}} \qquad (2.8)$$

is the generating function for the number of odd digits in the 1st, $(k+1)$th, $(2k+1)$th places of k. So with $k = 10$, for example

$$\sum_{n=0}^{\infty} \frac{A_{10}(n)}{10^n} = \frac{10}{99} + \varepsilon_n \qquad (2.9)$$

where $0 < |\varepsilon_n| < \frac{10}{9} \cdot 10^{-10^{10}}$, and the above approximation is correct to over a billion digits. ∎

Example 2.3. The number of times the digit $i > 0$ occurs in n has generating function

$$\sum_{n=0}^{\infty} g(n) q^n = \frac{1}{1-q} \sum_{n=0}^{\infty} \frac{q^{i \cdot 10^n}}{1 + q^{10^n} + \cdots + q^{9 \cdot 10^n}}.$$

So the generating function for $c(n)$ in Sum 4 is just

$$\sum_{n=0}^{\infty} c(n) q^n = \frac{1}{1-q} \sum_{n=0}^{\infty} \frac{32 q^{9 \cdot 10^n} - \dfrac{q^{8 \cdot 10^n}}{32}}{1 + q^{10^n} + \cdots + q^{9 \cdot 10^n}}.$$

At $q := \frac{1}{2}$, the second term vanishes to give

$$\sum_{n=0}^{\infty} \frac{c(n)}{2^n} = \frac{1}{1-q} \left(\frac{32 q^9 - \dfrac{q^{8 \cdot 10^n}}{32}}{1 + \cdots + q^9} \right) + O(q^{800})$$

$$= \frac{511}{8184} + \varepsilon$$

where $\varepsilon < 10^{-241}$.

Example 2.4. The generating function which reverses digits, as in Sum 6, is

$$\sum_{n=0}^{\infty} x^{e(n)} q^n = \prod_{n=0}^{\infty} \left(1 + x^{1/10^{n+1}} q^{10^n} + \cdots + x^{9/10^{n+1}} q^{9 \cdot 10^n} \right). \qquad (2.10)$$

So

$$\sum_{n=0}^{\infty} e(n) q^n = \frac{1}{1-q} \sum_{n=0}^{\infty} \frac{\dfrac{1}{10^{n+1}} q^{10^n} + \cdots + \dfrac{9}{10^{n+1}} q^{9 \cdot 10^n}}{1 + q^{10^n} + \cdots + q^{9 \cdot 10^n}} \qquad (2.11)$$

and as in Lemma 2.1

$$\sum_{n=1}^{\infty} \frac{e(n)}{n(n+1)} = \frac{10}{99} \log 10.$$

There are very many analogues of these results. All have variations in different bases. The binary digit counting functions δ has generating function

$$\sum_{n=0}^{\infty} x^{\delta(n)} q^n = \prod_{n=0}^{\infty} \left(1 + xq^{2^n}\right) \qquad (2.12)$$

and

$$\sum_{n=0}^{\infty} \delta(n) q^n = \frac{1}{1-q} \sum_{n=0}^{\infty} \frac{q^{2^n}}{1+q^{2^n}} \qquad (2.13)$$

whence

$$\sum_{n=1}^{\infty} \frac{\delta(n)}{n(n+1)} = 2 \log 2. \qquad (2.14)$$

(See the Putnam examinations of 1981, 1984 and 1987.) As in Example 2.1 we have Sum 14.

Sum 14. Let $\delta(n)$ denote the sum of the binary digits of n. Then

$$\sum_{n=1}^{\infty} \delta(n) \left(\frac{1}{n^k} - \frac{1}{(n+1)^k} \right) = \left(\frac{2^k}{2^k - 1} \right) \alpha(k)$$

where $\alpha(k)$ is the alternating zeta function.

The sum of the decimal digits of n denoted $s(n)$ has generating function

$$\sum_{n=0}^{\infty} x^{s(n)} q^n = \prod_{n=0}^{\infty} \left(1 + xq^{10^n} + x^2 q^{2 \cdot 10^n} + \cdots + x^9 q^{9 \cdot 10^n}\right) \qquad (2.15)$$

from which we deduce that

$$\sum_{n=1}^{\infty} \frac{s(n)}{n(n+1)} = \frac{10}{9} \log 10. \qquad (2.16)$$

Loxton and van der Poorten [10] and Mahler [11] treat transcendence questions for functions, with power series expansions at zero which satisfy functional equations. From these results, one knows that if f, holomorphic at zero and not an algebraic function, satisfies a function equation of the form

$$f(q^m) = f(q) + R(q) \qquad (2.17)$$

where m is an integer and R is a rational function, then $f(\alpha)$ is transcendental for algebraic α. From this we deduce that the exact answers in Sum 2, Sum 4 and Sum 8, are transcendental. This can also be deduced easily from Roth's Theorem [8].

GENERATING FUNCTIONS—PART TWO. A second type of digit function arises as follows.

Example 3.1. Let $\delta(n)$ as before, denote the sum of the binary digits of n, and let $\rho(n) := \prod\{S_i: i\text{th binary digit of } n \neq 0\}$ and $\rho(0) := 1$, where S_i is a given sequence and the product is taken over those binary digits of n which equal one. Then formally

$$\sum_{n=0}^{\infty} \frac{x^{\delta(n)} q^n}{\rho(n)} = \prod_{n=0}^{\infty} \left(1 + \frac{x}{S_{n+1}} q^{2^n}\right) \qquad (3.1)$$

and

$$\sum_{n=0}^{\infty} \frac{x^{\delta(n)}}{\rho(n)} = \prod_{n=0}^{\infty} \left(1 + \frac{x}{S_{n+1}}\right).$$

Example 3.2. Let $\delta(n)$ denote the sum of the binary digits of n, and let

$$D(n) = \prod i$$

where the product is taken over those i where the ith binary digit of n is non-zero (as in Sum 5). So, if $0 < n_1 < n_2 < \cdots < n_k$,

$$D(2^{n_1} + 2^{n_2} + \cdots + 2^{n_k}) = (n_1 + 1)(n_2 + 1) \cdots (n_k + 1).$$

Then as in Example 3.1, starting with

$$F_q(x) := x \prod_{n=1}^{\infty} \left(1 - \frac{x^2}{n^2} q^{2^{n-1}}\right) = x \prod_{n=0}^{\infty} \left(1 - \frac{x^2}{(n+1)^2} q^{2^n}\right) \quad (3.2)$$

we have, for $|x| < 1$,

$$F_1(x) = \frac{\sin \pi x}{\pi} = \sum_{n=0}^{\infty} \frac{(-1)^{\delta(n)} x^{2\delta(n)+1}}{[D(n)]^2} \quad (3.3)$$

and at $x := \frac{1}{2}$

$$\frac{2}{\pi} = \sum_{n=0}^{\infty} \frac{(-1)^{\delta(n)}}{4^{\delta(n)} [D(n)]^2}. \quad (3.4)$$

Similarly, starting with

$$\frac{(\sin \pi x)(\sinh \pi x)}{\pi^2} = x^2 \prod_{n=1}^{\infty} \left(1 - \frac{x^4}{n^4}\right) \quad (3.5)$$

$$= \sum_{n=0}^{\infty} \frac{(-1)^{\delta(n)} x^{4\delta(n)+2}}{[D(n)]^4},$$

we have, at $x := \frac{1}{2}$,

$$2\left(\frac{e^{\pi/2} - e^{-\pi/2}}{\pi^2}\right) = \sum_{n=0}^{\infty} \frac{(-1)^{\delta(n)}}{16^{\delta(n)} [D(n)]^4}, \quad (3.6)$$

which is Sum 5.

Example 3.3. Let $t(n) := \Sigma i$, where the sum is taken over the non-zero digits on n base 2. So $t(1011_2) = 4 + 0 + 2 + 1 = 7$. Then

$$\prod_{n=0}^{\infty} (1 - x^{n+1} q^{2^n}) = \sum_{n=0}^{\infty} (-1)^{\delta(n)} x^{t(n)} q^n. \quad (3.7)$$

So

$$\sum_{n=0}^{\infty} (-1)^{\delta(n)} x^{t(n)} = \prod_{n=1}^{\infty} (1 - x^n) = \sum_{-\infty}^{\infty} (-1)^n x^{(3n+1)n/2} \quad (3.8)$$

on using Euler's pentagonal number theorem [2] and on integrating, from zero to one,

$$\sum_{n=0}^{\infty} \frac{(-1)^{\delta(n)}}{t(n)+1} = \sum_{-\infty}^{\infty} \frac{2(-1)^n}{3n^2+n+2}. \qquad (3.9)$$

4. CONTINUED FRACTION EXPANSIONS. The identities of this section are based on the two functions

$$G_\alpha(z,w) := \sum_{n=1}^{\infty} z^n w^{\lfloor n\alpha \rfloor} \qquad (4.1)$$

and

$$F_\alpha(z,w) := \sum_{n=1}^{\infty} z^n \sum_{m=1}^{\lfloor n\alpha \rfloor} w^m \qquad (4.2)$$

where α is a non-negative real number and $\lfloor n\alpha \rfloor$ is the integer part of $n\alpha$, while z and w are complex with modulus so as to ensure convergence. The function F_α was studied by Mahler [11] and is obviously related to G_α by

$$F_\alpha(z,w) + \frac{w}{1-w} G_\alpha(z,w) = \frac{zw}{(1-z)(1-w)} \qquad (4.3)$$

for $|z|, |w| < 1$. Van der Poorten [10] comments that Mahler's paper has been largely overlooked. In [3] we explore these matters further. Note that for positive z and w, F_α is strictly increasing as a function of α.

For irrational α we will use the infinite continued fraction approximations generated by

(a) $\qquad p_{n+1} := p_n a_{n+1} + p_{n-1} \qquad p_0 := a_0 = \lfloor \alpha \rfloor, \quad p_{-1} := 1$
(b) $\qquad q_{n+1} := q_n a_{n+1} + q_{n-1} \qquad q_0 := 1, \qquad q_{-1} := 0 \qquad (4.4)$

for $n \geq 0$ where

$$\alpha = [a_0, a_1, \ldots, a_n, a_{n+1}, \ldots]$$
$$= a_0 + \cfrac{1}{a_1 + \cfrac{1}{a_2 + \cfrac{1}{a_3 + \cdots}}}$$

so that each a_i is integral, $a_0 \geq 0$ and $a_n \geq 1$ for $n \geq 1$. Then for $n \geq 0$ p_{2n}/q_{2n} increases to α while p_{2n+1}/q_{2n+1} decreases to α and

$$\frac{1}{q_n(q_n + q_{n+1})} < \left| \alpha - \frac{p_n}{q_n} \right| < \frac{1}{q_n q_{n+1}}. \qquad (4.5)$$

All of this is standard and may be found in [8], [9], or [16]. We will avoid using finite continued fractions which arise only for rational α. Let us write $q_n \alpha - p_n$ as ε_n. By (4.5) and (4.4)

$$|\varepsilon_{n+1}| < \frac{1}{q_n + q_{n+1}} < |\varepsilon_n| < \frac{1}{q_{n+1}} \leq 1.$$

A key lemma is:

Lemma 4.1. *For irrational $\alpha > 0$ and n, N in $\mathbf{N}$*

(a) $\qquad \lfloor n\alpha + \varepsilon_N \rfloor = \lfloor n\alpha \rfloor \qquad$ for $n < q_{N+1}$

(b) $\qquad \lfloor n\alpha + \varepsilon_N n \rfloor = \lfloor n\alpha \rfloor + (-1)^N \qquad$ for $n = q_{N+1}$.

Proof: Suppose N is even (the odd case is entirely parallel). Then $\varepsilon_N > 0$ and (a) fails when
$$n\alpha + \varepsilon_N > m > n\alpha \quad \text{for some } m \text{ in } \mathbf{N}. \qquad (4.6)$$
As $\alpha > p_N/q_N$, we have an integer k with
$$(n + q_N)\varepsilon_N > mq_N - np_N = k > 0.$$
If $k \geq 2$ then $n + q_n > 2/\varepsilon_N > 2q_{N+1}$ and $n > q_{N+1}$.

If $k = 1$ we have
$$p_N q_N - q_N p_N = 0, \qquad p_{N+1} q_N - q_{N+1} p_N = 1,$$
so that the linear Diophantine equation $mq_N - np_N = 1$ has general solution $m = p_{N+1} + sp_N$, $n = q_{N+1} + sq_N$ for s integer. However, $n + q_N > 1/\varepsilon_N > q_{N+1}$ so that s is non-negative. This establishes (a). For $n = q_{N+1}$ we have
$$q_{N+1}\alpha < p_{N+1} < q_{N+1}\alpha + \varepsilon_N < p_{N+1} + 1$$
since $p_{N+1} > q_{N+1}\alpha$ and $0 < \varepsilon_{N+1} + \varepsilon_N < 1$. This yields (b). ∎

Theorem 4.1.

(a) *For rational $\alpha = p/q$ (reducible or irreducible)*
$$(1 - z^q w^p) G_\alpha(z, w) = \sum_{j=1}^{q} z^j w^{\lfloor jp/q \rfloor}.$$

(b) *For irrational α and $N > 0$*
$$(1 - z^{q_N} w^{p_N}) G_\alpha(z, w)$$
$$= \sum_{n=1}^{q_n} z^n w^{\lfloor n\alpha \rfloor} + (-1)^N \left(\frac{w-1}{w}\right) z^{q_N} w^{p_N} z^{q_{N+1}} w^{p_{N+1}} + R_N(z, w)$$

with
$$|R_N(z, w)| \leq |1 - w| \frac{|z|^{q_{N+1} + q_N + 1}}{1 - |z|}.$$

Proof:

(a) $\qquad G_\alpha(z, w) = \sum_{k=0}^{\infty} \sum_{j=1}^{q} z^{qk+j} w^{kp + \lfloor j(p/q) \rfloor}$
$$= \sum_{k=0}^{\infty} (z^q w^p)^k \sum_{j=1}^{q} z^j w^{\lfloor j(p/q) \rfloor},$$

(b) $\qquad (1 - z^{q_N} w^{p_N}) G_\alpha(z, w) - \sum_{n=1}^{q_N} z^n w^{\lfloor n\alpha \rfloor}$
$$= \sum_{n=1}^{\infty} z^{n+q_N} \{ w^{\lfloor (n+q_N)\alpha \rfloor} - w^{p_N + \lfloor n\alpha \rfloor} \}$$
$$= \sum_{n=1}^{\infty} z^{n+q_N} w^{\lfloor n\alpha \rfloor + p_N} \{ w^{\lfloor n\alpha + \varepsilon_N \rfloor - \lfloor n\alpha \rfloor} - 1 \}.$$

By the proof of Lemma 4.1, the first non-zero term in this last expression is $(-1)^N(w-1)/w)z^{q_N+q_{N+1}}w^{p_N+p_{N+1}}$ while the other terms are dominated by $|z|^n|1-w|$ with $n > q_N + q_{N+1}$. ∎

For fixed $\alpha > 0$ we write
$$P_N := \sum_{n=1}^{q_N} z^n w^{\lfloor n\alpha \rfloor}, \qquad Q_N := 1 - z^{q_N}w^{p_N}$$

and observe that Theorem 4.1 shows that

$$G_\alpha - \frac{P_N}{Q_N} = (-1)^N \left(\frac{w-1}{w}\right) \frac{z^{q_N}w^{p_N}z^{q_{N+1}}w^{p_{N+1}}}{Q_N} + O(z^{q_N+q_{N+1}} + 1) \quad (4.7)$$

for α irrational (while $G_\alpha = P_N/Q_N$ for rational α). Thus as a function of z P_N/Q_N is the main diagonal Padé approximation to G_α of order q_N.

Corollary 4.1. *For irrational $\alpha > 0$*

$$G_\alpha(z,w) = \frac{zw^{p_0}}{1-zw^{p_0}} - \frac{1-w}{w}\sum_{n=0}^{\infty}(-1)^n \frac{z^{q_n}w^{p_n}z^{q_{n+1}}w^{p_{n+1}}}{(1-z^{q_n}w^{p_n})(1-z^{q_{n+1}}w^{p_{n+1}})}. \quad (4.8)$$

Proof: Let $A_N := P_{N+1}Q_N - Q_{N+1}P_N$. Then A_N is a polynomial of degree at most $q_{N+1} + q_N$ in z. From (4.7) we see that

$$\frac{P_{N+1}}{Q_{N+1}} - \frac{P_N}{Q_N} = \frac{A_N}{Q_N Q_{N+1}} = (-1)^N \left(\frac{w-1}{w}\right) \left\{\frac{z^{q_N}w^{p_N}z^{q_{N+1}}w^{p_{N+1}}}{Q_N Q_{N+1}}\right\}.$$

On summing from zero to infinity we produce (4.8). ∎

This is derived by Mahler for $\alpha \in (0,1)$ in [11].

Corollary 4.2. *For irrational $\alpha > 0$ and for $w \neq 1$*

$$F_\alpha(z,w) = \frac{zw}{(1-z)(1-w)}\frac{1-w^{p_0}}{1-zw^{p_0}} + \sum_{n=0}^{\infty}\frac{(-1)^n z^{q_n}w^{p_n}z^{q_{n+1}}w^{p_{n+1}}}{(1-z^{q_n}w^{p_n})(1-z^{q_{n+1}}w^{p_{n+1}})}. \quad (4.9)$$

In particular, for $w = 1$, the spectrum of α [7] is generated by

$$\sum_{n=1}^{\infty}\lfloor n\alpha\rfloor z^n = \frac{p_0 z}{(1-z)^2} + \sum_{n=0}^{\infty}(-1)^n \frac{z^{q_n}z^{q_{n+1}}}{(1-z^{q_n})(1-z^{q_{n+1}})}. \quad (4.10)$$

Proof: Equation (4.9) follows from (4.8) and (4.3). Equation (4.10) is now obtained by letting w tend to 1. ∎

If F_N denotes the truncation of the right-hand side of (4.9)

$$\frac{zw}{(1-z)(1-zw^{p_0})}\left(\frac{1-w^{p_0}}{1-w}\right) + \sum_{n=0}^{N-1}(-1)^n \frac{z^{q_n}w^{p_n}z^{q_{n+1}}w^{p_{n+1}}}{(1-z^{q_n}w^{p_n})(1-z^{q_{n+1}}w^{p_{n+1}})}$$

we observe that (4.7) and (4.3) show that

$$F_N = \frac{\left(\frac{w}{1-w}\right)[zQ_N - (1-z)P_N]}{(1-z)(1-z^{q_N}w^{p_N})} \quad (4.11)$$

and some manipulation shows that, for $q_N > 1$, the numerator may be rewritten as

$$B_N := wz \sum_{n=1}^{q_N} z^n \left(\frac{w^{\lfloor (n+1)\alpha \rfloor} - w^{\lfloor n\alpha \rfloor}}{1-w} \right) + wz \left(\frac{1-w^{p_0}}{1-w} \right)(1 - z^{q_N} w^{p_N}) \quad (4.12)$$

so that B_N is a very simple integer polynomial in w and z (of degree $q_N + 1$ in z), while

$$F_\alpha - F_N = O(z^{q_N + q_{N+1}}).$$

Note that F_N is especially simple for $w := 1$ and $0 < \alpha < 1$.

Example 4.1. (a) Let $\alpha := \pi/2$ in (4.11) or (4.10). As

$$\frac{\pi}{2} = [1, 1, 1, 31, \ldots]$$

we have $p_0 = 1$, $p_1 = 2$, $p_2 = 3$, $p_3 = 11$, $p_4 = 344$ and $q_0 = 1$, $q_1 = 1$, $q_2 = 2$, $q_3 = 7$, $q_4 = 219$. Thus

$$F_{\pi/2}(z, 1) = \sum_{n=1}^\infty \left\lfloor \frac{\pi}{2} n \right\rfloor z^n$$

$$= \frac{z}{(1-z)^2} + \frac{z^2}{(1-z)^2} - \frac{z^3}{(1-z)(1-z^2)}$$

$$+ \frac{z^9}{(1-z^2)(1-z^7)} - \frac{z^{226}}{(1-z^7)(1-z^{219})} + \cdots$$

and the approximation F_4 is also expressible as

$$\frac{z(z^7 + z^6 + 2z^5 + z^4 + 2z^3 + z^2 + 2z + 1)}{(1-z^7)(1-z)}$$

and has an error like z^{226}. In particular

$$\sum_{n=1}^\infty \frac{\left\lfloor \frac{\pi}{2} n \right\rfloor}{2^n} \doteq \frac{339}{127}$$

with error less than 10^{-68}.

(b) Sum 9 follows from using (4.10) for $\tanh(\pi) = [0, 1, 267, \ldots]$. This produces

$$\sum_{n=1}^\infty \lfloor n \tanh \pi \rfloor z^n = \frac{z^2}{(1-z)^2} - \frac{z^{269}}{(1-z)(1-z^{268})} + \cdots.$$

(c) Sum 10 follows similarly from (4.10) with one of our favorite transcendental numbers $\alpha := e^{\pi\sqrt{163/9}} = [640320, 1653264929, \ldots]$.

(d) Let $\alpha := \log_{10}(2) = [0, 3, 3, 9, \ldots]$. Then (4.11) with $N := 3$, $z := \frac{1}{2}$ and $w := 1$ gives

$$\sum_{n=1}^\infty \frac{\lfloor n \log_{10}(2) \rfloor}{2^n} \doteq \frac{146}{1023}$$

to 30 places since $q_0 = 1$, $q_1 = 3$, $q_2 = 10$, $q_3 = 93$. Thus, as the number of even digits in 2^n is $\lfloor n \log_{10}(2) \rfloor + 1$ less the number of odd digits in 2^n, the "false" Sum 8 follows from Sum 7 and this "false" identity. In fact, see below, Sum 8 is transcendental while Sum 7 is rational. ∎

Other lovely approximations follow from

$$\log_{10}(6) = [0, 1, 3, 1, 1, 32, \dots]$$
$$\tanh(1) = [1, 3, 7, 9, 11, \dots]$$
$$\frac{e-1}{2} = [0, 1, 6, 10, 14, \dots]$$

and other simple transcendental numbers. Thus

$$\sum_{n=1}^{\infty} \frac{\lfloor n\zeta(3) \rfloor}{2^n} \doteq \frac{64}{31}$$

to 30 places.

Example 4.2. Many other related sums can be derived from (4.8) and (4.9). We indicate some classes.

(a) For irrational $\alpha > 0$

$$G_\alpha(1, w) = \sum_{n=1}^{\infty} w^{\lfloor n\alpha \rfloor} = \left(\frac{1-w}{w}\right) F_{1/\alpha}(w, 1),$$

and more generally

$$G_\alpha(z, w) = \left(\frac{1-w}{w}\right) F_{1/\alpha}(w, z).$$

This follows either from the elementary identity in [11]

$$F_\alpha(z, w) + F_{\alpha-1}(w, z) = \frac{zw}{(1-z)(1-w)} \tag{4.13}$$

or from Theorem 2 in [13], when $z = 1$.

(b) Letting $w := -1$ in (4.9) produces a Lambert-like series for $\sum_{\lfloor n\alpha \rfloor \text{ odd}} z^n$. As an example,

$$\sum \left\{ \frac{1}{2^n} \Big| \text{length}(2^n) \text{ even} \right\} \doteq \frac{114}{1025}$$

to 30 places.

(c) Observe that

$$\sum_{k=0}^{M} \frac{(-1)^k \binom{M}{R} G_\alpha(z, w^k)}{(1-w)^M} = \sum_{n=1}^{\infty} \left(\frac{1 - w^{\lfloor n\alpha \rfloor}}{1-w}\right)^M z^n$$

so that on letting w tend to unity we obtain the approximation

$$\sum_{n=1}^{\infty} \lfloor n\alpha \rfloor^M z^n = \frac{\Delta_N^M(z)}{(1-z)(1-z^{q_N})^M} + O(z^{q_N + q_{N+1}})$$

where Δ_N^M is an integer polynomial in z of degree $MqN + 1$. In particular

$$\sum_{n=1}^{\infty} \lfloor n\alpha \rfloor^2 z^n = \sum_{n=0}^{\infty} \frac{z^{q_n + q_{n+1}}}{(1-z^{q_n})^2 (1-z^{q_{n+1}})^2}$$
$$\times \{(2p_n + 2p_{n+1} - 1) - z^{q_n} z^{q_{n+1}}$$
$$- (2p_n - 1) z^{q_{n+1}} - (2p_{n+1} - 1) z^{q_n}\}$$

for $0 < \alpha < 1$, α irrational. Thus

$$\sum_{n=1}^{\infty} \frac{(\text{length}(6^n))^2}{6^n} \doteq \frac{196669}{37303}$$

to 88 places.

(d) Similarly, if w is a primitive Nth root of unity

$$\frac{1}{N} \sum_{k=1}^{N} G_\alpha(z, w^k) \overline{w}^{Mk} = \sum_{\lfloor n\alpha \rfloor \equiv M \pmod{N}} z^n$$

[compare (b)]. Thus

$$\sum_{3 \mid \lfloor n \log_{10} 2 \rfloor} \frac{1}{3^n} \doteq \frac{3554}{7381}$$

to 50 places.

(e) Let $w := e^{i\theta}$ (θ real) in (4.9). We obtain

$$\sum_{n=1}^{\infty} \cos(\lfloor n\alpha \rfloor) z^n = \frac{\sum_{n=1}^{q_N} \cos(\lfloor n\alpha \rfloor \theta) z^n - \sum_{n=1}^{q_N} \cos(p_N - \lfloor n\alpha \rfloor \theta) z^{n+q_N}}{1 - 2 z^{q_N} \cos(p_N \theta) + z^{2q_N}}$$

$$+ O(z^{q_N + q_{N+1}}),$$

with a similar expression for sin replacing cos. ∎

The rational counterpart to (4.13) is

$$F_{p/q}(z, w) + F_{q/p}(w, z) = \frac{zw}{(1 - z)(1 - w)} + \frac{z^q w^p}{1 - z^q w^p}, \quad (4.14)$$

for p and q relatively prime.

We consider $F(\alpha) := F_\alpha(z, w)$ as a function of α, and observe that $F(\alpha)$ is continuous at each irrational. Moreover, $\lim_{\alpha \downarrow p/q} F(\alpha) = F(p/q)$. Thus, on using (4.13) and (4.14) $\lim_{\alpha \uparrow p/q} F(\alpha) = F(p/q) - z^q w^p / (1 - z^q w^p)$. In consequence, F is discontinuous at every rational and $F(1) - F(0) = \sum_{0 < p/q < 1} \{F(p/q) - F(\frac{p}{q} -)\}$ so that dF is a "pure jump measure" on the rationals in $[0, 1]$. [This observation was made by H. Diamond.] Explicitly the jumps are expressed as

$$J := \sum_{s=1}^{\infty} \sum_{\substack{1 \le r \le s \\ (r,s)=1}} \frac{z^s w^r}{1 - z^s w^r}$$

$$= \sum_{k=1}^{\infty} \sum_{s=1}^{\infty} z^{sk} \sum_{\substack{(r,s)=1 \\ 1 \le r \le s}} w^{rk}. \quad (4.15)$$

Now, on setting $n = sk$, this yields

$$\sum_{n=1}^{\infty} z^k \left\{ \sum_{s/n} \sum_{\substack{(r,s)=1 \\ 1 \le r \le s}} w^{(r/s)n} \right\}.$$

Equation (16.2.3) in [8] applies with $F(w) := w^n$ and shows that the bracketed term is just $\sum_{m=1}^{n} w^m$. Hence $J = \sum_{n=1}^{\infty} z^n \sum_{m=1}^{n} w^m = F_1(z, w)$ as claimed. This is valid for $|z| < 1$, $|w| \le 1$. ∎

We have also shown, using Theorem 4.1(a) and $\lfloor n\alpha \rfloor = \lfloor n(p_N/q_N) \rfloor$ for $n < q_N$, that for $0 < \alpha < 1$

$$F_N = \begin{cases} F_{p_N/q_N} & N \text{ even} \\ F_{p_N/q_N} - \dfrac{z^{q_N} w^{p_N}}{1 - z^{q_N} w^{p_N}} & N \text{ odd}. \end{cases} \quad (4.16)$$

Clearly $F: Q \to Q$. In [10], [11] (4.9) is used to obtain transcendence estimates by functional equation methods. For $w := \pm 1$ and $z := 1/b$, $b = 2, 3, 4, \ldots$ we can get very accessible estimates for F_α or G_α from Roth's theorem [2], [9], [15].

First, observe that Corollary 4.2 shows F_α is irrational when α is irrational and w, z are rational. It is convenient to introduce

$$s := s(\alpha) = \limsup_{n \to \infty} a_n.$$

Thus s is infinite when α has unbounded continued fraction coefficients. For b and w as above, we have from (4.12)

$$0 < \left| F(\alpha) - \frac{P_N}{Q_N} \right| \leq O\left(\frac{1}{b^{q_N + q_{N+1}}} \right) \leq O\left(\frac{1}{Q_N^{(1 + q_{N+1}/q_N)}} \right) \quad (4.17)$$

for integers P_N and $Q_N := (b - 1)(b^{q_n} - w^{p_N})$. Hence, Roth's theorem shows $F(\alpha)$ is transcendental when

$$\limsup_{n \to \infty} \frac{q_{N+1}}{q_N} > 1,$$

and clearly α is Liouville when $s(\alpha) = \infty$. Since almost all numbers have unbounded coefficients, $F(\alpha)$ is Liouville in almost all cases and F maps Liouville numbers to Liouville numbers as they have $s = $ infinity. When $s(\alpha)$ is finite, we have $q_{N+1} \leq sq_N + q_{N-1} \leq (s + 1)q_N$ eventually and so infinitely often

$$q_{N+1} \geq sq_N + q_{N-1} \geq \frac{s^2 + s + 1}{s + 1} q_N$$

and (4.17) shows $F(\alpha)$ is approximable to order at least $(s + 1) + (1/(s + 1)) \geq 5/2$. If $s = 1$ then α is equivalent to $(\sqrt{5} + 1)/2$. In every other case $F(\alpha)$ is approximable to order $10/3$. In summary $F(\alpha)$ is never algebraic, indeed never has the expected rate of rational approximation and is usually Liouville ([2], [8], [15]). In fact almost all irrationals have only finitely many solutions to

$$\left| \alpha - \frac{p}{q} \right| < \frac{1}{q^2 (\log q)^{1+}}.$$

Example 4.3. (a) Arguing similarly from Example 4.2 we see that for almost all α,

$$\sum_{n=1}^{\infty} \frac{p(\lfloor n\alpha \rfloor)}{b^n}$$

is a Liouville number, for any integer polynomial p.

It is hard to find explicit numbers with unbounded continued fraction coefficients but e and $\tanh(1)$ are two examples:

$$\sum_{n=1}^{\infty} \frac{p(\lfloor ne \rfloor)}{b^n}$$

is Liouville for all p and b.

(b) Correspondingly, $\sum_{n=1}^{\infty} p(\lfloor n\alpha \rfloor)/b^n$ is approximable to order at least
$$\frac{1 + s(\alpha)}{\deg(p)}.$$
∎

For irrational $0 < \alpha < 1$, $F_\alpha(z, w)$ may be computed entirely from the continued fraction expansion via
$$F_\alpha(z, w) = \sum_{n=0}^{\infty} (-1)^n \frac{z_n z_{n+1}}{(1 - z_n)(1 - z_{n+1})}$$
where $z_{n+1} := z_n^{a_{n+1}} z_{n-1}$, $z_0 := z$, $z_{-1} := w$. This follows from (4.9) and an easy induction.

We conclude with some remarks about iterates of $F(\alpha) := \sum_{n=1}^{\infty} \lfloor n\alpha \rfloor 2^{-n}$. For $\alpha = p/q$ ($0 < \alpha < 1$) we have
$$F_\alpha(z, w) = zw \frac{\sum_{n=1}^{q} \left(\left\lfloor (n+1)\frac{p}{q} \right\rfloor - \left\lfloor n\frac{p}{q} \right\rfloor \right) w^{\lfloor n(p/q) \rfloor} z^n}{(1-z)(1-z^q w^p)} \quad (4.18)$$
either by direct computation or from (4.11) and (4.16). We now set $z := \frac{1}{2}$, $w := 1$ and observe that
$$F\left(\frac{p}{q}\right) + F\left(1 - \frac{p}{q}\right) = 1 + \frac{1}{2^q - 1}.$$
In particular $F(\frac{1}{2}) = \frac{2}{3}$. Moreover, (4.18) shows that
$$F\left(1 - \frac{1}{q}\right) = 1 - \frac{1}{2^q - 1}.$$
Let $q_0 := 2$ and $q_{n+1} := 1/(2^{q_n} - 1)$ to deduce that
$$F^{(n)}\left(\frac{1}{2}\right) = 1 - \frac{1}{q_{n+1}}$$
and so converges to 1. Similar analysis shows that
$$F\left(\frac{1}{q}\right) = \frac{2}{2^q - 1} < \frac{1}{2^{q-2}},$$
and so that
$$F^{(n)}\left(\frac{1}{3}\right) \to 0, \quad \text{because } F^{(2)}\left(\frac{1}{3}\right) = \frac{18}{127} < \frac{1}{7}.$$
Note that $\alpha \geq \frac{1}{2}$ implies $F^{(n)}(\alpha) \geq F^{(n)}(\frac{1}{2})$ and $\alpha < \frac{1}{2}$ implies $F^{(n+1)}(\alpha) \to 0$ for $0 \leq \alpha < \frac{1}{2}$. For rational α, the entire sequence is rational, otherwise it is entirely transcendental, usually Liouville.

5. RATIONAL DIGIT SUMS. This section is based on the following Lemma whose proof we owe to A. C. Thompson.

Lemma 5.1. *For $0 < q < 1$ and integer $m > 1$*
$$q = \sum_{n=1}^{\infty} \frac{\lfloor m^n q \rfloor \,(\bmod\, m)}{m^n}. \quad (5.1)$$

Proof: Consider the base m expansion of q

$$q = \sum_{k=1}^{\infty} \frac{a_k}{m^k} \qquad 0 \le a_k < m$$

where when ambiguous we take the terminating expansion. Then

$$m^n q = \sum_{k=1}^{n-1} m^{n-k} a_k + a_n + \theta_n$$

for some θ_n in $[0, 1[$. Thus a_n is the remainder of $\lfloor m^n q \rfloor$ modulo m, and (5.1) follows. ∎

Let $F(q) := \sum_{n=1}^{\infty} c_n q^n$ be any formal power series.

Theorem 5.1. *For $0 < q < 1/\limsup_{n \to \infty} |c_n|^{1/n}$,*

$$F(q) = \sum_{n=1}^{\infty} \frac{f(n)}{m^n}$$

where

$$f(n) = \sum_{k \ge 1} c_k (\lfloor m^n q^k \rfloor \bmod m).$$

Proof: From Lemma 5.1

$$F(q) = \sum_{k=1}^{\infty} c_k q^k = \sum_{k=1}^{\infty} c_k \sum_{n=1}^{\infty} \frac{\lfloor m^n q^k \rfloor \bmod m}{m^n}$$

$$= \sum_{n=1}^{\infty} \frac{f(n)}{m^n}$$

on exchanging order of summation, as is valid within the radius of convergence of F. ∎

Theorem 5.1 can be extended so as to replace m^n by $\prod_{k=1}^{n} r_k$ where r_k are integers ≥ 2, and where the remainder is computed modulo r_n.

If we specialize Theorem 5.1 to the case where $q := 1/b$ and b is an integer divisible by m we may observe that $\lfloor m^n/b^k \rfloor \bmod m$ coincides with the coefficient (mod m) of b^k in the base b expansion of m^n (the $(k+1)^{th}$ digit).

Specializing further so that $m := 2$ and b is even we have

$$F\left(\frac{1}{b}\right) = \sum_{n=1}^{\infty} \frac{f_b(n)}{2^n} \qquad (5.2)$$

where

$$f_b(n) := \sum \{c_k | 2^n \text{ has } (k+1)^{th} \text{ digit odd base } b\}.$$

Example 5.1. (a) Let $F(q) := q/(1-q)$. Then $f_b(n)$ counts the number of odd digits in 2^n base b. Sum 7 is established on setting $b := 10$.

(b) Sum 1 corresponds to taking $F(q) = q^2/(1-q^2)$ and $q = 1/10$.

(c) Let $F(q) = q/(1-q-q^2)$. Now F is the generating function of the Fibonacci numbers ($F_1 = 1$, $F_2 = 1$, $F_{n+1} = F_n + F_{n-1}$). Again with $q := 1/10$, we

obtain for
$$f(n) := \sum \{F_k | 2^n \text{ has } (k+1)^{\text{th}} \text{ digit odd}\},$$
as in Bowman and White [4], that
$$\sum_{n=1}^{\infty} \frac{f(n)}{2^n} = \frac{10}{89}.$$

The generating function for F_k^2 is $\dfrac{q - q^2}{1 - 2q - 2q^2 + q^3}$ and so for
$$f(n) := \sum \{F_k^2 | 2^n \text{ has } (k+1)^{\text{th}} \text{ digit odd}\}$$
$$\sum_{n=1}^{\infty} \frac{f(n)}{2^n} = \frac{90}{781}.$$

(d) Let
$$F(q) = \sum_{n=1}^{\infty} q^{n^2} = \frac{\theta_3(q) - 1}{2}.$$

Then
$$\sum_{n=1}^{\infty} \frac{f(n)}{2^n} = \frac{\theta_3\left(\frac{1}{10}\right) - 1}{2}$$
where $f(n)$ counts the number of odd digits of 2^n in square positions (the second, fifth, tenth digits etc.).

(e) If we apply Theorem 5.1 to $F(q) := q/(1-q)$ with $b := 10$ and $m := 5$ we deduce that again
$$\sum_{n=1}^{\infty} \frac{f(n)}{5^n} = \frac{1}{9}$$
where $f(n)$ sums the digits (mod 5) of 5^n base 10 (e.g. $f(3125) = 6$). ∎

6. THETA FUNCTION EXAMPLES. The underlying identity for this section is really just a modular transformation of $\theta_3(q) := \sum_{n=-\infty}^{\infty} q^{n^2}$. (See [2].)

Lemma 6.1. *For $\alpha, \beta > 0$ with $\alpha\beta = 2\pi$*
$$\sqrt{\alpha}\left[\sum_{n=-\infty}^{\infty} e^{-\alpha^2 n^2/2}\right] = \sqrt{\beta}\left[\sum_{n=-\infty}^{\infty} e^{-\beta^2 n^2/2}\right].$$

Example 6.1. From the Lemma, with $s = 2/\beta^2$ so $\alpha^2 = 2\pi^2 s$
$$\sqrt{\pi s} - \sum_{n=-\infty}^{\infty} e^{-n^2/s} = 2\sqrt{\pi s}\, e^{-\pi^2 s} + O(e^{-\pi^2 4s}) \qquad (6.1)$$
$$\sim 2\sqrt{\pi s}\, 10^{-(4.2863\ldots)s}.$$

Now with $s := 10^{10}$ we get
$$\left|\sqrt{\pi} - \left(\frac{1}{10^5}\sum_{n=-\infty}^{\infty} e^{-n^2/10^{10}}\right)\right| \leq 10^{-4.2 \cdot 10^{10}}, \qquad (6.2)$$
which is Sum 12.

If we set
$$s = \frac{1}{\log 10^{1/N}} = \frac{N}{\log 10}$$
we get
$$\sqrt{\frac{N\pi}{\log 10}} - \sum_{-\infty}^{\infty} \frac{1}{10^{n^2/N}} \sim 2 \cdot \sqrt{\frac{N\pi}{\log 10}} \, 10^{-(1.861\cdots)N} \qquad (6.3)$$
and with $N := 10^4$ we get Sum 11.

Similarly we have
$$\sqrt{\frac{q\pi}{\log q}} - \sum_{-\infty}^{\infty} \frac{1}{q^{n^2/q}} \sim 2\sqrt{\frac{q\pi}{\log q}} \, e^{-\pi^2 q/\log q}. \qquad (6.4)$$

REFERENCES

1. P. E. Böhmer, Über die Transzendenz gewisser dyadischer Brüche, *Math. Ann.* 96 (1927), 367–377.
2. J. M. Borwein and P. B. Borwein, *Pi and the AGM—A Study in Analytic Number Theory and Computational Complexity*, Wiley, N.Y. N.Y., 1987.
3. J. M. Borwein and P. B. Borwein, Generating functions of integer parts, *J. Number Theory*, In Press.
4. D. Bowman and T. White, private communication.
5. J. L. Davison, A series and its associated continued fraction, *Proc. Amer. Math. Soc.* 63 (1977), 26–32.
6. H. M. Diamond, Elementary Problem #3353, This MONTHLY, 96 (1989) 838.
7. R. L. Graham, D. E. Knuth and O. Patashnik, *Concrete Mathematics*, Addison-Wesley, Reading, Mass., 1989.
8. G. H. Hardy and E. M. Wright, *An Introduction to the Theory of Numbers*, 4th ed. Oxford University Press, London, 1960.
9. W. J. LeVeque, *Fundamentals of Number Theory*, Addison-Wesley, Reading, Mass., 1977.
10. J. H. Loxton and A. J. van der Poorten, *Transcendence and Algebraic Independence by a Method of Mahler*, Transcendence Theory–Advances and Applications, ed. A. Baker and D. W. Masser, Academic Press, 1977, 211–226.
11. Kurt Mahler, Arithmetische Eigenschaften der Lösungen einer Klasse von Funktionalgleichungen, *Math. Ann.* 101 (1929), 342–366.
12. D. J. Newman, On the number of binary digits in a multiple of three, *Proc. Amer. Math. Soc.* 21 (1969), 719–722.
13. M. Newman, Irrational power series, *Proc. Amer. Math. Soc.* 11 (1960), 699–702.
14. J. O. Shallit, On infinite products associated with sums of digits, *J. Number Theory* 21 (1985), 128–134.
15. M. Waldschmidt, Nombres transcendents, *Lecture Notes in Mathematics* 402 Springer, New York, 1974.
16. H. S. Wall, *Analytic Theory of Continued Fractions*, Van Nostrand, Toronto, New York, London, 1948.
17. Rolf Wallisser, Eine Bemerkung über irrationale Werte und Nichtfortsetzbarkeit von Potenzreihen mit ganzzahligen Koeffizienten, *Colloq. Math.* 23 (1971), 141–144.
18. J.-P. Allouche and J. Shallit, Sums of Digits and the Hurwitz Zeta Function, *Lecture Notes in Mathematics* 1434, Springer-Verlag, New York, 1990, 19–30.

Department of Mathematics, Statistics and Computing Science
Dalhousie University
Halifax, Nova Scotia, B3H 3J5
Canada

6. The amazing number π

Discussion

As mentioned below, this text accompanies an address given at the celebration to replace the lost tombstone of Ludolph van Ceulen at the Pieterskerk (St. Peter's Church) in Leiden on the fifth of July, 2000.

The authors are reasonably sure that they will never give another lecture in a packed cathedral from their pulpit. Another singular feature was that the very urbane crown prince of the Netherlands was in the audience. This reflects one of the "amazing" features of the number pi. It has entered the public imagination in a way that few pieces of mathematics have or for that matter ever can.

One of the things we hope to illustrate in this article, and in others in this collection, is that there is a meaningful and mathematically interesting story to be told. But the cultural side of the story holds its own interest and has found itself in unusual places such as in movies (*Pi the movie*) and novels (*Contact* and *Life of Pi*) to name a few.

Physics has its quarks and black holes to capture the public imagination. There are few places where mathematics has a similar popular appeal. We would suggest that while the average educated person knows no more about black holes than about *Pi*, it is possible to have a metaphorical understanding of physics in a way that is very hard in mathematics. The story of pi is an exception.

As a forcible example, imagine the following excerpt from Eli Mandel's 2002 Booker Prize winning novel *Life of Pi* being written about another transcendental number:

> "My name is
> Piscine Molitor Patel
> known to all as Pi Patel.
>
> For good measure I added '$\pi = 3.14$' and I then drew a large circle which I sliced in two with a diameter, to evoke that basic lesson of geometry."

At the end of the novel, Piscine (Pi) Patel writes

> I am a person who believes in form, in harmony of order. Where we can, we must give things a meaningful shape. For example—I wonder—could you tell my jumbled story in exactly one hundred chapters, not one more, not one less? I'll tell you, that's one thing I hate about my nickname, the way that number runs on forever. It's important in life to conclude things properly. Only then can you let go.

We may well not share the irrational sentiment, but we should celebrate that Pi knows π to be irrational.

Source

P. Borwein, "The amazing number π," *Nieuw Arch. Wiskd.*, (5)**1** (2000), 254–258.

Peter Borwein
Department of Mathematics and Statistics
Simon Fraser University, Burnaby, B.C., Canada V5A 1S6
pborwein@cecm.sfu.ca

Beeger lecture 2000

The amazing number π

This text accompanies an address given at the celebration to replace the lost tombstone of Ludolph van Ceulen at the Pieterskerk (St. Peter's Church) in Leiden on the fifth of July, 2000. It honours the particular achievements of Ludolph as well as the long and important tradition of intellectual inquiry associated with understanding the number π and numbers generally.

The history of π parallels virtually the entire history of Mathematics. At times it has been of central interest and at times the interest has been quite peripheral (no pun intended). Certainly Lindemann's proof of the transcendence of π was one of the highlights of nineteenth century mathematics and stands as one of the seminal achievements of the millennium (very loosely this result says that π is not an easy number). One of the low points was the Indiana State legislature's attempt to legislate a value of π in 1897; an attempt as plausible as repealing the law of gravity.

Why π?

The amount of human ingenuity that has gone into understanding the nature of π and computing its digits is quite phenomenal and begs the question "why π?". After all there are more numbers than one can reasonably contemplate that could get a similar treatment. And π is just one of the very infinite firmament of numbers. Part of the answer is historical. It is the earliest and the most naturally occurring hard number (technically, hard means transcendental which means not the solution of a simple equation). Even the choice of label 'transcendental' gives it something of a mystical aura.

What is pi? First and foremost it is a number, between 3 and 4 (3.14159...). It arises in any computations involving circles: the area of a circle of radius 1 or equivalently, though not obviously, the perimeter of a circle of radius 1/2. The nomenclature π is presumably the Greek letter 'p' in periphery. The most basic properties of π were understood in the period of classical Greek mathematics by the time of the death of Archimedes in 212 BC.

Ruler and compass

The Greek notion of number was quite different from ours, so the Greek numbers were our whole numbers: 1, 2, 3... In Greek geometry the essential idea was not number but continuous magnitude, e.g. line segments. It was based on the notion of multiplicity of units and, in this

---

Bill No. 246, 1897. State of Indiana

"Be it enacted by the General Assembly of the State of Indiana: It has been found that the circular area is to the quadrant of the circumference, as the area of an equilateral rectangle is to the square on one side."

Pi-henge by Lisa Hakesley

sense, numbers that existed were numbers that could be drawn with just an unmarked ruler and compass. The rules allowed for starting with a fixed length of 1 and seeing what could be constructed with straight edge and compass alone. (Our current notion is much more based on counting.) The question of whether π is a constructible magnitude had been explicitly raised as a question by the sixth century BC and the time of the Pythagoreans. Unfortunately π is not constructible, though a proof of this would not be available for several thousand years. In this context there isn't a more basic question than "is π a number?" Of course, our more modern notion of number embraces the Greek notion of constructible and doesn't depend on construction. The existence of π as a number given by an infinite (albeit unknown) decimal expansion poses little problem.

Very early on the Greeks had hypothesized that π wasn't constructible, Aristophanes already makes fun of "circle squarers" in his fifth century BC play "The Birds."

Lindemann's proof of the transcendence of π in 1882 settles the issue that π is not constructible by the Greek rules and a truly marvelous proof was given a few years later by Hilbert. Not that this has stopped cranks from still trying to construct π.

Does this tell us everything we wish to know about π? No, our ignorance is still much more profound than our knowledge! For example, the second most natural hard number is e which is provably transcendental. But what about $\pi + e$? This embarrassingly easy question is currently totally intractable (we don't even know how to show that $\pi + e$ is irrational). The number π is a mathematical apple and e is a mathematical orange and we have no idea how to mix them.

The need for π

Why compute the digits of π? Sometimes it is necessary to do so, though hardly ever more than the 6 or so digits that Archimedes computed several thousand years ago are needed for physical applications. Even far fetched computations like the volume of a spherical universe only require a few dozen digits. There is also the 'Everest Hypothesis' ('because it's there'). Probably the number of people involved and the effort in time has been similar in the two quests. A few thousand people have reached the computational level that requires the carrying of oxygen — though so far I know of no π related fatalities. There has been significant knowledge accumulated in this slightly quixotic pursuit. But this knowledge could have been derived from computing a host of other numbers in a variety of different bases. Once again the answer to "why π" is largely historical and cultural. These are good but not particularly scientific reasons. Pi was first, pi is hard and pi has captured the educated imagination. (Have you ever seen a cartoon about log 2 — a number very similar to π?)

Whatever the personal motivations, π has been much computed and a surprising amount has been learnt along the way.

The mathematics involved

In constructing the all star hockey team of great mathematicians, there seems to

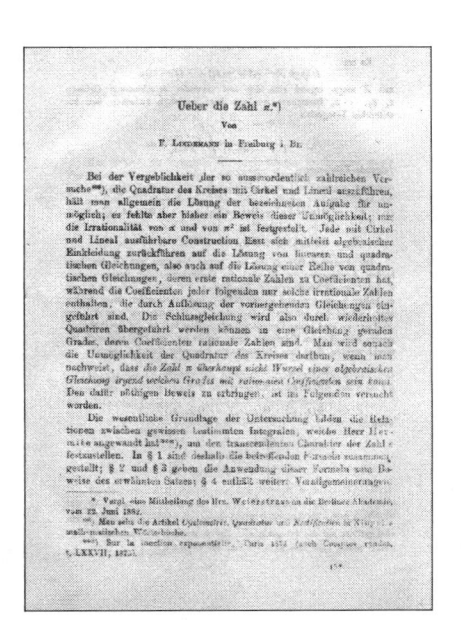

The title page of Lindemann's proof of the irrationality of π

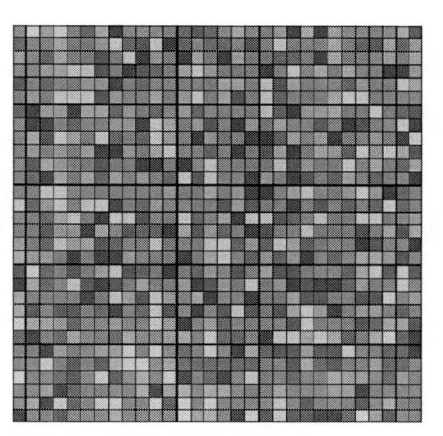

Digits of π represented by grey tones

be pretty wide agreement that the front line consists of Archimedes, Newton, and Gauss. Both Archimedes and Newton invented methods for computing π. In Newton's case this was an application of his newly invented calculus. I know of no such calculation from Gauss though his exploration of the Arithmetic-Geometric Mean iteration laid the foundation of the most successful methods for doing such calculations. There is less consensus about who comes next. I might add Hilbert and Euler next (on defense). Both of these mathematicians also contribute to the story of π. Perhaps von Neumann is in goal — certainly he is a candidate for the most versatile and smartest mathematician of the twentieth century. One of the first calculations done on ENIAC (one of the first real computers) was the computations of roughly a thousand digits of π and von Neumann was part of the team that did the calculation.

One doesn't often think of a problem like this having economic benefits. But as is often the case with pure mathematics and curiosity driven research the rewards can be surprising. Large recent records depend on three things: better algorithms for π; larger and faster computers; and an understanding of how to do arithmetic with numbers that are billions of digits long

The better algorithms are due to a variety of people including Ramanujan, Brent, Salamin, the Chudnovsky brothers and ourselves (the Borwein brothers, NAW). Some of the mathematics is both beautiful and subtle. (The Ramanujan type series listed in the appendix are, for me, of this nature.)

The better computers are, of course, the most salient technological advance of the

Cow Pi by Lisa Hakesley

second half of the twentieth century.

Understanding arithmetic is an interesting and illuminating story in its own right. A hundred years ago we knew how to add and multiply — do it the way we all learned in school. Now we are not so certain except that we now know that the "high school method" is a disaster for multiplying really big numbers. The mathematical technology that allows for multiplying very large numbers together is essentially the same as the mathematical technology that allows image processing devices like CAT scanners to work (FFTs). In making the record setting algorithms work, David Bailey tuned the FFT algorithms in several of the standard implementations and saved the US economy millions of dollars annually. Most recent records are set when new computers are being installed and tested. (Recent records are more or less how many digits can be computed in a day — a reasonable amount of test time on a costly machine.) The computation of π seems to stretch the machine and there is a history of uncovering subtle and sometimes not so subtle bugs at this stage.

Patterns in π

What do the calculations of π reveal and what does one expect? One expects that the digits of π should look random — that roughly one out of each ten digits should be a 7 et cetera. This appears to be true at least for the first few hundred billion. But this is far from a proof — an actual proof of this is way out of the reach of current mathematics. As is so often the case in mathematics some of the most basic questions are some of the most intractable. What mathematicians believe is that every pattern possible eventually occurs in the digits of π — with a suitable encoding the Bible is written in entirety in the digits, as is the New York phone book and everything else imaginable.

The question of whether there are subtle patterns in the digits is an interesting one. (Perhaps every billionth digit is a seven after a while. While unlikely this is not provably impossible. Or perhaps π is buried within π in some predictable way.) Looking for subtle patterns in long numbers is exactly the kind of problem one needs to tackle in handling the human genome (a chromosome is just a large number written in base 4, at least to a mathematician).

Acknowledgment
The author would like pay tribute to the mathematical community of the Netherlands on the occasion of its honouring one of its founding fathers, the mathematician Ludolph van Ceulen (1540–1610).

Appendices
I have included two appendices. One is from David H. Bailey, Jonathan M. Borwein, Peter B. Borwein, and Simon Plouffe, "The Quest for Pi," (June, 1996) *The Mathematical Intelligencer*. It is a chronology of the computation of digits of π. The second is taken from: Lennert Berggren, Jonathan M. Borwein and Peter B. Borwein, *Pi: A Source Book*, Springer–Verlag 1988. It is a list of significant mathematical formulae related to π. These are reproduced with permission from Springer–Verlag New York.

The previously mentioned chronology is of the problem of computing all of the initial digits of π. There is also a shorter chronology of computing just a few very distant bits of π. The record here is 40 trillion and is due to Colin Percival using the methods described in the last reference above. It is surprising that this is possible at all.

Appendix Ia. A computational chronology for pi
History of π calculations (pre 20th century)

Babylonians	2000? BCE	1	3.125 ($3\frac{1}{8}$)
Egyptians	2000? BCE	1	3.16045 ($4(\frac{8}{9})^2$)
China	1200? BCE	1	3
Bible (1 Kings 7:23)	550? BCE	1	3
Archimedes	250? BCE	3	3.1418 (ave.)
Hon Han Shu	130 AD	1	3.1622 ($=\sqrt{10}$?)
Ptolemy	150	3	3.14166
Chung Hing	250?	1	3.16227 ($\sqrt{10}$)
Wang Fau	250?	1	3.15555 ($\frac{142}{45}$)
Liu Hui	263	5	3.14159
Siddhanta	380	3	3.1416
Tsu Ch'ung Chi	480?	7	3.1415926
Aryabhata	499	4	3.14156
Brahmagupta	640?	1	3.162277 ($=\sqrt{10}$)
Al-Khowarizmi	800	4	3.1416
Fibonacci	1220	3	3.141818
Al-Kashi	1429	14	
Otho	1573	6	3.1415929
Viète	1593	9	3.1415926536 (ave.)
Romanus	1593	15	
Van Ceulen	1596	20	
Van Ceulen	1610	35	
Newton	1665	16	
Sharp	1699	71	
Seki	1700?	10	
Kamata	1730?	25	
Machin	1706	100	
De Lagny	1719	127	(112 correct)
Takebe	1723	41	
Matsunaga	1739	50	
Vega	1794	140	
Rutherford	1824	208	(152 correct)
Strassnitzky and Dase	1844	200	
Clausen	1847	248	
Lehmann	1853	261	
Rutherford	1853	440	
Shanks	1874	707	(527 correct)

Appendix Ib.
History of π calculations (20th century)

Ferguson	1946	620
Ferguson	Jan. 1947	710
Ferguson and Wrench	Sep. 1947	808
Smith and Wrench	1949	1,120
Reitwiesner et al. (ENIAC)	1949	2,037
Nicholson and Jeenel	1954	3,092
Felton	1957	7,480
Genuys	Jan. 1958	10,000
Felton	May 1958	10,021
Guilloud	1959	16,167
Shanks and Wrench	1961	100,265
Guilloud and Filliatre	1966	250,000
Guilloud and Dichampt	1967	500,000
Guilloud and Bouyer	1973	1,001,250
Miyoshi and Kanada	1981	2,000,036
Guilloud	1982	2,000,050
Tamura	1982	2,097,144
Tamura and Kanada	1982	8,388,576
Kanada, Yoshino and Tamura	1982	16,777,206
Ushiro and Kanada	Oct. 1983	10,013,395
Gosper	1985	17,526,200
Bailey	Jan. 1986	29,360,111
Kanada and Tamura	Sep. 1986	33,554,414
Kanada and Tamura	Oct. 1986	67,108,839
Kanada, Tamura, Kubo, et al.	Jan. 1987	134,217,700
Kanada and Tamura	Jan. 1988	201,326,551
Chudnovskys	May 1989	480,000,000
Chudnovskys	Jun. 1989	525,229,270
Kanada and Tamura	Jul. 1989	536,870,898
Kanada and Tamura	Nov. 1989	1,073,741,799
Chudnovskys	Aug. 1989	1,011,196,691
Chudnovskys	Aug. 1991	2,260,000,000
Chudnovskys	May 1994	4,044,000,000
Takahashi and Kanada	Jun. 1995	3,221,225,466
Kanada	Aug. 1995	4,294,967,286
Kanada	Oct. 1995	6,442,450,938
Kanada	Jun. 1997	51,539,600,000
Kanada	Sep. 1999	206,158,430,000

Appendix II. Selected formulae for pi

Archimedes (ca. 250 BC)

Let $a_0 := 2\sqrt{3}$, $b_0 := 3$ and

$$a_{n+1} := \frac{2a_n b_n}{a_n + b_n} \quad \text{and} \quad b_{n+1} := \sqrt{a_{n+1} b_n}.$$

Then a_n and b_n converge linearly to π (with an error $O(4^{-n})$.)

François Viète (ca. 1579)

$$\frac{2}{\pi} = \sqrt{\frac{1}{2}} \sqrt{\frac{1}{2} + \frac{1}{2}\sqrt{\frac{1}{2}}} \sqrt{\frac{1}{2} + \frac{1}{2}\sqrt{\frac{1}{2} + \frac{1}{2}\sqrt{\frac{1}{2}}}} \cdots$$

John Wallis (ca. 1650)

$$\frac{\pi}{2} = \frac{2 \cdot 2 \cdot 4 \cdot 4 \cdot 6 \cdot 6 \cdot 8 \cdot 8 \cdots}{1 \cdot 3 \cdot 3 \cdot 5 \cdot 5 \cdot 7 \cdot 7 \cdot 9 \cdots}$$

William Brouncker (ca. 1650)

$$\pi = \cfrac{4}{1 + \cfrac{1}{2 + \cfrac{9}{2 + \cfrac{25}{2 + \cdots}}}}$$

Madhava, James Gregory, Gottfried Wilhelm Leibnitz (1450–1671)

$$\frac{\pi}{4} = 1 - \frac{1}{3} + \frac{1}{5} - \frac{1}{7} + \cdots$$

Isaac Newton (ca. 1666)

$$\pi = \frac{3\sqrt{3}}{4} + 24\left(\frac{2}{3 \cdot 2^3} - \frac{1}{5 \cdot 2^5} - \frac{1}{28 \cdot 2^7} - \frac{1}{72 \cdot 2^9} - \cdots\right)$$

Machin Type Formulae (1706–1776)

$$\frac{\pi}{4} = 4\arctan(\frac{1}{5}) - \arctan(\frac{1}{239})$$
$$\frac{\pi}{4} = \arctan(\frac{1}{2}) + \arctan(\frac{1}{3})$$
$$\frac{\pi}{4} = 2\arctan(\frac{1}{2}) - \arctan(\frac{1}{7})$$
$$\frac{\pi}{4} = 2\arctan(\frac{1}{3}) + \arctan(\frac{1}{7})$$

Leonard Euler (ca. 1748)

$$\frac{\pi^2}{6} = 1 + \frac{1}{2^2} + \frac{1}{3^2} + \frac{1}{4^2} + \frac{1}{5^2} + \cdots$$
$$\frac{\pi^4}{90} = 1 + \frac{1}{2^4} + \frac{1}{3^4} + \frac{1}{4^4} + \frac{1}{5^4} + \cdots$$
$$\frac{\pi^2}{6} = 3\sum_{m=1}^{\infty}\frac{1}{m^2\binom{2m}{m}}$$

Srinivasa Ramanujan (1914)

$$\frac{1}{\pi} = \sum_{n=0}^{\infty}\binom{2n}{n}^3 \frac{42n+5}{2^{12n+4}}.$$

$$\frac{1}{\pi} = \frac{\sqrt{8}}{9801}\sum_{n=0}^{\infty}\frac{(4n)!}{(n!)^4}\frac{[1103+26390n]}{396^{4n}}.$$

Each additional term of the latter series adds roughly 8 digits.

Louis Comtet (1974)

$$\frac{\pi^4}{90} = \frac{36}{17}\sum_{m=1}^{\infty}\frac{1}{m^4\binom{2m}{m}}$$

Eugene Salamin, Richard Brent (1976)

Set $a_0 = 1$, $b_0 = 1/\sqrt{2}$ and $s_0 = 1/2$. For $k = 1, 2, 3, \cdots$ compute

$$a_k = \frac{a_{k-1}+b_{k-1}}{2}$$
$$b_k = \sqrt{a_{k-1}b_{k-1}}$$
$$c_k = a_k^2 - b_k^2$$
$$s_k = s_{k-1} - 2^k c_k$$
$$p_k = \frac{2a_k^2}{s_k}$$

Then p_k converges *quadratically* to π.

Jonathan Borwein and Peter Borwein (1991)

Set $a_0 = 1/3$ and $s_0 = (\sqrt{3}-1)/2$. Iterate

$$r_{k+1} = \frac{3}{1+2(1-s_k^3)^{1/3}}$$
$$s_{k+1} = \frac{r_{k+1}-1}{2}$$
$$a_{k+1} = r_{k+1}^2 a_k - 3^k(r_{k+1}^2 - 1)$$

Then $1/a_k$ converges *cubically* to π.

[1985] Set $a_0 = 6 - 4\sqrt{2}$ and $y_0 = \sqrt{2} - 1$. Iterate

$$y_{k+1} = \frac{1-(1-y_k^4)^{1/4}}{1+(1-y_k^4)^{1/4}}$$
$$a_{k+1} = a_k(1+y_{k+1})^4 - 2^{2k+3}y_{k+1}(1+y_{k+1}+y_{k+1}^2)$$

Then a_k converges *quartically* to $1/\pi$.

David Chudnovsky and Gregory Chudnovsky (1989)

$$\frac{1}{\pi} = 12\sum_{n=0}^{\infty}(-1)^n\frac{(6n)!}{(n!)^3(3n)!}\frac{13591409+n545140134}{(640320^3)^{n+1/2}}.$$

Each additional term of the series adds roughly 15 digits.

Jonathan Borwein and Peter Borwein (1989)

$$\frac{1}{\pi} = 12\sum_{n=0}^{\infty}\frac{(-1)^n(6n)!}{(n!)^3(3n)!}\frac{(A+nB)}{C^{n+1/2}}$$

where

$A := 212175710912\sqrt{61} + 1657145277365$
$B := 13773980892672\sqrt{61} + 107578229802750$
$C := [5280(236674 + 30303\sqrt{61})]^3.$

Each additional term of the series adds roughly 31 digits.

[1985] The following is not an identity but is correct to over 42 billion digits:

$$\left(\frac{1}{10^5}\sum_{n=-\infty}^{\infty}e^{-\frac{n^2}{10^{10}}}\right)^2 \doteq \pi.$$

Roy North (1989)

Gregory's series for π, truncated at 500,000 terms gives to forty places

$$4\sum_{k=1}^{500,000}\frac{(-1)^{k-1}}{2k-1}$$
$$= 3.14159\underline{0}65358979324\underline{0}46264338326\underline{9}502884197.$$

Only the underlined digits are incorrect.

David Bailey, Peter Borwein and Simon Plouffe (1996)

$$\pi = \sum_{i=0}^{\infty}\frac{1}{16^i}\left(\frac{4}{8i+1} - \frac{2}{8i+4} - \frac{1}{8i+5} - \frac{1}{8i+6}\right)$$

7. Experimental mathematics: Recent developments and future outlook

Discussion

This article is the first of several in the collection—notable selections 10 and 13—which at the invitation of various editors have attempted to place in context and to popularize the methods of experimental mathematics. Over the past twenty years, the term has gone from sounding oxymoronic to being a commonplace. That said, we remain mindful of sentiments, such as the following two given by Richard Feynman in 1964:

> Another thing I must point out is that you cannot prove a vague theory wrong.
>
> ...
>
> Also, if the process of computing the consequences is indefinite, then with a little skill any experimental result can be made to look like the expected consequences.[1]

In similar fashion, it is refreshing to read David Avnir's assessment that:

> The common situation is this: An experimentalist performs a resolution analysis and finds a limited-range power law with a value of D smaller than the embedding dimension. Without necessarily resorting to special underlying mechanistic arguments, the experimentalist then often chooses to label the object for which she or he finds this power law a "fractal". This is the fractal geometry of nature.[2]

Source

D.H. Bailey and J.M. Borwein, "Experimental Mathematics: Recent Developments and Future Outlook," pp. 51–66 in Volume I of *Mathematics Unlimited—2∞1 and Beyond,* B. Engquist and W. Schmid (Eds.), Springer-Verlag, 2000.

---

[1] Quoted by Gary Taubes in "The (Political) Science of Salt," *Science* August 14, 1998, 898–907.

[2] Avnir et al in "Is the geometry of nature fractal?" *Science* January 2, 1998, 39–40. Their review of all articles from 1990 to 1996 in *Physical Reviews* suggested very little substance for claims of 'fractility'.

Experimental Mathematics: Recent Developments and Future Outlook

David H. Bailey[1] and Jonathan M. Borwein[2]

[1] Lawrence Berkeley Laboratory, Berkeley, CA 94720, USA,
   `dhbailey@lbl.gov`.[***]
[2] Gordon M. Shrum Professor of Science, Centre for Experimental and Constructive Mathematics, Simon Fraser University, Burnaby, BC, Canada, `jborwein@cecm.sfu.ca`.[†]

1 Introduction

While extensive usage of high-performance computing has been a staple of other scientific and engineering disciplines for some time, research mathematics is one discipline that has heretofore not yet benefited to the same degree. Now, however, with sophisticated mathematical computing tools and environments widely available on desktop computers, a growing number of remarkable new mathematical results are being discovered partly or entirely with the aid of these tools. With currently planned improvements in these tools, together with substantial increases expected in raw computing power, due both to Moore's Law and the expected implementation of these environments on parallel supercomputers, we can expect even more remarkable developments in the years ahead.

This article briefly discusses the nature of mathematical experiment. It then presents a few instances primarily of our own recent computer-aided mathematical discoveries, and sketches the outlook for the future. Additional examples in diverse fields and broader citations to the literature may be found in [16] and its references.

2 Preliminaries

The crucial role of high performance computing is now acknowledged throughout the physical, biological and engineering sciences. Numerical experimentation, using increasingly large-scale, three-dimensional simulation programs, is now a staple of fields such as aeronautical and electrical engineering, and research scientists heavily utilize computing technology to collect and analyze data, and to explore the implications of various physical theories.

[***] Bailey's work supported by the Director, Office of Computational and Technology Research, Division of Mathematical, Information, and Computational Sciences of the U.S. Department of Energy, under contract number DE-AC03-76SF00098.

[†] Borwein's work supported by the Natural Sciences and Engineering Research Council of Canada and the Networks of Centres of Excellence programme.

However, "pure" mathematics (and closely allied areas such as theoretical physics) only recently has begun to capitalize on this new technology. This is ironic, because the basic theoretical underpinnings of modern computer technology were set out decades ago by mathematicians such as Alan Turing and John Von Neumann. But only in the past decade, with the emergence of powerful mathematical computing tools and environments, together with the growing availability of very fast desktop computers and highly parallel supercomputers, as well as the pervasive presence of the Internet, has this technology reached the level where the research mathematician can enjoy the same degree of intelligent assistance that has graced other technical fields for some time.

This new approach is often termed *experimental mathematics*, namely the utilization of advanced computing technology to explore mathematical structures, test conjectures and suggest generalizations. And there is now a thriving journal of *Experimental Mathematics*. In one sense, there is nothing new in this approach — mathematicians have used it for centuries. Gauss once confessed, "I have the result, but I do not yet know how to get it." [2]. Hadamard declared, "The object of mathematical rigor is to sanction and legitimize the conquests of intuition, and there was never any other object for it." [34]. In recent times Milnor has stated this philosophy very clearly:

> If I can give an abstract proof of something, I'm reasonably happy. But if I can get a concrete, computational proof and actually produce numbers I'm much happier. I'm rather an addict of doing things on computer, because that gives you an explicit criterion of what's going on. I have a visual way of thinking, and I'm happy if I can see a picture of what I'm working with. [35]

What is really meant by an *experiment* in the context of mathematics? In *Advice to a Young Scientist*, Peter Medawar [31] identifies four forms of experiment:

> 1. The *Kantian* experiment is one such as generating "the classical non-Euclidean geometries (hyperbolic, elliptic) by replacing Euclid's axiom of parallels (or something equivalent to it) with alternative forms."
> 2. The *Baconian* experiment is a contrived as opposed to a natural happening, it "is the consequence of 'trying things out' or even of merely messing about."
> 3. The *Aristotelian* experiment is a demonstration: "apply electrodes to a frog's sciatic nerve, and lo, the leg kicks; always precede the presentation of the dog's dinner with the ringing of a bell, and lo, the bell alone will soon make the dog dribble."
> 4. The *Galilean* experiment is "a critical experiment – one that discriminates between possibilities and, in doing so, either gives us confidence in the view we are taking or makes us think it in need of correction."

The first three are certainly common in mathematics. However, as discussed in detail in [15], the Galilean experiment is the only one of the four forms which can make experimental mathematics a truly serious enterprise.

3 Tools of the Trade

The most obvious development in mathematical computing technology has been the growing availability of powerful symbolic computing tools. Back in the 1970s, when the first symbolic computing tools became available, their limitations were quite evident — in many cases, these programs were unable to handle operations that could be done by hand. In the intervening years these programs, notably the commercial products such as Maple and Mathematica, have greatly improved. While numerous deficiencies remain, they nonetheless routinely and correctly dispatch many operations that are well beyond the level that a human could perform with reasonable effort.

Another recent development that has been key to a number of new discoveries is the emergence of practical integer relation detection algorithms. Let $x = (x_1, x_2, \cdots, x_n)$ be a vector of real or complex numbers. x is said to possess an integer relation if there exist integers a_i, not all zero, such that $a_1 x_1 + a_2 x_2 + \cdots + a_n x_n = 0$. By an *integer relation algorithm*, we mean a practical computational scheme that can recover the vector of integers a_i, if it exists, or can produce bounds within which no integer relation exists. The problem of finding integer relations was studied by numerous mathematicians, including Euclid and Euler. The first general integer relation algorithm was discovered in 1977 by Ferguson and Forcade [24]. There is a close connection between integer relation detection and finding small vectors in an integer lattice, and thus one common solution to the integer relation problem is to apply the Lenstra-Lenstra-Lovasz (LLL) lattice reduction algorithm [30]. At the present time, the most effective scheme for integer relation detection is Ferguson's "PSLQ" algorithm [23,6].

Integer relation detection, as well as a number of other techniques used in modern experimental mathematics, relies heavily on very high precision arithmetic. The most advanced tools for performing high precision arithmetic utilize fast Fourier transforms (FFTs) for multiplication operations. Armed with one of these programs, a researcher can often effortlessly evaluate mathematical constants and functions to precision levels in the many thousands of decimal digits. The software products Maple and Mathematica include relatively complete and well-integrated multiple precision arithmetic facilities, although until very recently they did not utilize FFTs, or other accelerated multiplication techniques. One may also use any of several freeware multiprecision software packages [3,22] and for many purposes tools such as Matlab, MuPAD or more specialized packages like Pari-GP are excellent.

4   David H. Bailey and Jonathan M. Borwein

High precision arithmetic, when intelligently used with integer relation detection programs, allows researchers to discover heretofore unknown mathematical identities. It should be emphasized that these numerically discovered "identities" are only approximately established. Nevertheless, in the cases we are aware of, the results have been numerically verified to hundreds and in some cases thousands of decimal digits beyond levels that could reasonably be dismissed as numerical artifacts. Thus while these "identities" are not firmly established in a formal sense, they are supported by very compelling numerical evidence. After all, which is more compelling, a formal proof that in its full exposition requires hundreds of difficult pages of reasoning, fully understood by only two or three colleagues, or the numerical verification of a conjecture to 100,000 decimal digit accuracy, subsequently validated by numerous subsidiary computations? In the same way, these tools are often even more useful as a way of *excluding* the possibility of hoped for relationships, as in equation (1) below.

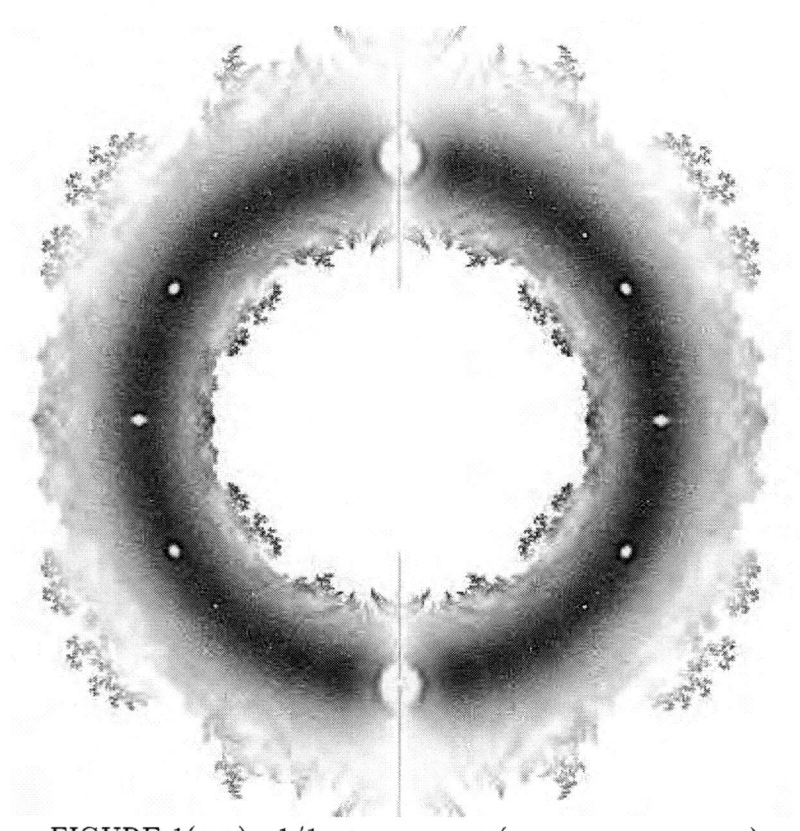

FIGURE 1(A-D): -1/1 POLYNOMIALS (TO BE SET IN COLOR)

We would be remiss not to mention the growing power of visualization especially when married to high performance computation. The pictures

Experimental Mathematics   5

in FIGURE 1 represents the zeroes of all polynomials with ±1 coefficients of degree at most 18. One of the most striking features of the picture, its fractal nature excepted, is the appearance of different sized "holes" at what transpire to be roots of unity. This observation which would be very hard to make other than pictorially led to a detailed and rigorous analysis of the phenomenon and more [17,27]. They were lead to this analysis by the interface which was built for Andrew Odlyzko's seminal online paper [32].

One additional tool that has been utilized in a growing number of studies is Sloane and Plouffe's *Encyclopedia of Integer Sequences* [36]. As the title indicates, it identifies many integer sequences based on the first few terms. A very powerful on-line version is also available and is a fine example of the changing research paradigm. Another wonderful resource is Stephen Finch's "Favorite Mathematical Constants," which contains a wealth of frequently updated information, links and references on 125 constants, [25], such as the *hard hexagon constant* $\kappa \approx 1.395485972$ for which Zimmermann obtained a minimal polynomial of degree 24 in 1996.[1]

In the following, we illustrate this – both new and old – approach to mathematical research using a handful of examples with which we are personally familiar. We will then sketch some future directions in this emerging methodology. We have focussed on the research of our own circle of direct collaborators. We do so for resaons of fmailiarity and because we believe it is representative of broad changes in the way mathematics is being done rather than to claim primacy for our own skills or expertise.

4 A New Formula for Pi

Through the centuries mathematicians have assumed that there is no shortcut to determining just the n-th digit of π. Thus it came as no small surprise when such a scheme was recently discovered [5]. In particular, this simple algorithm allows one to calculate the n-th hexadecimal (or binary) digit of π without computing any of the first $n-1$ digits, without the need for multiple-precision arithmetic software, and requiring only a very small amount of memory. The one millionth hex digit of π can be computed in this manner on a current-generation personal computer in only about 30 seconds run time.

This scheme is based on the following remarkable formula, whose formal proof involves nothing more sophisticated than freshman calculus:

$$\pi = \sum_{k=0}^{\infty} \frac{1}{16^k} \left[\frac{4}{8k+1} - \frac{2}{8k+4} - \frac{1}{8k+5} - \frac{1}{8k+6} \right]$$

This formula was found using months of PSLQ computations, after corresponding but simpler n-th digit formulas were identified for several

---
[1] See http://www.mathsoft.com/asolve/constant/square/square.html.

other constants, including log(2). This is likely the first instance in history that a significant new formula for π was discovered by a computer.

Similar base-2 formulas are given in [5,21] for a number of other mathematical constants. In [20] some base-3 formulas were obtained, including the identity

$$\pi^2 = \frac{2}{27} \sum_{k=0}^{\infty} \frac{1}{729^k} \left[\frac{243}{(12k+1)^2} - \frac{405}{(12k+2)^2} - \frac{81}{(12k+4)^2} \right.$$
$$- \frac{27}{(12k+5)^2} - \frac{72}{(12k+6)^2} - \frac{9}{(12k+7)^2}$$
$$\left. - \frac{9}{(12k+8)^2} - \frac{5}{(12k+10)^2} + \frac{1}{(12k+11)^2} \right]$$

In [8], it is shown that the question of whether π, log 2 and certain other constants are normal can be reduced to a plausible conjecture regarding dynamical iterations of the form $x_0 = 0$,

$$x_n = (bx_{n-1} + r_n) \bmod 1$$

where b is an integer and $r_n = p(n)/q(n)$ is the ratio of two nonzero polynomials with $\deg(p) < \deg(q)$. The conjecture is that these iterates either have a finite set of attractors or else are equidistributed in the unit interval. In particular, it is shown that the question of whether π is normal base 16 (and hence base 2) can be reduced to the assertion that the dynamical iteration

$$x_n = \left(16x_{n-1} + \frac{120n^2 - 89n + 16}{512n^4 - 1024n^3 + 712n^2 - 206n + 21} \right) \bmod 1$$

is equidistributed in $[0, 1)$. There are also connections between the question of normality for certain constants and the theory of linear congruential pseudorandom number generators. All of these results derive from the discovery of the individual digit-calculating formulas mentioned above. For details, see [8].

5 Identities for the Riemann Zeta Function

Another application of computer technology in mathematics is to determine whether or not a given constant α, whose value can be computed to high precision, is algebraic of some degree n or less. This can be done by first computing the vector $x = (1, \alpha, \alpha^2, \cdots, \alpha^n)$ to high precision and then applying an integer relation algorithm. If a relation is found for x, then this relation vector is precisely the set of integer coefficients of a polynomial satisfied by α. Even if no relation is found, integer relation detection programs can produce bounds within which no relation can exist. In fact, exclusions of this type are solidly established by integer relation calculations, whereas "identities" discovered in this fashion are only approximately established, as noted above.

Consider, for example, the following identities, with that for $\zeta(3)$ due to Apéry [10,14]:

$$\zeta(2) = 3\sum_{k=1}^{\infty} \frac{1}{k^2\binom{2k}{k}}$$

$$\zeta(3) = \frac{5}{2}\sum_{k=1}^{\infty} \frac{(-1)^{k-1}}{k^3\binom{2k}{k}}$$

$$\zeta(4) = \frac{36}{17}\sum_{k=1}^{\infty} \frac{1}{k^4\binom{2k}{k}}$$

where $\zeta(n) = \sum_k k^{-n}$ is the Riemann zeta function at n. These results have led many to hope that

$$Z_5 = \zeta(5)/\sum_{k=1}^{\infty} \frac{(-1)^{k-1}}{k^5\binom{2k}{k}} \tag{1}$$

might also be a simple rational or algebraic number. However, computations using PSLQ established, for instance, that if Z_5 satisfies a polynomial of degree 25 or less, then the Euclidean norm of the coefficients must exceed 2×10^{37}. Given these results, there is no "easy" identity, and researchers are licensed to investigate the possibility of multi-term identities for $\zeta(5)$. One recently discovered [14], using a PSLQ computation, was the polylogarithmic identity

$$\sum_{k=1}^{\infty} \frac{(-1)^{k+1}}{k^5\binom{2k}{k}} = 2\zeta(5) + 80\sum_{k=1}^{\infty}\left[\frac{1}{(2k)^5} - \frac{L}{(2k)^4}\right]\rho^{2k}$$
$$- \frac{4}{3}L^5 + \frac{8}{3}L^3\zeta(2) + 4L^2\zeta(3)$$

where $L = \log(\rho)$ and $\rho = (\sqrt{5} - 1)/2$. This illuastrates neatly that one can only find a closed form if one knows where to look.

Other earlier evaluations involving the central binomial coefficient suggested general formulas [12], which were pursued by a combination of PSLQ and heavy-duty symbolic manipulation. This led, most unexpectedly, to the identity

$$\sum_{k=1}^{\infty}\zeta(4k+3)z^{4k} = \sum_{k=1}^{\infty} \frac{1}{k^3(1-z^4/k^4)}$$
$$= \frac{5}{2}\sum_{k=1}^{\infty} \frac{(-1)^{k-1}}{k^3\binom{2k}{k}(1-z^4/k^4)} \prod_{m=1}^{k-1} \frac{1+4z^4/m^4}{1-z^4/m^4}.$$

Experimental analysis of the first ten terms showed that the rightmost above series necessarily had the form

$$\frac{5}{2}\sum_{k=1}^{\infty} \frac{(-1)^{k-1}P_k(z)}{k^3\binom{2k}{k}(1-z^4/k^4)}$$

David H. Bailey and Jonathan M. Borwein

where

$$P_k(z) = \prod_{j=1}^{k-1} \frac{1 + 4z^4/j^4}{1 - z^4/j^4}.$$

Also discovered in this process was the intriguing *equivalent* combinatorial identity

$$\binom{2n}{n} = \sum_{k=1}^{\infty} \frac{2n^2 \prod_{i=1}^{n-1}(4k^4 + i^4)}{k^2 \prod_{i=1, i \neq k}^{n}(k^4 - i^4)}.$$

This evaluation was discovered as the result of an serendipitous error in an input to Maple[2]— the computational equivalent of discovering penicillin after a mistake in a Petri dish.

With the recent proof of this last conjectured identity, by Almkvist and Granville [1], the above identities have now been rigorously established. But other numerically discovered "identities" of this type appear well beyond the reach of current formal proof methods. For example, in 1999 British physicist David Broadhurst used a PSLQ program to recover an explicit expression for $\zeta(20)$ involving 118 terms. The problem required 5,000 digit arithmetic and over six hours computer run time. The complete solution is given in [6].

6 Identification of Multiple Sum Constants

Numerous identities were experimentally discovered in some recent research on multiple sum constants. After computing high-precision numerical values of these constants, a PSLQ program was used to determine if a given constant satisfied an identity of a conjectured form. These efforts produced empirical evaluations and suggested general results [4]. Later, elegant proofs were found for many of these specific and general results [13], using a combination of human intuition and computer-aided symbolic manipulation. Three examples of experimentally discovered re-

---

[2] Typing 'infty' for 'infinity' revealed that the program had an algorithm when a formal variable was entered.

sults that were subsequently proven are:

$$\sum_{k=1}^{\infty}\left(1+\frac{1}{2}+\cdots+\frac{1}{k}\right)^2 (k+1)^{-4} = \frac{37}{22680}\pi^6 - \zeta^2(3)$$

$$\sum_{k=1}^{\infty}\left(1+\frac{1}{2}+\cdots+\frac{1}{k}\right)^3 (k+1)^{-6} = \zeta^3(3) + \frac{197}{24}\zeta(9) + \frac{1}{2}\pi^2\zeta(7)$$
$$- \frac{11}{120}\pi^4\zeta(5) - \frac{37}{7560}\pi^6\zeta(3)$$

$$\sum_{k=1}^{\infty}\left(1-\frac{1}{2}+\cdots+(-1)^{k+1}\frac{1}{k}\right)^2 (k+1)^{-3} = 4\operatorname{Li}_5(\tfrac{1}{2}) - \frac{1}{30}\ln^5(2)$$
$$- \frac{17}{32}\zeta(5) - \frac{11}{720}\pi^4\ln(2)$$
$$+ \frac{7}{4}\zeta(3)\ln^2(2) + \frac{1}{18}\pi^2\ln^3(2)$$
$$- \frac{1}{8}\pi^2\zeta(3)$$

where again $\zeta(n) = \sum_{j=1}^{\infty} j^{-n}$ is a value of the Riemann zeta function, and $\operatorname{Li}_n(x) = \sum_{j=1}^{\infty} x^j j^{-n}$ denotes the classical polylogarithm function.

More generally, one may define *multi-dimensional Euler sums* (or *multiple zeta values*) by

$$\zeta\begin{pmatrix} s_1, s_2 \cdots s_r \\ \sigma_1, \sigma_2 \cdots \sigma_r \end{pmatrix} := \sum_{k_1 > k_2 > \cdots > k_r > 0} \frac{\sigma_1^{k_1}}{k_1^{s_1}} \frac{\sigma_2^{k_2}}{k_2^{s_2}} \cdots \frac{\sigma_r^{k_r}}{k_r^{s_r}}$$

where $\sigma_j = \pm 1$ are signs and $s_j > 0$ are integers. When all the signs are positive, one has a multiple zeta value. The integer r is the sum's depth and $s_1 + s_2 + \cdots + s_r$ is the weight. These sums have connections with diverse fields such as knot theory, quantum field theory and combinatorics. Constants of this form with alternating signs appear in problems such as computation of the magnetic moment of the electron.

Multi-dimensional Euler sums satisfy many striking identities. The discovery of the more recondite identities was facilitated by the development of Hölder convolution algorithms that permit very high precision numerical values to be rapidly computed. See [13] and a computational interface at www.cecm.sfu.ca/projects/ezface+/. One beautiful general identity discovered by Zagier [37] in the course of similar research is

$$\zeta(3,1,3,1,\ldots,3,1) = \frac{1}{2n+1}\zeta(2,2,\ldots,2) = \frac{2\pi^{4n}}{(4n+2)!}$$

where there are n instances of '(3,1)' and '2' in the arguments to $\zeta(\cdot)$. This has now been proven in [13] and the proof, while entirely conventional, was obtained by guided experimentation. A related conjecture for which overwhelming evidence but no hint of a proof exists is the

10   David H. Bailey and Jonathan M. Borwein

"identity"
$$8^n \zeta \begin{pmatrix} 2, & 1, & 2, & 1, & \cdots, & 2, & 1 \\ -1, & 1, & -1, & 1, & \cdots, & -1, & 1 \end{pmatrix} = \zeta(2,1,2,1,\ldots,2,1).$$

Along this line, Broadhurst conjectured, based on low-degree numerical results, that the dimension of the space of Euler sums with weight w is the Fibonacci number $F_{w+1} = F_w + F_{w-1}$, with $F_1 = F_2 = 1$. In testing this conjecture, complete reductions of all Euler sums to a basis of size F_{w+1} were obtained with PSLQ at weights $w \leq 9$. At weights $w = 10$ and $w = 11$ the conjecture was stringently tested by application of PSLQ in more than 600 cases. At weight $w = 11$ such tests involve solving integer relations of size $n = F_{12} + 1 = 145$. In a typical case, each of the 145 constants was computed to more than 5,000 digit accuracy, and a working precision level of 5,000 digits was employed in an advanced "multi-pair" PSLQ program. In these problems the ratios of adjacent coefficients in the recovered integer vector usually have special values, such as $11! = 39916800$. These facts, combined with confidence ratios typically on the order of 10^{-300} in the detected relations, render remote the chance that these identities are spurious numerical artifacts, and lend substantial support to this conjecture [6].

7 Mathematical Computing Meets Parallel Computing

The potential future power of highly parallel computing technology has been underscored in some recent results. Not surprisingly, many of these computations involve the constant π, underscoring the enduring interest in this most famous of mathematical constants. In 1997 Fabrice Bellard of INRIA used a more efficient formula, similar to the one mentioned in section three, programmed on a network of workstations, to compute 150 binary digits of π starting at the *trillionth* position. Not to be outdone, 17-year-old Colin Percival of Simon Fraser University in Canada organized a computation of 80 binary digits of π beginning at the five trillionth position, using a network of 25 laboratory computers. He an many others are presently computing binary digits at the quadrillionth position on the web [33]. As we write, the most recent computational result was Yasumasa Kanada's calculation (September 1999) of the first 206 billion decimal digits of π. This spectacular computation was made on a Hitachi parallel supercomputer with 128 processors, in little over a day, and employed the Salamin-Brent algorithm [10], with a quartically convergent algorithm from [10] as an independent check.

Several large-scale parallel integer relation detection computations have also been performed in the past year or two. One arose from the discovery by Broadhurst that

$$\alpha^{630} - 1 = \frac{(\alpha^{315}-1)(\alpha^{210}-1)(\alpha^{126}-1)^2(\alpha^{90}-1)(\alpha^3-1)^3(\alpha^2-1)^5(\alpha-1)^3}{(\alpha^{35}-1)(\alpha^{15}-1)^2(\alpha^{14}-1)^2(\alpha^5-1)^6\alpha^{68}}$$

where $\alpha = 1.176280818\ldots$ is the largest real root of Lehmer's polynomial [29]

$$0 = 1 + \alpha - \alpha^3 - \alpha^4 - \alpha^5 - \alpha^6 - \alpha^7 + \alpha^9 + \alpha^{10}.$$

The above cyclotomic relation was first discovered by a PSLQ computation, and only subsequently proven. Broadhurst then conjectured that there might be integers a, b_j, c_k such that

$$a\,\zeta(17) = \sum_{j=0}^{8} b_j\, \pi^{2j} (\log \alpha)^{17-2j} + \sum_{k \in D(\mathcal{S})} c_k\, \mathrm{Li}_{17}(\alpha^{-k})$$

where the 115 indices k are drawn from the set, $D(\mathcal{S})$, of positive integers that divide at least one element of

$$\mathcal{S} = \{29, 47, 50, 52, 56, 57, 64, 74, 75, 76, 78, 84, 86, 92, 96, 98, 108, 110, 118,$$
$$124, 130, 132, 138, 144, 154, 160, 165, 175, 182, 186, 195, 204, 212, 240,$$
$$246, 270, 286, 360, 630\}.$$

Indeed, such a relation was found, using a parallel multi-pair PSLQ program running on a SGI/Cray T3E computer system at Lawrence Berkeley Laboratory. The run employed 50,000 decimal digit arithmetic and required approximately 44 hours on 32 processors. The resulting integer coefficients are as large as 10^{292}, but the "identity" nonetheless was confirmed to 13,000 digits beyond the level of numerical artifact [7].

8 Connections to Quantum Field Theory

In another surprising recent development, David Broadhurst has found, using these methods, that there is an intimate connection between Euler sums and constants resulting from evaluation of Feynman diagrams in quantum field theory [18,19]. In particular, the renormalization procedure (which removes infinities from the perturbation expansion) involves multiple zeta values. As before, a fruitful theory has emerged, including a large number of both specific and general results [13].

Some recent quantum field theory results are even more remarkable. Broadhurst has now shown [20], using PSLQ computations, that in each of ten cases with unit or zero mass, the finite part the scalar 3-loop tetrahedral vacuum Feynman diagram reduces to 4-letter "words" that represent iterated integrals in an alphabet of seven "letters" comprising the one-forms $\Omega := dx/x$ and $\omega_k := dx/(\lambda^{-k} - x)$, where $\lambda := (1 + \sqrt{-3})/2$ is the primitive sixth root of unity, and k runs from 0 to 5. A 4-letter word is a 4-dimensional iterated integral, such as

$$U := \zeta(\Omega^2 \omega_3 \omega_0) =$$
$$\int_0^1 \frac{dx_1}{x_1} \int_0^{x_1} \frac{dx_2}{x_2} \int_0^{x_2} \frac{dx_3}{(-1-x_3)} \int_0^{x_3} \frac{dx_4}{(1-x_4)} = \sum_{j>k>0} \frac{(-1)^{j+k}}{j^3 k}.$$

12   David H. Bailey and Jonathan M. Borwein

There are 7^4 such four-letter words. Only two of these are primitive terms occurring in the 3-loop Feynman diagrams: U, above, and

$$V := \mathrm{Real}[\zeta(\Omega^2 \omega_3 \omega_1)] = \sum_{j>k>0} \frac{(-1)^j \cos(2\pi k/3)}{j^3 k}.$$

The remaining terms in the diagrams reduce to products of constants found in Feynman diagrams with fewer loops. These ten cases as shown in Figure 1. In these diagrams, dots indicate particles with nonzero rest mass. The formulas that have been found, using PSLQ, for the corresponding constants are given in Table 2. In the table the constant $C = \sum_{k>0} \sin(\pi k/3)/k^2$.

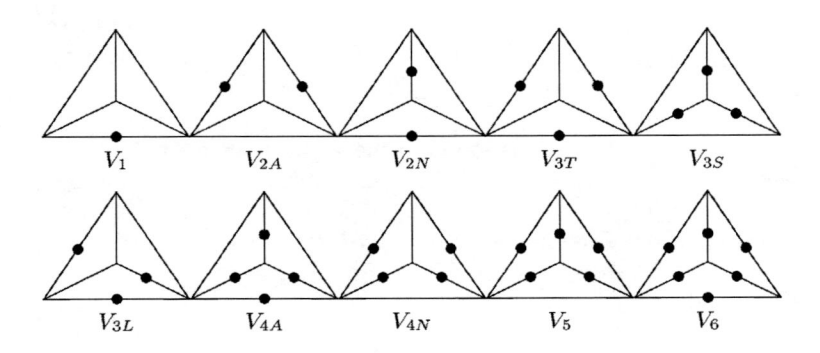

Fig. 1. The ten tetrahedral cases

V_1	$= 6\zeta(3) + 3\zeta(4)$
V_{2A}	$= 6\zeta(3) - 5\zeta(4)$
V_{2N}	$= 6\zeta(3) - \frac{13}{2}\zeta(4) - 8U$
V_{3T}	$= 6\zeta(3) - 9\zeta(4)$
V_{3S}	$= 6\zeta(3) - \frac{11}{2}\zeta(4) - 4C^2$
V_{3L}	$= 6\zeta(3) - \frac{15}{4}\zeta(4) - 6C^2$
V_{4A}	$= 6\zeta(3) - \frac{77}{12}\zeta(4) - 6C^2$
V_{4N}	$= 6\zeta(3) - 14\zeta(4) - 16U$
V_5	$= 6\zeta(3) - \frac{469}{27}\zeta(4) + \frac{8}{3}C^2 - 16V$
V_6	$= 6\zeta(3) - 13\zeta(4) - 8U - 4C^2$

Table 1. Formulas found by PSLQ for the ten cases of Figure 1

9 A Note of Caution

In spite of the remarkable successes of this methodology, some caution is in order. First of all, the fact that an identity is established to high precision is *not* a guarantee that it is indeed true. One example is

$$\sum_{n=1}^{\infty} \frac{[n \tanh \pi]}{10^n} \approx \frac{1}{81}$$

which holds to 267 digits, yet is not an exact identity, failing in the 268'th place. Several other such bogus "identities" are exhibited and explained in [11].

More generally speaking, caution must be exercised when extrapolating results true for small n to all n. For example,

$$\int_0^\infty \frac{\sin(x)}{x} \, dx = \frac{\pi}{2}$$

$$\int_0^\infty \frac{\sin(x)}{x} \frac{\sin(x/3)}{x/3} \, dx = \frac{\pi}{2}$$

$$\cdots$$

$$\int_0^\infty \frac{\sin(x)}{x} \frac{\sin(x/3)}{x/3} \cdots \frac{\sin(x/13)}{x/13} \, dx = \frac{\pi}{2}$$

yet

$$\int_0^\infty \frac{\sin(x)}{x} \frac{\sin(x/3)}{x/3} \cdots \frac{\sin(x/15)}{x/15} \, dx = \frac{467807924713440738696537864469}{935615849440640907310521750000} \pi.$$

When this fact was recently observed by a researcher using a mathematical software package, he concluded that there must be a "bug" in the software. Not so. What is happening here is that

$$\int_0^\infty \frac{\sin(x)}{x} \frac{\sin(x/h_1)}{x/h_1} \cdots \frac{\sin(x/h_n)}{x/h_n} \, dx = \frac{\pi}{2}$$

only so long as $1/h_1 + 1/h_2 + \cdots + 1/h_n < 1$. In the above example, $1/3 + 1/5 + \cdots + 1/13 < 1$, but with the addition of $1/15$, the sum exceeds 1 and the identity no longer holds [9]. Changing the h_n lets this pattern persist indefinitely but still fail in the large.

10 Future Outlook

Computer mathematics software is now becoming a staple of university departments and government research laboratories. Many university departments now offer courses where the usage of one of these software packages is an integral part of the course. But further expansion of these facilities into high schools has been inhibited by a number of factors, including the fairly high cost of such software, the lack of appropriate computer equipment, difficulties in standardizing such coursework at a regional or national level, a paucity of good texts incorporating such tools into a realistic curriculum, lack of trained teachers and many other demands on their time.

But computer hardware continues its downward spiral in cost and its upward spiral in power. It thus appears that within a very few years, moderately powerful symbolic computation facilities can be incorporated into relatively inexpensive hand calculators, at which point it will be much easier to successfully integrate these tools into high school curricula. Thus it seems that we are poised to see a new generation of students coming into university mathematics and science programs who are completely comfortable using such tools. This development is bound to have a profound impact on the future teaching, learning and doing of mathematics.

A likely and fortunate spin-off of this development is that the commercial software vendors who produce these products will likely enjoy a broader financial base, from which they can afford to further enhance their products geared at serious researchers. Future enhancements are likely to include more efficient algorithms, more extensive capabilities mixing numerics and symbolics, more advanced visualization facilities, and software optimized for emerging symmetric multiprocessor and highly parallel, distributed memory computer systems. When combined with expected increases in raw computing power due to Moore's Law — improvements which almost certainly will continue unabated for at least ten years and probably much longer — we conclude that enormously more powerful computer mathematics systems will be available in the future.

We only now are beginning to experience and comprehend the potential impact of computer mathematics tools on mathematical research. In ten more years, a new generation of computer-literate mathematicians, armed with significantly improved software on prodigiously powerful computer systems, are bound to make discoveries in mathematics that we can only dream of at the present time. Will computer mathematics eventually replace, in near entirety, the solely human form of research, typified by Andrew Wiles' recent proof of Fermat's Last Theorem? Will computer mathematics systems eventually achieve such intelligence that they discover deep new mathematical results, largely or entirely without human assistance? Will new computer-based mathematical discovery techniques enable mathematicians to explore the realm, proved to exist by Gödel, Chaitin and others, that is fundamentally beyond the limits of formal reasoning?

11 Conclusion

We have shown a small but we hope convincing selection of what the present allows and what the future holds in store. We have hardly mentioned the growing ubiquity of web based computation, or of pervasive access to massive data bases, both public domain and commercial. Neither have we raised the human/computer interface or intellectual property issues and the myriad other not-purely-technical issues these raise.

Whatever the outcome of these developments, we are still persuaded that mathematics is and will remain a uniquely human undertaking. One could even argue that these developments confirm the fundamentally human nature of mathematics. Indeed, Reuben Hersh's arguments [26] for a humanist philosophy of mathematics, as paraphrased below, become more convincing in our setting:

1. *Mathematics is human.* It is part of and fits into human culture. It does not match Frege's concept of an abstract, timeless, tenseless, objective reality.
2. *Mathematical knowledge is fallible.* As in science, mathematics can advance by making mistakes and then correcting or even re-correcting them. The "fallibilism" of mathematics is brilliantly argued in Lakatos' *Proofs and Refutations* [28].
3. *There are different versions of proof or rigor.* Standards of rigor can vary depending on time, place, and other things. The use of computers in formal proofs, exemplified by the computer-assisted proof of the four color theorem in 1977, is just one example of an emerging nontraditional standard of rigor.
4. *Empirical evidence, numerical experimentation and probabilistic proof all can help us decide what to believe in mathematics.* Aristotelian logic isn't necessarily always the best way of deciding.
5. *Mathematical objects are a special variety of a social-cultural-historical object.* Contrary to the assertions of certain post-modern detractors, mathematics cannot be dismissed as merely a new form of literature or religion. Nevertheless, many mathematical objects can be seen as shared ideas, like Moby Dick in literature, or the Immaculate Conception in religion.

Certainly the recognition that "quasi-intuitive" analogies can be used to gain insight in mathematics can assist in the learning of mathematics. And honest mathematicians will acknowledge their role in discovery as well.

We look forward to what the future will bring.

References

1. G. Almkvist and A. Granville, "Borwein and Bradley's Apéry-like formulae for $\zeta(4n+3)$", *Experimental Mathematics* **8** (1999), 197–204.

16   David H. Bailey and Jonathan M. Borwein

2. Issac Asimov and J. A. Shulman, ed., *Isaac Asimov's Book of Science and Nature Quotations*, Weidenfield and Nicolson, New York, 1988, pg. 115.
3. David H. Bailey, "A Fortran-90 Based Multiprecision System", *ACM Transactions on Mathematical Software*, **21** (1995), pg. 379-387. Available from http://www.nersc.gov/~dhbailey.
4. David H. Bailey, Jonathan M. Borwein and Roland Girgensohn, "Experimental Evaluation of Euler Sums", *Experimental Mathematics*, 4 (1994), 17–30.
5. David H. Bailey, Peter B. Borwein and Simon Plouffe, "On The Rapid Computation of Various Polylogarithmic Constants", *Mathematics of Computation*, **66**,(1997, 903–913.
6. David H. Bailey and David Broadhurst, "Parallel Integer Relation Detection: Techniques and Applications". Available from http://www.nersc.gov/~dhbailey.
7. David H. Bailey and David Broadhurst, "A Seventeenth-Order Polylogarithm Ladder". Available from http://www.nersc.gov/~dhbailey.
8. David H. Bailey and Richard E. Crandall, "On the Random Character of Fundamental Constant Expansions", manuscript (2000). Available from http://www.nersc.gov/~dhbailey.
9. David Borwein and Jonathan M. Borwein, "Some Remarkable Properties of Sinc and Related Integrals", CECM Preprint 99:142, available from http://www.cecm.sfu.ca/preprints.
10. Jonathan M. Borwein and Peter B. Borwein, *Pi and the AGM: A Study in Analytic Number Theory and Computational Complexity*, John Wiley and Sons, New York, 1987.
11. J. M. Borwein and P. B. Borwein, "Strange Series and High Precision Fraud", *American Mathematical Monthly*, **99** (1992), 622–640.
12. J.M. Borwein and D.M. Bradley, "Empirically determined Apéry–like formulae for zeta($4n+3$)," *Experimental Mathematics*, **6** (1997), 181–194.
13. Jonathan M. Borwein, David M. Bradley, David J. Broadhurst and Peter Lisonek, "Special Values of Multidimensional Polylogarithms", *Trans. Amer. Math. Soc.*, in press. CECM Preprint 98:106, available from http://www.cecm.sfu.ca/preprints.
14. Jonathan M. Borwein, David J. Broadhurst and Joel Kamnitzer, "Central binomial sums and multiple Clausen values," preprint, November 1999. CECM Preprint 99:137, , available from http://www.cecm.sfu.ca/preprints.
15. J.M. Borwein, P.B. Borwein, R. Girgensohn and S. Parnes, "Making Sense of Experimental Mathematics," *Mathematical Intelligencer*, **18**, Number 4 (Fall 1996), 12–18.
16. Jonathan M. Borwein and Robert Corless, "Emerging tools for experimental mathematics," *MAA Monthly*, **106**(1999), 889–909. CECM Preprint 98:110, , available from http://www.cecm.sfu.ca/preprints.
17. Peter. B. Borwein and Christopher Pinner, "Polynomials with $\{0, +1, -1\}$ Coefficients and Root Close to a Given Point," *Canadian J. Mathematics* **49** (1998), 887–915.

18. David J. Broadhurst, John A. Gracey and Dirk Kreimer, "Beyond the Triangle and Uniqueness Relations: Non-zeta Counterterms at Large N from Positive Knots", *Zeitschrift für Physik*, **C75** (1997), 559–574.
19. David J. Broadhurst and Dirk Kreimer, "Association of Multiple Zeta Values with Positive Knots via Feynman Diagrams up to 9 Loops", *Physics Letters*, **B383** (1997), 403–412.
20. David J. Broadhurst, "Massive 3-loop Feynman Diagrams Reducible to SC* Primitives of Algebras of the Sixth Root of Unity", preprint, March 1998, to appear in *European Physical Journal C*. Available from http://xxx.lanl.gov/abs/hep-th/9803091 .
21. David J. Broadhurst, "Polylogarithmic Ladders, Hypergeometric Series and the Ten Millionth Digits of $\zeta(3)$ and $\zeta(5)$", preprint, March 1998. Available from http://xxx.lanl.gov/abs/math/9803067.
22. Sid Chatterjee and Herman Harjono, "MPFUN++: A Multiple Precision Floating Point Computation Package in C++", University of North Carolina, Sept. 1998. Available from http://www.cs.unc.edu/Research/HARPOON/mpfun++.
23. Helaman R. P. Ferguson, David H. Bailey and Stephen Arno, "Analysis of PSLQ, An Integer Relation Finding Algorithm", *Mathematics of Computation*, **68** (1999), 351–369.
24. Helaman R. P. Ferguson and Rodney W. Forcade, "Generalization of the Euclidean Algorithm for Real Numbers to All Dimensions Higher Than Two", *Bulletin of the American Mathematical Society*, **1** (1979), 912–914.
25. Stphen Finch, "Favorite Mathematical Constants", http://www.mathsoft.com/asolve/constant/constant.html.
26. Reuben Hersh, "Fresh Breezes in the Philosophy of Mathematics", the *American Mathematical Monthly*, August-September 1995, 589–594.
27. Loki Jörgenson, "Zeros of Polynomials with Constrained Roots", http//www.cecm.sfu.ca/personal/loki/Projects/Roots/Book.
28. Imre Lakatos, *Proofs and Refutations: The Logic of Mathematical Discovery*, Cambridge University Press, 1977.
29. Derrick H. Lehmer, "Factorization of Certain Cyclotomic Functions", *Annals of Mathematics*, **34** (1933), 461–479.
30. A. K. Lenstra, H. W. Lenstra, Jr. and L. Lovasz, "Factoring Polynomials with Rational Coefficients", *Mathematische Annalen*, **261** (1982), 515-534.
31. P. B. Medawar, *Advice to a young Scientist*, Harper Colophon, New York, 1981.
32. Andrew Odlyzko, "Zeros of polynomials with 0,1 coefficients", http://www.cecm.sfu.ca/organics/authors/odlyzko/ and /organics/papers/odlyzko/support/polyform.html.
33. Colin Percival, "Pihex: A Distributed Effort To Calculate Pi", http://www.cecm.sfu.ca/projects/pihex/.
34. George Polya, *Mathematical Discovery: On Understanding, Learning, and Teaching Problem Solving*, Combined Edition, New York, Wiley and Sons, 1981, pg. 129.

35. Ed Regis, *Who Got Einstein's Office?*, Addison-Wesley, 1986, pg. 78.
36. N.J.A. Sloane and Simon Plouffe, *The Encyclopedia of Integer Sequences*, Academic Press, 1995. The on-line version can be accessed at
`http://www.research.att.com/~njas/sequences/Seis.html`.
37. Don Zagier, *Values of zeta functions and their applications*, First European Congress of Mathematics, Volume II, Birkhäuser, Boston, 1994, 497–512.

8. Visible structures in number theory

Discussion

Videre est credere—Seeing is believing. This ancient dictum is, like most folk wisdom, both right and wrong. Seeing is both believing and a caution against believing. At best, like Figure 11 of the accompanying paper, subtle and interesting phenomena are suggested and patterns are seen that can't be found by paper and pencil. However, patterns and pictures can also be highly misleading. (See Figures 9 and 10.)

Mathematics is a highly logical field, and the process of presentation and proof is very structured with well-defined rules. But few if any mathematicians discover mathematics in a highly structured way, and many mathematicians (as Hadamard suggests) approach mathematics in an unstructured visual way. It is only recently with sophisticated mathematical tools like *Mathematica* or *Maple* or *Matlab* that it has become possible to visually explore the complicated worlds of things like fractals.

Learning how to "see mathematics" is, we would suggest, an exciting new direction in mathematics. With the sophisticated computational and visualization tools now available, the reemergence of the experimental side of mathematics is given extraordinary currency and vibrancy.

Additionally, as John Edensor Littlewood wrote long before the current fine graphic, visualization and geometric tools were available:

> A heavy warning used to be given [by lecturers] that pictures are not rigorous; this has never had its bluff called and has permanently frightened its victims into playing for safety. Some pictures, of course, are not rigorous, but I should say most are (and I use them whenever possible myself).[1] (1885-1977).

Source

P. Borwein and L. Jörgenson, "Visible structures in number theory," *MAA Monthly*, **108** (2001), 897–910. Featured in the Math Forum newsletter (Feb. 26, 2001).

---

[1] From *Littlewood's Miscellany* (p. 35 in 1953 edition).

Visible Structures in Number Theory

Peter Borwein and Loki Jörgenson

1. INTRODUCTION.

> I see a confused mass.   —Jacques Hadamard (1865–1963)

These are the words the great French mathematician used to describe his initial thoughts when he proved that there is a prime number greater than 11 [**11**, p. 76]. His final mental image he described as "...a place somewhere between the confused mass and the first point". In commenting on this in his fascinating but quirky monograph, he asks "What may be the use of such a strange and cloudy imagery?".

Hadamard was of the opinion that mathematical thought is visual and that words only interfered. And when he inquired into the thought processes of his most distinguished mid-century colleaugues, he discovered that most of them, in some measure, agreed (a notable exception being George Pólya).

For the non-professional, the idea that mathematicians "see" their ideas may be surprising. However the history of mathematics is marked by many notable developments grounded in the visual. Descartes' introduction of "cartesian" co–ordinates, for example, is arguably the most important advance in mathematics in the last millenium. It fundamentally reshaped the way mathematicians thought about mathematics, precisely because it allowed them to "see" better mathematically.

Indeed, mathematicians have long been aware of the significance of visualization and made great effort to exploit it. Carl Friedrich Gauss lamented, in a letter to Heinrich Christian Schumacher, how hard it was to draw the pictures required for making accurate conjectures. Gauss, whom many consider the greatest mathematician of all time, in reference to a diagram that accompanies his first proof of the fundamental theorem of algebra, wrote

> It still remains true that, with negative theorems such as this, transforming personal convictions into objective ones requires deterringly detailed work. To visualize the whole variety of cases, one would have to display a large number of equations by curves; each curve would have to be drawn by its points, and determining a single point alone requires lengthy computations. You do not see from Fig. 4 in my first paper of 1799, how much work was required for a proper drawing of that curve.   —Carl Friedrich Gauss (1777–1855)

The kind of pictures Gauss was looking for would now take seconds to generate on a computer screen.

Newer computational environments have greatly increased the scope for visualizing mathematics. Computer graphics offers magnitudes of improvement in resolution and speed over hand–drawn images and provides increased utility through color, animation, image processing, and user interactivity. And, to some degree, mathematics has evolved to exploit these new tools and techniques. We explore some subtle uses of interactive graphical tools that help us "see" the mathematics more clearly. In particular, we focus on cases where the right picture suggests the "right theorem", or where it indicates structure where none was expected, or where there is the possibility of "visual proof".

For all of our examples, we have developed Internet-accessible interfaces. They allow readers to interact and explore the mathematics and possibly even discover new results of their own—visit www.cecm.sfu.ca/projects/numbers/.

2. IN PURSUIT OF PATTERNS.

> Computers make it easier to do a lot of things, but most of the things they make it easier to do don't need to be done.
> —Andy Rooney

Mathematics can be described as the science of patterns, relationships, generalized descriptions, and recognizable structure in space, numbers, and other abstracted entities. This view is borne out in numerous examples such as [16] and [15]. Lynn Steen has observed [19]:

> Mathematical theories explain the relations among patterns; functions and maps, operators and morphisms bind one type of pattern to another to yield lasting mathematical structures. Application of mathematics use these patterns to "explain" and predict natural phenomena that fit the patterns. Patterns suggest other patterns, often yielding patterns of patterns.

This description conjures up images of cycloids, Sierpinski gaskets, "cowboy hat" surfaces, and multi-colored graphs. However it isn't immediately apparent that this patently visual reference to patterns applies throughout mathematics. Many of the higher order relationships in fields such as number theory defy pictorial representation or, at least, they don't immediately lend themselves intuitively to a graphic treatment. Much of what is "pattern" in the knowledge of mathematics is instead encoded in a linear textual format born out of the logical formalist practices that now dominate mathematics.

Within number theory, many problems offer large amounts of "data" that the human mind has difficulty assimilating directly. These include classes of numbers that satisfy certain criteria (e.g., primes), distributions of digits in expansions, finite and infinite series and summations, solutions to variable expressions (e.g., zeroes of polynomials), and other unmanageable masses of raw information. Typically, real insight into such problems has come directly from the mind of the mathematician who ferrets out their essence from formalized representations rather than from the data. Now computers make it possible to "enhance" the human perceptual/cognitive systems through many different kinds of visualization and patterns of a new sort emerge in the morass of numbers.

However the epistemological role of computational visualization in mathematics is still not clear, certainly not any clearer than the role of intuition where mental visualization takes place. However, it serves several useful functions in current practice. These include inspiration and discovery, informal communication and demonstration, and teaching and learning. Lately though, the area of experimental mathematics has expanded to include exploration and experimentation and, perhaps controversially, formal exposition and proof. Some carefully crafted questions have been posed about how experiment might contribute to mathematics [5]. Yet answers have been slow to come, due in part to general resistance and, in some cases, alarm [11] within the mathematical community. Moreover, experimental mathematics finds only conditional support from those who address the issues formally [7], [9].

The value of visualization hardly seems to be in question. The real issue seems to be what it can be used for. Can it contribute directly to the body of mathematical knowledge? Can an image act as a form of "visual proof"? Strong cases can be made to the affirmative [7], [3] (including in number theory), with examples typically in

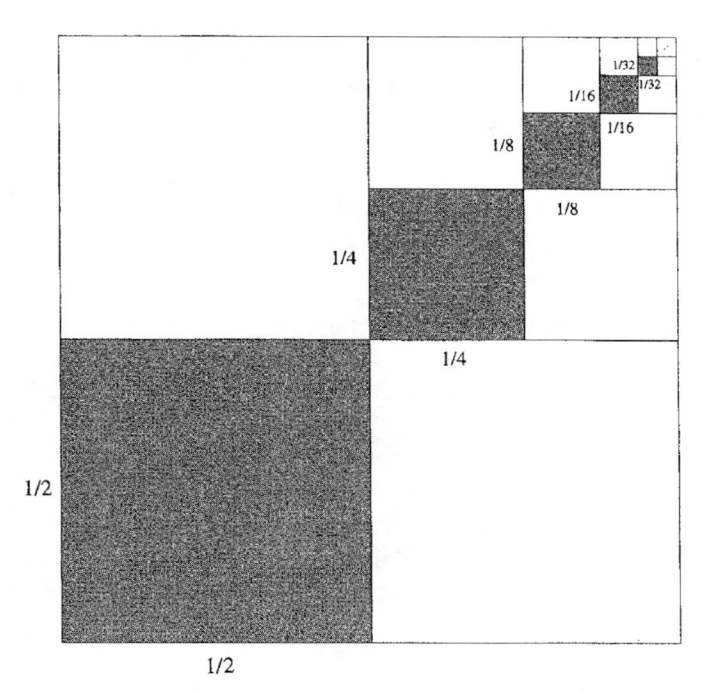

Figure 1. A simple "visual proof" of $\sum_{n=1}^{\infty}(\frac{1}{2})^{2n} = \frac{1}{3}$

the form of simplified, heuristic diagrams such as Figure 1. These carefully crafted examples call into question the epistemological criteria of an acceptable proof.

Establishing adequate criteria for mathematical proof is outside the scope of this paper; interested readers can visit [20], a repository for information related to reasoning with visual representations. Its authors suggest that three necessary, but perhaps not sufficient, conditions may be:

- *reliability*: the underlying means of arriving at the proof are reliable and the result is unvarying with each inspection
- *consistency*: the means and end of the proof are consistent with other known facts, beliefs, and proofs
- *repeatability*: the proof may be confirmed by or demonstrated to others

Each requirement is difficult to satisfy in a single, static visual representation. Most criticisms of images as mathematical knowledge or tools make this clear [8], [13].

Traditional exposition differs significantly from that of the visual. In the logical formal mode, proof is provided in linearly connected sentences composed of words that are carefully selected to convey unambiguous meaning. Each sentence follows the previous, specifying an unalterable path through the sequence of statements. Although error and misconception are still possible, the tolerances are extremely demanding and follow the strict conventions of deductivist presentation [12].

In graphical representations, the same facts and relationships are often presented in multiple modes and dimensions. For example, the path through the information is usually indeterminate, leaving the viewer to establish what is important (and what is not) and in what order the dependencies should be assessed. Further, unintended information and relationships may be perceived, either due to the unanticipated interaction of the complex array of details or due to the viewer's own perceptual and cognitive processes.

As a consequence, successful visual representations tend to be spartan in their detail. And the few examples of visual proof that withstand close inspection are limited in their scope and generalizability. The effort to bring images closer to conformity with the prevailing logical modes of proof has resulted in a loss of the richness that is intrinsic to the visual.

3. IN SUPPORT OF PROOF.

> Computers are useless. They can only give you answers. —Pablo Picasso (1881–1973)

In order to offer the reliability, consistency, and repeatability of the written word and still provide the potential inherent in the medium, visualization needs to offer more than just the static image. It too must guide, define, and relate the information presented. The logical formalist conventions for mathematics have evolved over many decades, resulting in a mode of discourse that is precise in its delivery. The order of presentation of ideas is critical, with definitions preceding their usage, proofs separated from the general flow of the argument for modularity, and references to foundational material listed at the end.

To do the same, visualization must include additional mechanisms or conventions beyond the base image. It isn't appropriate simply to ape the logical conventions and find some visual metaphor or mapping that works similarly (this approach is what limits existing successful visual proofs to very simple diagrams). Instead, an effective visualization needs to offer several key features

- *dynamic*: the representation should vary through some parameter(s) to demonstrate a range of behaviours (instead of the single instance of the static case)
- *guidance*: to lead the viewer through the appropriate steps in the correct order, the representation should offer a "path" through the information that builds the case for the proof
- *flexibility*: it should support the viewer's own exploration of the ideas presented, including the search for counterexamples or incompleteness
- *openness*: the underlying algorithms, libraries, and details of the programming languages and hardware should be available for inspection and confirmation

With these capabilities available in an interactive representation, the viewer could then follow the argument being made visually, explore all the ramifications, check for counterexamples, special cases, and incompleteness, and even confirm the correctness of the implementation. In fact, the viewer should be able to inspect a visual representation and a traditional logical formal proof with the same rigor.

Although current practice does not yet offer any conclusions as to how images and computational tools may impact mathematical methodologies or the underlying epistemology, it does indicate the direction that subsequent work may take. Examples offer some insight into how emerging technologies may eventually provide an unambiguous role for visualization in mathematics.

4. THE STRUCTURE OF NUMBERS. Numbers may be generated by a myriad of means and techniques. Each offers a very small piece of an infinitely large puzzle. Number theory identifies patterns of relationship between numbers, sifting for the subtle suggestions of an underlying fundamental structure. The regularity of observable features belies the seeming abstractness of numbers.

Chapter 8

Figure 2. The first 1600 decimal digits of π mod 2.

Binary Expansions. In the 17th century, Gottfried Wilhelm Leibniz asked in a letter to one of the Bernoulli brothers if there might be a pattern in the binary expansion of π. Three hundred years later, his question remains unanswered. The numbers in the expansion appear to be completely random. In fact, the most that can now be said of any of the classical mathematical constants is that they are largely non-periodic.

With traditional analysis revealing no patterns of interest, generating images from the expansions offers intriguing alternatives. Figures 2 and 3 show 1600 decimal digits of π and 22/7 respectively, both taken mod 2. The light pixels are the even digits and the dark ones are the odd. The digits read from left to right, top to bottom, like words in a book.

What does one see? The even and odd digits of π in Figure 2 seem to be distributed randomly. And the fact that 22/7 (a widely used approximation for π) is rational appears clearly in Figure 3. Visually representing randomness is not a new idea; Pickover [18] and Voelcker [22] have previously examined the possibility of "seeing randomness". Rather, the intention here is to identify patterns where none has so far been seen, in this case in the expansions of irrational numbers.

These are simple examples but many numbers have structures that are hidden both from simple inspection of the digits and even from standard statistical analysis. Figure 4 shows the rational number 1/65537, this time as a binary expansion, with a period of 65536. Unless graphically represented with sufficient resolution, the presence of a regularity might otherwise be missed in the unending string of 0's and 1's.

Figures 5 a) and b) are based on similar calculations using 1600 terms of the simple continued fractions of π and e respectively. Continued fractions have the form

$$\cfrac{1}{a_1 + \cfrac{1}{a_2 + \cfrac{1}{a_3 + \cdots}}}$$

December 2001]   VISIBLE STRUCTURES IN NUMBER THEORY

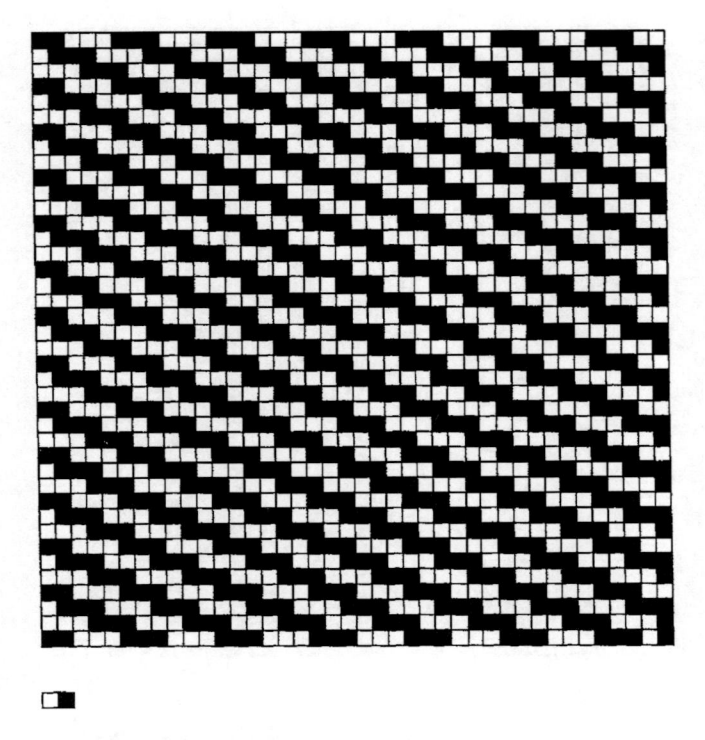

Figure 3. The first 1600 decimal digits of 22/7 mod 2.

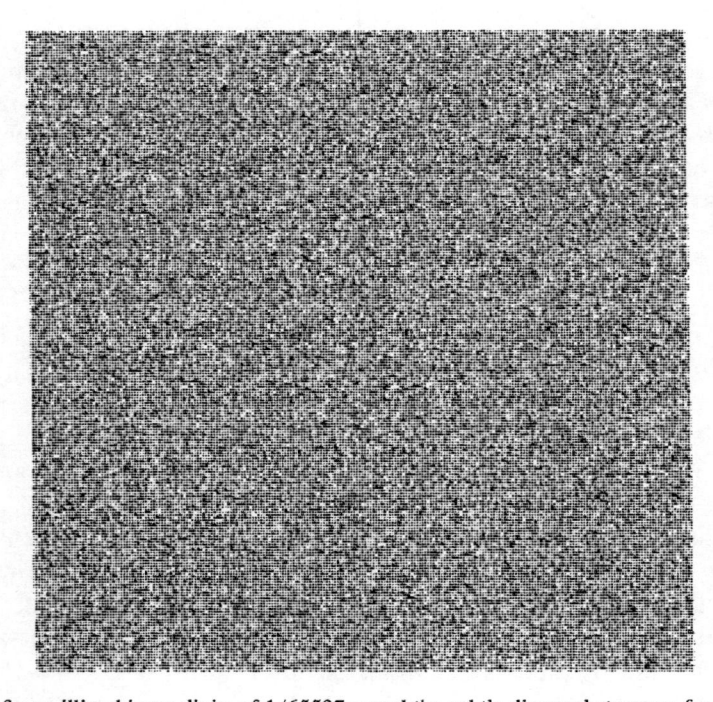

Figure 4. The first million binary digits of 1/65537 reveal the subtle diagonal structure from the periodicity.

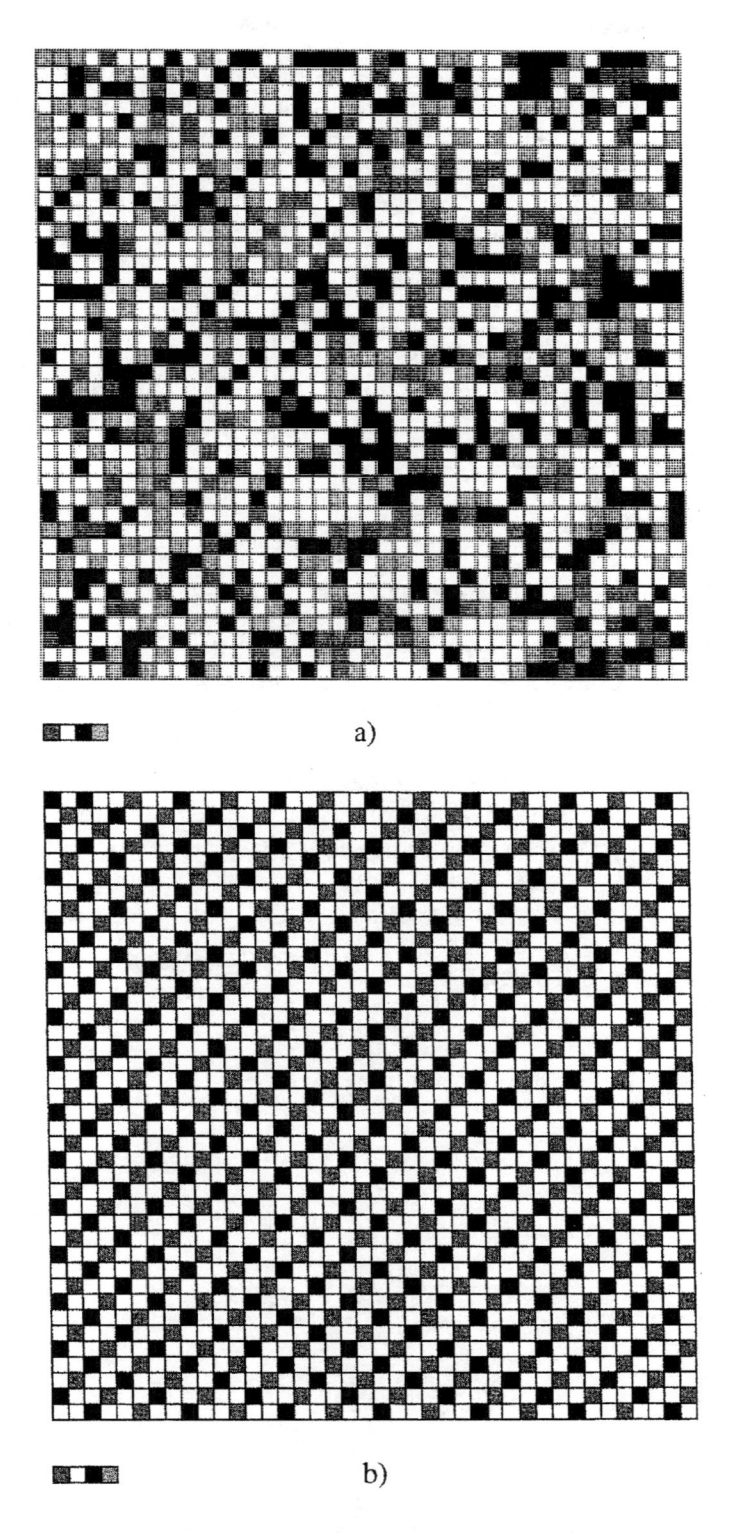

Figure 5. The first 1600 values of the continued fraction for a) π and b) e, both mod 4

In these images, the decimal values have been taken mod 4. Again the distribution of the a_i of π appears random though now, as one would expect, there are more odds than evens. However for e, the pattern appears highly structured. This is no surprise on closer examination, as the continued fraction for e is

$$[2, 1, 2, 1, 1, 4, 1, 1, 6, 1, 1, 8, 1, 1, 10, 1, 1, 12, \ldots]$$

and, if taken as a sequence of digits, is a rational number mod 4. It is apparent from the images that the natures of the various distributions are quite distinct and recognizable. In contrast no such simple pattern exists for exp(3) mod 4.

Presumably this particular visual representation offers a qualitative characterization of the numbers. It tags them in an instantly distinguishable fashion that would be almost impossible to do otherwise.

Sequences of Polynomials.

> Few things are harder to put up with than the annoyance of a good example.
> —Mark Twain (1835–1910)

In a similar vein, structures are found in the coefficients of sequences of polynomials. The first example in Figure 6 shows the binomial coefficients $\binom{n}{m}$ mod 3, or equivalently Pascal's Triangle mod 3. For the sake of what follows, it is convenient to think of the ith row as the coefficients of the polynomial $(1 + x)^i$ taken mod three. This apparently fractal pattern has been the object of much careful study [10].

Figure 7 shows the coefficients of the first eighty Chebyshev polynomials mod 3 laid out like the binomial coefficients of Figure 6. Recall that the nth Chebyshev polynomial T_n, defined by $T_n(x) := \cos(n \arccos x)$, has the explicit representation

$$T_n(x) = \frac{n}{2} \sum_{k=0}^{\lfloor n/2 \rfloor} (-1)^k \frac{(n-k-1)!}{k!(n-2k)!} (2x)^{n-2k},$$

and satisfies the recursion

$$T_n(x) = 2x T_{n-1}(x) - T_{n-2}(x), \quad n = 2, 3, \ldots.$$

Figure 6. Eighty rows of Pascal's Triangle mod 3

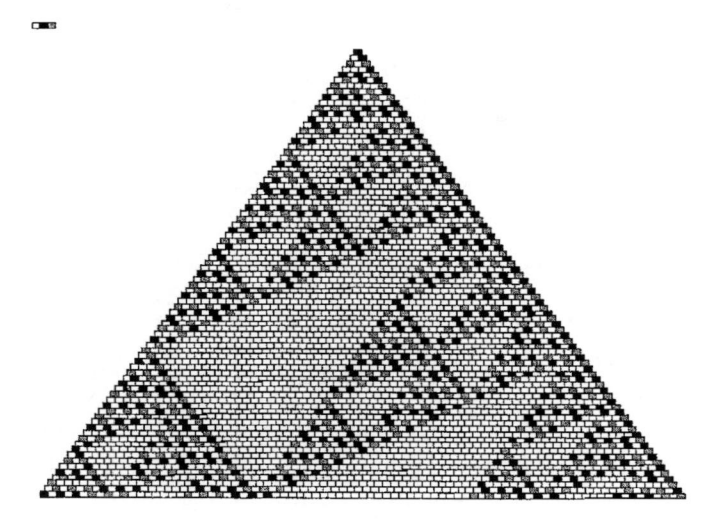

Figure 7. Eighty Chebyshev Polynomials mod 3

The expression for $T_n(x)$ resembles the $\binom{n}{m}$ form of the binomial coefficients and its recursion relation is similar to that for the Pascal's Triangle.

Figure 8 shows the Stirling numbers of the second kind mod 3, again organized as a triangle. They are defined by

$$S(n,m) := \frac{1}{m!} \sum_{k=0}^{m} \binom{m}{k} (-1)^{m-k} k^n$$

and give the number of ways of partitioning a set of n elements into m non-empty subsets. Once again the form of $\binom{n}{m}$ appears in its expression.

The well-known forms of the polynomials appear distinct. Yet it is apparent that the polynomials are graphically related to each other. In fact, the summations are variants of the binomial coefficient expression.

It is possible to find similar sorts of structure in virtually any sequence of polynomials: Legendre polynomials; Euler polynomials; sequences of Padé denominators to the exponential or to $(1-x)^\alpha$ with α rational. Then, selecting any modulus, a distinct

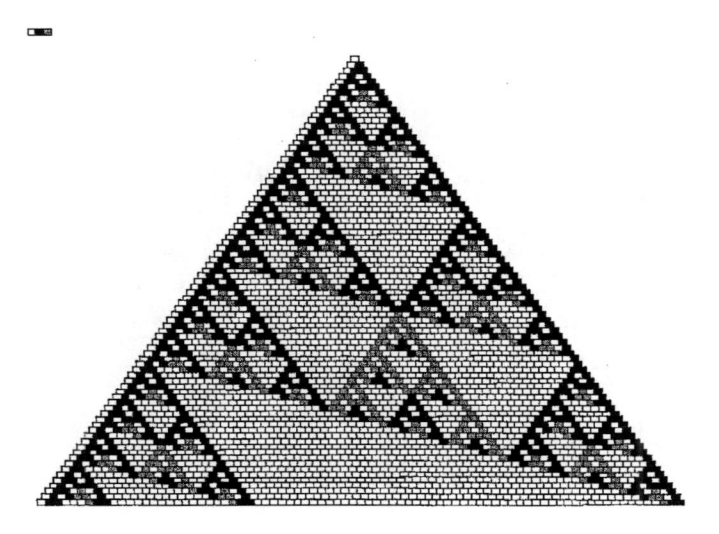

Figure 8. Eighty rows of Stirling Numbers of the second kind mod 3

pattern emerges. These images indicate an underlying structure within the polynomials themselves and demand some explanation. While conjectures exist for their origin, proofs for the theorems suggested by these pictures do not yet exist. And when there finally is a proof, might it be offered in some visual form?

Quasi-Rationals.

> For every problem, there is one solution which is simple, neat, and wrong.
> —H.L. Mencken (1880–1956)

Having established a visual character for irrationals and their expansions, it is interesting to note the existence of "quasi-rational" numbers. These are certain well-known irrational numbers whose images appear suspiciously rational. The sequences pictured in Figures 9 and 10 are $\{i\pi\}_{i=1}^{1600}$ mod 2 and $\{ie\}_{i=1}^{1600}$ mod 2, respectively. One way of thinking about these sequences is as binary expansions of the numbers

$$\sum_{n=1}^{\infty} \frac{[m\alpha] \bmod 2}{2^i},$$

where α is, respectively, π or e.

The resulting images are very regular. And yet these are transcendental numbers; having observed this phenomenon, we were subsequently able to explain this behavior rigorously from the study of

$$\sum_{n=1}^{\infty} \frac{[m\alpha]}{2^i},$$

Figure 9. Integer part of $\{i\pi\}_{i=1}^{1600}$ mod 2; note the slight irregularities in the pseudo-periodic pattern.

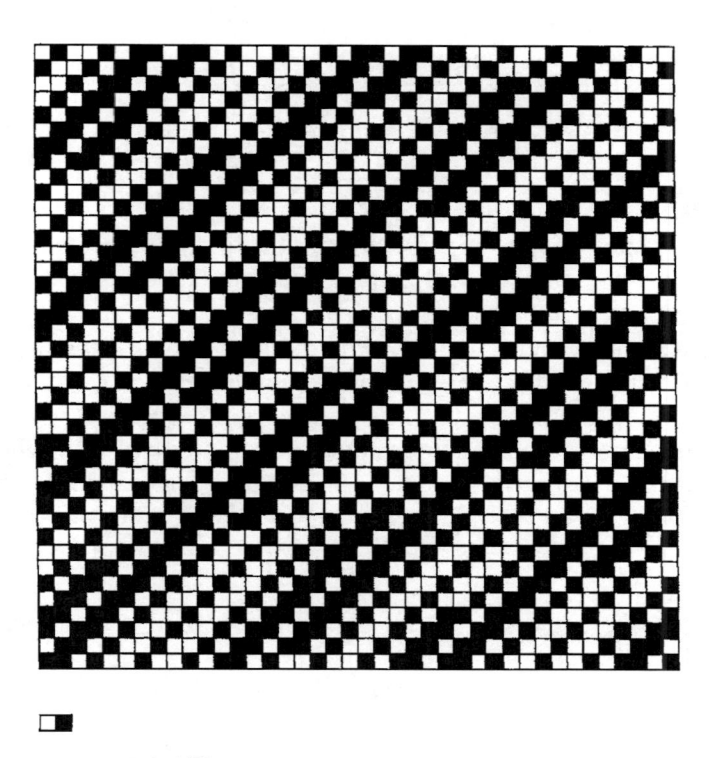

Figure 10. Integer part of $\{ie\}_{i=1}^{1600}$ mod 2; note the slight irregularities in the pseudo-periodic pattern.

which is transcendental for all irrational α. This follows from the remarkable continued fraction expansion of Böhmer [4]

$$\sum_{n=1}^{\infty} [m\alpha] z^n = \sum_{n=0}^{\infty} \frac{(-1)^{n+1}}{(1-z^{q_n})(1-z^{q_{n+1}})}.$$

Here (q_n) is the sequence of denominators in the simple continued fraction expansion of α.

Careful examination of Figures 9 and 10 show that they are only *pseudo*-periodic; slight irregularities appear in the pattern. Rational–like behaviour follows from the very good rational approximations evidenced by the expansions. Or put another way, there are very large terms in the continued fraction expansion. For example, the expansion of

$$\sum_{n=1}^{\infty} \frac{[m\pi] \bmod 2}{2^i}$$

is

[0, 1, 2, 42, 638816050508714029100700827905, 1, 126, . . .],

with a similar phenomenon for e.

This behaviour makes it clear that there is subtlety in the nature of these numbers. Indeed, while we were able to establish these results rigorously, many related phenomena exist whose proofs are not yet in hand. For example, there is no proof or explanation for the visual representation of

$$\sum_{n=1}^{\infty} \frac{[m\pi]\bmod 2}{3^i}$$

Proofs for these graphic results might well offer further refinements to their representations, leading to yet another critical graphic characterization.

Complex Zeros. Polynomials with constrained coefficients have been much studied [2], [17], [6]. They relate to the Littlewood conjecture and many other problems. Littlewood notes that "these raise fascinating questions" [14].

Certain of these polynomials demonstrate suprising complexity when their zeros are plotted appropriately. Figure 11 shows the complex zeros of all polynomials

$$P_n(z) = a_0 + a_1 z + a_2 z^2 + \cdots + a_n z^n$$

of degree $n \leq 18$, where $a_i = \{-1, +1\}$. This image, reminiscent of pictures for polynomials with all coefficients in the set $\{0, +1\}$ [17], does raise many questions: Is the set fractal and what is its boundary? Are there holes at infinite degree? How do the holes vary with the degree? What is the relationship between these zeros and those of polynomials with real coefficients in the neighbourhood of $\{-1, +1\}$?

Some, but definitely not all, of these questions have found some analytic answer [17], [6]. Others have been shown to relate subtly to standing problems of some significance in number theory. For example, the nature of the holes involves a old problem known as *Lehmer's conjecture* [1]. It is not yet clear how these images contribute to a solution to such problems. However they are provoking mathematicians to look at numbers in new ways.

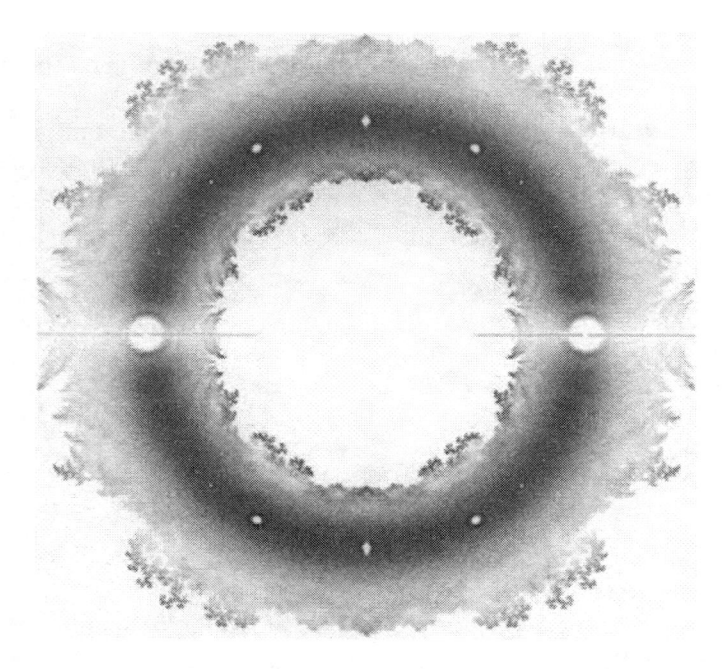

Figure 11. Roots of Littlewood Polynomials of degree at most 18 for coefficients ±1.

5. CONCLUSION. Visualization extends the natural capacity of the mathematician to envision his subject, to see the entities and objects that are part of his work with the aid of software and hardware. Since graphic representations are firmly rooted in verifiable algorithms and machines, the images and interfaces may also provide new forms of exposition and possibly even proof. Most important of all, like spacecraft, diving bells, and electron microscopes, visualization of mathematical structures takes the human mind to places it has never been and shows the mind's eye images from a realm previously unseen.

Readers are encouraged to review this paper in full color on-line [21].

ACKNOWLEDGEMENTS. This work has been supported in part by the NCE for Mathematics of Information Technology and Complex Systems (MITACS), and in part by research and equipment grants from the Natural Sciences and Engineering Research Council of Canada (NSERC). We thank the Centre for Experimental & Constructive Mathematics.

REFERENCES

1. F. Beaucoup, P. Borwein, D. Boyd, and C. Pinner, Multiple roots of $[-1, 1]$ power series. *London Math. Soc.* **57** (1998) 135–147.
2. A.T. Bharucha-Reid and M. Sambandham, *Random Polynomials*, Academic Press, Orlando, FL, 1986.
3. Alex Bogomolny, http://www.cut-the-knot.com/ctk/pww.html
4. J. Borwein and P. Borwein, On the generating function of the integer part: $[n\alpha + \gamma]$, *J. Number Theory* **43** (1993) 293–318.
5. J.M. Borwein, P.B. Borwein, R. Girgensohn, and S. Parnes, Making sense of experimental mathematics, www.cecm.sfu.ca/organics/vault/expmath/
6. P. Borwein and C. Pinner, Polynomials with $\{0, +1, -1\}$ coefficients and a root close to a given point, *Canadian J. Math.* **49** (1997) 887–915.
7. J.R. Brown, Proofs and pictures, *Brit. J. Phil. Sci.* **48** (1997) 161–180.
8. T. Eisenberg and T. Dreyfuss, On the reluctance to visualize in mathematics, in *Visualization in Teaching and Learning Mathematics*, Mathematical Association of America, Washington, DC, 1991, pp. 25–37.
9. Marcus Giaquinto, Epistemology of visual thinking in elementary real analysis, *Brit. J. Phil. Sci.* **45** (1994) 789–813.
10. A. Granville, The arithmetic properties of binomial coefficients, in *Proceedings of the Organic Mathematics Workshop, Dec. 12–14*, http://www.cecm.sfu.ca/organics/, 1995. IMpress, Simon Fraser University, Burnaby, BC.
11. John Horgan, The death of proof, *Scientific American* October, 1993, pp. 92–103.
12. Imre Lakatos, *Proofs and refutations: the logic of mathematical discovery*, Cambridge University Press, Cambridge, 1976.
13. Bruno Latour, Drawing things together, in *Representation in Scientific Practice*, Michael Lych and Steve Woolgar, eds., MIT Press, Cambridge, MA, 1990, pp. 25–37.
14. J.E. Littlewood, Some problems in real and complex analysis, in *Heath Mathematical Monographs*, D.C. Heath, Lexington, MA, 1968.
15. Roger A. Nelson, *Proofs Without Words II, More Exercises in Visual Thinking*, Mathematical Association of America, Washington, DC, 2000.
16. Roger A. Nelson, *Proofs Without Words: Exercises in Visual Thinking*, Mathematical Association of America, Washington, DC, 1993.
17. A. Odlyzko and B. Poonen, Zeros of polynomials with 0,1 coefficients, *Enseign. Math.* **39** (1993) 317–348.
18. C. Pickover, Picturing randomness on a graphics supercomputer, *IBM J. Res. Develop.* **35** (1991) 227–230.
19. Lynn Arthur Steen, The science of patterns, *Science* **240** (1988) 611–616.
20. http://www.hcrc.ed.ac.uk/gal/Diagrams/biblio.html
21. http://www.cecm.sfu.ca/~pborwein/
22. J. Voelcker, Picturing randomness, *IEEE Spectrum* **8** (1988) 13–16.

PETER BORWEIN is a Professor of Mathematics at Simon Fraser University, Vancouver, British Columbia. His Ph.D. is from the University of British Columbia under the supervision of David Boyd. After a postdoctoral year in Oxford and a dozen years at Dalhousie University in Halifax, Nova Scotia, he took up his current position. He has authored five books and over a hundred research articles. His research interests span diophantine and computational number theory, classical analysis, and symbolic computation. He is co-recipient of the Cauvenet Prize and the Hasse Prize, both for exposition in mathematics.
CECM, Simon Fraser University, Burnaby, BC, Canada V5A 1S6
pborwein@bb.cecm.sfu.ca

LOKI JÖRGENSON is an Adjunct Professor of Mathematics at Simon Fraser University, Vancouver, British Columbia. Previously the Research Manager for the Centre for Experimental and Constructive Mathematics, he is a senior scientist at Jaalam Research. He maintains his involvement in mathematics as the digital editor for the Canadian Mathematical Society. His Ph.D. is in computational physics from McGill University, and he has been active in visualization, simulation, and computation for over 15 years. His research has included forays into philosophy, graphics, educational technologies, high performance computing, statistical mechanics, high energy physics, logic, and number theory.
CECM, Simon Fraser University, Burnaby, BC, Canada V5A 1S6
loki@cecm.sfu.ca

9. The experimental mathematician: The pleasure of discovery and the role of proof

Discussion

This article was encouraged and nurtured by colleagues in mathematical education. Despite three decades of significant failures and only modest successes, we remain convinced of the power of the computer and mathematical software to revitalize and enliven the teaching of school and college mathematics. We do wish that prior attempts had been mindful of the good business advice that "one should always under-promise but over-deliver," rather than dumping mediocre software and unmaintained equipment in the classroom and ignoring the teacher.

Self-indulgent neglect by mathematical educators—aided and abetted by governments and computer vendors—is more than a little to blame. The related current semiotic malaise in mathematical education is cleanly dissected by Tony Brown:

> So to summarise, according to the citation count, in order of descent, the authors are listening to themselves, dead philosophers, other specialists in semiotic work in mathematics education research, other mathematics education research researchers and then just occasionally to social scientists but almost never to other education researchers, including mathematics teacher education researchers, school teachers and teacher educators. The engagement with Peirce is being understood primarily through personal engagements with the original material rather than as a result of working through the filters of history, including those evidenced within mathematics education research reports in the immediate area. The reports, and the hierarchy of power relations implicit in them, marginalise links to education, policy implementation or the broader social sciences.[1]

Source

J.M. Borwein, "The Experimental Mathematician: The Pleasure of Discovery and the Role of Proof," *International Journal of Computers for Mathematical Learning,* **10** (2005), 75–108. [CECM Preprint 02:178; 264]. A counterpart presentation online at **www.cecm.sfu.ca/personal/jborwein/proof.pdfs**, published in *CMESG25 Proceedings,* 2002.

---

[1] From "Signifying 'students', 'teachers' and 'mathematics' a reading of a special issue." Published online: at http://www.springerlink.com/content/x51838k6367w416g/, May 2008.

JONATHAN M. BORWEIN

THE EXPERIMENTAL MATHEMATICIAN: THE PLEASURE OF DISCOVERY AND THE ROLE OF PROOF

'...where almost one quarter hour was spent, each beholding the other with admiration before one word was spoken: at last Mr. Briggs began "My Lord, I have undertaken this long journey purposely to see your person, and to know by what wit or ingenuity you first came to think of this most excellent help unto Astronomy, viz. the Logarithms: but my Lord, being by you found out, I wonder nobody else found it out before, when now being known it appears so easy." '[1]

ABSTRACT. The emergence of powerful mathematical computing environments, the growing availability of correspondingly powerful (multi-processor) computers and the pervasive presence of the internet allow for mathematicians, students and teachers, to proceed heuristically and 'quasi-inductively'. We may increasingly use symbolic and numeric computation, visualization tools, simulation and data mining. The unique features of our discipline make this both more problematic and more challenging. For example, there is still no truly satisfactory way of displaying mathematical notation on the web; and we care more about the reliability of our literature than does any other science. The traditional role of proof in mathematics is arguably under siege – for reasons both good and bad.

AMS Classifications: 00A30, 00A35, 97C50

KEY WORDS: aesthetics, constructivism, experimental mathematics, humanist philosophy, insight, integer relations, proof

1. EXPERIMENTAL MATH: AN INTRODUCTION

"There is a story told of the mathematician Claude Chevalley (1909–1984), who, as a true Bourbaki, was extremely opposed to the use of images in geometric reasoning.

He is said to have been giving a very abstract and algebraic lecture when he got stuck. After a moment of pondering, he turned to the blackboard, and, trying to hide what he was doing, drew a little diagram, looked at it for a moment, then quickly erased it, and turned back to the audience and proceeded with the lecture...

76   JONATHAN M. BORWEIN

> ...The computer offers those less expert, and less stubborn than Chevalley, access to the kinds of images that could only be imagined in the heads of the most gifted mathematicians, ..." (Nathalie Sinclair[2])

For my coauthors and I, *Experimental Mathematics* (Borwein and Bailey, 2003) connotes the use of the computer for some or all of:

1. Gaining insight and intuition.
2. Discovering new patterns and relationships.
3. Graphing to expose math principles.
4. Testing and especially falsifying conjectures.
5. Exploring a possible result to see if it *merits* formal proof.
6. Suggesting approaches for formal proof.
7. Computing replacing lengthy hand derivations.
8. Confirming analytically derived results.

This process is studied very nicely by Nathalie Sinclair in the context of pre-service teacher training.[3] Limned by examples, I shall also raise questions such as:

What constitutes secure mathematical knowledge? When is computation convincing? Are humans less fallible? What tools are available? What methodologies? What about the 'law of the small numbers'? How is mathematics actually done? How should it be? Who cares for certainty? What is the role of proof?

And I shall offer some personal conclusions from more than twenty years of intensive exploitation of the computer as an adjunct to mathematical discovery.

1.1. *The Centre for Experimental Math*

About 12 years ago I was offered the signal opportunity to found the *Centre for Experimental and Constructive Mathematics* (CECM) at Simon Fraser University. On its web-site (www.cecm.sfu.ca) I wrote

> "At CECM we are interested in developing methods for exploiting mathematical computation as a tool in the development of mathematical intuition, in hypotheses building, in the generation of symbolically assisted proofs, and in the construction of a flexible computer environment in which researchers and research students can undertake such research. That is, in doing 'Experimental Mathematics.'"

The decision to build CECM was based on: (i) more than a decade's personal experience, largely since the advent of the personal computer, of the value of computing as an adjunct to mathematical insight and correctness; (ii) on a growing conviction that the future of mathematics would rely much more on collaboration and intelligent computation; (iii) that such developments needed to be enshrined in, and were equally valuable for, mathematical education; and (iv) that experimental mathematics is *fun*.

A decade or more later, my colleagues and I are even more convinced of the value of our venture – and the 'mathematical universe is unfolding' much as we anticipated. Our efforts and philosophy are described in some detail in the recent books (Borwein and Bailey, 2003; Borwein et al., 2004) and in the survey articles (Borwein et al., 1996; Borwein and Carless, 1999; Bailey and Borwein, 2000; Borwein and Borwein, 2001). More technical accounts of some of our tools and successes are detailed in (Borwein and Bradley (1997) and Borwein and Lisoněk (2000). About 10 years ago the term 'experimental mathematics' was often treated as an oxymoron. Now there is a highly visible and high quality journal of the same name. About 15 years ago, most self-respecting research pure mathematicians would not admit to using computers as an adjunct to research. Now they will talk about the topic whether or not they have any expertise. The centrality of information technology to our era and the growing need for concrete implementable answers suggests why we had attached the word 'Constructive' to CECM – and it motivated my recent move to Dalhousie to establish a new *Distributed Research Institute and Virtual Environment*, D-DRIVE (www.cs.dal.ca/ddrive).

While some things have happened much more slowly than we guessed (e.g., good character recognition for mathematics, any substantial impact on classroom parole) others have happened much more rapidly (e.g., the explosion of the world wide web[4], the quality of graphics and animations, the speed and power of computers). Crudely, the tools with broad societal or economic value arrive rapidly, those interesting primarily in our niche do not.

Research mathematicians for the most part neither think deeply about nor are terribly concerned with either pedagogy or the philosophy of mathematics. Nonetheless, aesthetic and philosophical notions have always permeated (pure and applied) mathematics. And the top researchers have always been driven by an aesthetic imperative:

> "We all believe that mathematics is an art. The author of a book, the lecturer in a classroom tries to convey the structural beauty of mathematics to his readers, to his listeners. In this attempt, he must always fail. Mathematics is logical to be sure,

each conclusion is drawn from previously derived statements. Yet the whole of it, the real piece of art, is not linear; worse than that, its perception should be instantaneous. We have all experienced on some rare occasions the feeling of elation in realizing that we have enabled our listeners to see at a moment's glance the whole architecture and all its ramifications." (Emil Artin, 1898–1962)[5]

Elsewhere, I have similarly argued for aesthetics before utility (Borwein, 2004, in press). The opportunities to tie research and teaching to aesthetics are almost boundless – at all levels of the curriculum.[6] This is in part due to the increasing power and sophistication of visualization, geometry, algebra and other mathematical software. That said, in my online lectures and resources,[7] and in many of the references one will find numerous examples of the utility of experimental mathematics.

In this article, my primary concern is to explore the relationship between proof (deduction) and experiment (induction). I borrow quite shamelessly from my earlier writings.

There is a disconcerting pressure at all levels of the curriculum to derogate the role of proof. This is in part motivated by the aridity of some traditional teaching (e.g., of Euclid), by the alternatives now being offered by good software, by the difficulty of teaching and learning the tools of the traditional trade, and perhaps by laziness.

My own attitude is perhaps best summed up by a cartoon in a book on learning to program in APL (a very high level language). The blurb above reads *Remember 10 minutes of computation is worth 10 hours of thought*. The blurb below reads *Remember 10 minutes of thought is worth 10 hours of computation*. Just as the unlived life is not much worth examining, proof and rigour should be in the service of things worth proving. And equally foolish, but pervasive, is encouraging students to 'discover' fatuous generalizations of uninteresting facts. As an antidote, In Section 2, I start by discussing and illustrating a few of George Polya's views. Before doing so, I review the structure of this article.

Section 2 discusses some of George Polya's view on heuristic mathematics, while Section 3 visits opinions of various eminent mathematicians. Section 4 discusses my own view and their genesis. Section 5 contains a set of mathematical examples amplifying the prior discussion. Sections 6 and 7 provide two fuller examples of computer discovery, and in Section 8 I return to more philosophical matters – in particular, a discussion of proof versus truth and the nature of secure mathematical knowledge.

2. POLYA ON PICTURE-WRITING

"[I]ntuition comes to us much earlier and with much less outside influence than formal arguments which we cannot really understand unless we have reached a relatively high level of logical experience and sophistication." (Geroge Polya)[8]

Polya, in his engaging eponymous 1956 *American Mathematical Monthly* article on picture writing, provided three provoking examples of converting pictorial representations of problems into generating function solutions:

1. *In how many ways can you make change for a dollar?*
   This leads to the (US currency) generating function

$$\sum_{k=1}^{\infty} P_k x^k = \frac{1}{(1-x^1)(1-x^5)(1-x^{10})(1-x^{25})(1-x^{50})},$$

which one can easily expand using a *Mathematica* command, Series $[1/((1-x) \times (1-x\hat{\ }5) \times (1-x\hat{\ }10) \times (1-x\hat{\ }25) \times (1-x\hat{\ }50)), \{x, 0, 100\}]$
to obtain $P_{100} = 292$ (243 for Canadian currency, which lacks a 50 cent piece but has a dollar coin). Polya's diagram is shown in Figure 1.[9]
To see why we use geometric series and consider the so-called *ordinary generating function*

$$\frac{1}{1-x^{10}} = 1 + x^{10} + x^{20} + x^{30} + \cdots$$

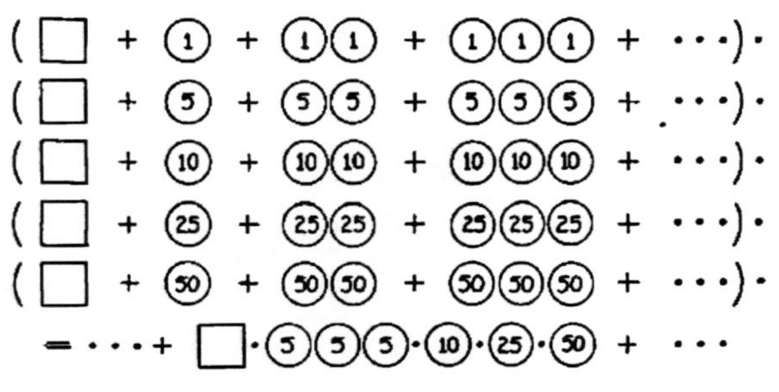

Figure 1. Polya's illustration of the change solution.

for dimes and
$$\frac{1}{1-x^{25}} = 1 + x^{25} + x^{50} + x^{75} + \cdots$$

for quarters, etc. If we multiply these two together and compare coefficients, we get

$$\frac{1}{1-x^{10}} \times \frac{1}{1-x^{25}} = 1 + x^{10} + x^{20} + x^{25} + x^{30} + x^{35}$$
$$+ x^{40} + x^{45} + 2x^{50} + x^{55} + 2x^{60} + \cdots$$

and can argue that the coefficient of x^{60} on the right is precisely the number of ways of making 60 cents out of identical dimes and quarters.

This is easy to check with a handful of change or a calculator and the more general question with more denominations is handled similarly. I leave it to the reader to decide whether it is easier to decode the generating function from the picture or vice versa. In any event, symbolic and graphic experiment can provide abundant and mutual reinforcement and assistance in concept formation.

2. *Dissect a polygon with n sides into n − 2 triangles by n − 3 diagonals and compute D_n, the number of different dissections of this kind.*
This leads to the fact that the generating function for $D_3 = 1$, $D_4 = 2$, $D_5 = 5$, $D_6 = 14$, $D_7 = 42$, ...

$$D(x) = \sum_{k=1}^{\infty} D_k x^k$$

satisfies

$$D(x) = x[1 + D(x)]^2,$$

whose solution is therefore

$$D(x) = \frac{1 - 2x - \sqrt{1-4x}}{x}$$

and D_{n+2} turns out to be the *n*-th Catalan number $\binom{2n}{n}/(n+1)$.

3. *Compute T_n, the number of different (rooted) trees with n knots.* The generating function of the T_n becomes a remarkable result due to Cayley:

$$T(x) = \sum_{k=1}^{\infty} T_k x^k = x \prod_{k=1}^{\infty} (1-x^k)^{-T_k}, \qquad (1)$$

where remarkably the product and the sum share their coefficients. This produces a recursion for T_n in terms of $T_1, T_2, \ldots, T_{n-1}$, which starts: $T_1 = 1$, $T_2 = 1$, $T_3 = 2$, $T_4 = 4$, $T_5 = 9$, $T_6 = 20, \ldots$

In each case, Polya's main message is that one can usefully draw pictures of the component elements – (a) in pennies, nickels dimes and quarters (plus loonies in Canada and half dollars in the US), (b) in triangles and (c) in the simplest trees (e.g., those with the fewest branches).

> "In the first place, the beginner must be convinced that proofs deserve to be studied, that they have a purpose, that they are interesting." (George Polya)[10]

While by 'beginner' George Polya largely intended young school students, I suggest that this is equally true of anyone engaging for the first time with an unfamiliar topic in mathematics.

3. GAUSS, HADAMARD AND HARDY'S VIEWS

Three of my personal mathematical heroes, very different men from different times, all testify interestingly on these points and on the nature of mathematics.

3.1. *Carl Friedrich Gauss*

Carl Friedrich Gauss (1777–1855) wrote in his diary[11]

> "I have the result, but I do not yet know how to get it."

Ironically I have been unable to find the precise origin of this quote.

One of Gauss's greatest discoveries, in 1799, was the relationship between the lemniscate sine function and the arithmetic-geometric mean iteration. This was based on a purely computational observation. The young Gauss wrote in his diary that the result *"will surely open up a whole new field of analysis."*

He was right, as it prised open the whole vista of 19th century elliptic and modular function theory. Gauss's specific discovery, based on tables of integrals provided by Stirling (1692–1770), was that the reciprocal of the integral

$$v \frac{2}{\pi} \int_0^1 \frac{dt}{\sqrt{1-t^4}}$$

agreed numerically with the limit of the rapidly convergent iteration given by $a_0 := 1$, $b_0 := \sqrt{2}$ and computing

$$a_{n+1} := \frac{a_n + b_n}{2}, \quad b_{n+1} := \sqrt{a_n b_n}.$$

The sequences a_n, b_n have a common limit 1.1981402347355922074...

Which object, the integral or the iteration, is more familiar, which is more elegant – then and now? Aesthetic criteria change: 'closed forms' have yielded centre stage to 'recursion', much as biological and computational metaphors (even 'biology envy') have replaced Newtonian mental images with Richard Dawkin's 'blind watchmaker'.

This experience of 'having the result' is reflective of much research mathematics. Proof and rigour play the role described next by Hadamard. Likewise, the back-handed complement given by Briggs to Napier underscores that is often harder to discover than to explain or digest the new discovery.

3.2. *Jacques Hadamard*

A constructivist, experimental and aesthetic driven rationale for mathematics could hardly do better than to start with:

> "The object of mathematical rigor is to sanction and legitimize the conquests of intuition, and there was never any other object for it." (J. Hadamard[12])

Jacques Hadamard (1865–1963) was perhaps the greatest mathematician to think deeply and seriously about cognition in mathematics[13]. He is quoted as saying "... *in arithmetic, until the seventh grade, I was last or nearly last"* which should give encouragement to many young students.

Hadamard was both the author of "The psychology of invention in the mathematical field" (1945), a book that still rewards close inspection, and co-prover of the *Prime Number Theorem* (1896):

> "The number of primes less than n tends to ∞ as does $n/\log n$."

This was one of the culminating results of 19th century mathematics and one that relied on much preliminary computation and experimentation.

One rationale for experimental mathematics and for heuristic computations is that one generally does not know during the course of research how it will pan out. Nonetheless, one must frequently prove all the pieces along the way as assurance that the project remains on course. The methods of experimental mathematics, alluded to below, allow one to maintain the necessary level of assurance without nailing down all the lemmas. At the end of the day, one can decide if the result merits proof. It may not be the answer one sought, or it may just not be interesting enough.

3.3. *Hardy's Apology*

Correspondingly, G. H. Hardy (1877–1947), the leading British analyst of the first half of the 20th century was also a stylish author who wrote compellingly in defense of pure mathematics. He noted that

> "All physicists and a good many quite respectable mathematicians are contemptuous about proof."

in his apologia, "A Mathematician's Apology". The Apology is a spirited defense of beauty over utility:

> "Beauty is the first test. There is no permanent place in the world for ugly mathematics."

That said, his comment that

> "Real mathematics...is almost wholly 'useless'."

has been over-played and is now to my mind very dated, given the importance of cryptography and other pieces of algebra and number theory devolving from very pure study. But he does acknowledge that

84   JONATHAN M. BORWEIN

> "If the theory of numbers could be employed for any practical and obviously honourable purpose, ..."

even Gauss would be persuaded.

The Apology is one of Amazon's best sellers. And the existence of Amazon, or Google, means that I can be less than thorough with my bibliographic details without derailing a reader who wishes to find the source.

Hardy, on page 15 of his tribute to Ramanujan entitled *Ramanujan, Twelve Lectures ...,* gives the so-called 'Skewes number' as a *"striking example of a false conjecture"*. The integral

$$\operatorname{li} x = \int_0^x \frac{dt}{\log t}$$

is a very good approximation to $\pi(x)$, the number of primes not exceeding x. Thus, $\operatorname{li} 10^8 = 5,762,209.375\ldots$ while $\pi(10^8) = 5,761,455$.

It was conjectured that

$$\operatorname{li} x > \pi(x)$$

holds for all x and indeed it so for many x. Skewes in 1933 showed the first explicit crossing at $10^{10^{10^{34}}}$. This has by now been now reduced to a relatively tiny number, a mere 10^{1167}, still vastly beyond direct computational reach or even insight.

Such examples show forcibly the limits on numeric experimentation, at least of a naive variety. Many will be familiar with the 'Law of large numbers' in statistics. Here we see what some number theorists call the 'Law of small numbers': *all small numbers are special*, many are primes and direct experience is a poor guide. And sadly or happily depending on one's attitude even 10^{1166} may be a small number. In more generality one never knows when the initial cases of a seemingly rock solid pattern are misleading. Consider the classic sequence counting the maximal number of regions obtained by joining n points around a circle by straight lines:

$$1, 2, 4, 8, 16, \mathbf{31}, \mathbf{57}, \ldots$$

(see entry A000127 in Sloane's Encyclopedia).

4. RESEARCH GOALS AND MOTIVATIONS

As a computational and experimental pure mathematician my main goal is: *insight*. Insight demands speed and increasingly parallelism as described in Borwein and Borwein (2001). Extraordinary speed and enough space are prerequisite for rapid verification and for validation and falsification ('proofs *and* refutations'). One can not have an 'aha' when the 'a' and 'ha' come minutes or hours apart.

What is 'easy' changes as computers and mathematical software grow more powerful. We see an exciting merging of disciplines, levels and collaborators. We are more and more able to marry theory & practice, history & philosophy, proofs & experiments; to match elegance and balance to utility and economy; and to inform all mathematical modalities computationally – analytic, algebraic, geometric & topological.

This has lead us to articulate an *Experimental Mathodology*,[14] as a philosophy (Borwein et al., 1996; Borwein and Bailey, 2003) and in practice (Borwein and Corless, 1999), based on: (i) meshing computation and mathematics (intuition is often acquired not natural, notwithstanding the truth of Polya's observations above); (ii) visualization (even three is a lot of dimensions). Nowadays we can exploit pictures, sounds and other haptic stimuli; and on (iii) 'caging' and 'monster-barring' (Imre Lakatos' and my terms for how one rules out exceptions and refines hypotheses). Two particularly useful components are:

- *Graphic checks.* comparing $y-y^2$ and y^2-y^4 to $-y^2 \ln(y)$ for $0 < y < 1$ pictorially (as in Figure 2) is a much more rapid way to divine which is larger than traditional analytic methods. It is clear that in the later case they cross, it is futile to try to prove one majorizes the other. In the first case, evidence is provided to motivate a proof.
- *Randomized checks.* of equations, linear algebra, or primality can provide enormously secure knowledge or counter-examples when deterministic methods are doomed.

All of these are relevant at every level of learning and research. My own methodology depends heavily on: (i) (*High Precision*) computation of object(s) for subsequent examination; (ii) *Pattern Recognition* of *Real Numbers* (e.g., using CECM's Inverse Calculator and

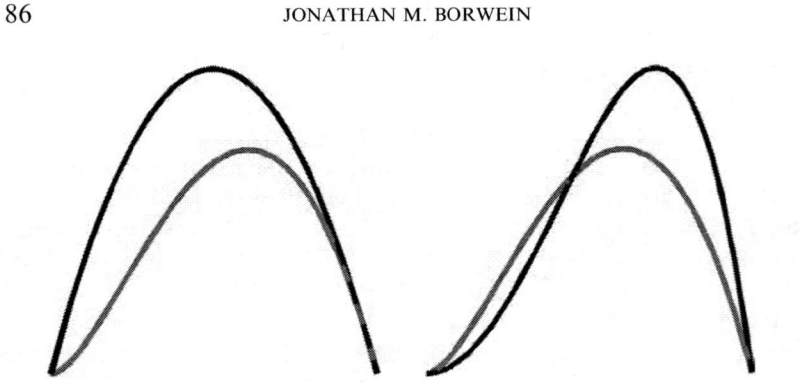

Figure 2. Graphical comparison of $y-y^2$ and y^2-y^4 to $-y^2 \ln(y)$ (red).

'RevEng'.[15]) or *Sequences* (e.g., using Salvy & Zimmermann's 'gfun' or Sloane and Plouffe's Online Encyclopedia); and (iii) extensive use of *Integer Relation Methods*: *PSLQ & LLL* and FFT.[16] Exclusion bounds are especially useful and such methods provide a great test bed for 'Experimental Mathematics'. All these tools are accessible through the listed CECM websites and those at www.expmath.info. To make more sense of this it is helpful to discuss the nature of experiment.

4.1. *Four Kinds of Experiment*

Peter Medawar usefully distinguishes four forms of scientific experiment.

1. The Kantian example: Generating "the classical non-Euclidean geometries (hyperbolic, elliptic) by replacing Euclid's axiom of parallels (or something equivalent to it) with alternative forms."
2. The Baconian experiment is a contrived as opposed to a natural happening, it "is the consequence of 'trying things out' or even of merely messing about."
3. Aristotelian demonstrations: "apply electrodes to a frog's sciatic nerve, and lo, the leg kicks; always precede the presentation of the dog's dinner with the ringing of a bell, and lo, the bell alone will soon make the dog dribble."
4. The most important is Galilean: "a critical experiment – one that discriminates between possibilities and, in doing so, either gives us

confidence in the view we are taking or makes us think it in need of correction."

The first three forms are common in mathematics, the fourth is not. It is also the only one of the four forms which has the promise to make Experimental Mathematics into a serious replicable scientific enterprise.[17]

5. FURTHER MATHEMATICAL EXAMPLES

The following suite of examples aims to make the case that modern computational tools can assist both by encapsulating concepts and by unpacking them as needs may be.

5.1. *Two Things About* $\sqrt{2}$...

Remarkably one can still find new insights in the oldest areas:

5.1.1. *Irrationality*

We present graphically, Tom Apostol's lovely new geometric proof[18] of the irrationality of $\sqrt{2}$. Earlier variants have been presented, but I like very much that this was published in the present millennium.

PROOF. To say $\sqrt{2}$ is rational is to draw a right-angled isoceles triangle with integer sides. Consider the *smallest* right-angled isoceles

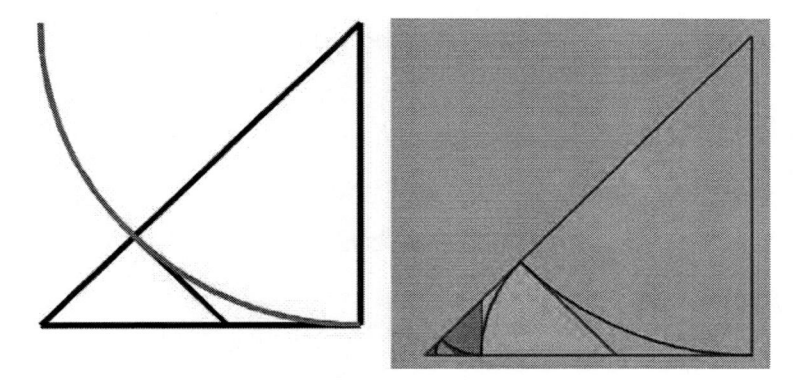

Figure 3. Root two is irrational (static and dynamic pictures).

triangle with integer sides – that is with shortest hypotenuse. Circumscribe a circle of radius the vertical side and construct the tangent on the hypotenuse, as in the picture in Figure 3. Repeating the process once more produces an even smaller such triangle in the same orientation as the initial one.

The *smaller* right-angled isoceles triangle again has integer sides... **QED**

This can be beautifully illustrated in a dynamic geometry package such as *Geometer's SketchPad, Cabri* or *Cinderella*, as used here. We can continue to draw smaller and smaller integer-sided similar triangles until the area palpably drops below 1/2. But I give it here to emphasize the ineffably human component of the best proofs.

A more elaborate picture can be drawn to illustrate the irrationality of $\sqrt{n}$ for $n = 3, 5, 6, \ldots$

5.1.2. *Rationality*
$\sqrt{2}$ also makes things rational:

$$\left(\sqrt{2}^{\sqrt{2}}\right)^{\sqrt{2}} = \sqrt{2}^{(\sqrt{2} \cdot \sqrt{2})} = \sqrt{2}^2 = 2.$$

Hence by *the principle of the excluded middle*

$$\text{Either } \sqrt{2}^{\sqrt{2}} \in \mathbb{Q} \text{ or } \sqrt{2}^{\sqrt{2}} \notin \mathbb{Q}.$$

In either case we can deduce that there are irrational numbers α and β with α^β rational. But how do we know which ones? This is not an adequate proof for an Intuitionist or a Constructivist. We may build a whole mathematical philosophy project around this. Compare the assertion that

$$\alpha := \sqrt{2} \quad \text{and} \quad \beta := 2\ln_2(3) \text{ yield } \alpha^\beta = 3$$

as *Maple* confirms. This illustrates nicely that verification is often easier than discovery (similarly the fact multiplication is easier than

factorization is at the base of secure encryption schemes for e-commerce).
There are eight possible (ir)rational triples:

$$\alpha^\beta = \gamma$$

and finding examples of all cases is now a fine student project.

5.2. *Exploring Integrals and Products*

Even *Maple* 'knows' $\pi \neq 22/7$ since

$$0 < \int_0^1 \frac{(1-x)^4 x^4}{1+x^2} dx = \frac{22}{7} - \pi,$$

though it would be prudent to ask 'why' it can perform the evaluation and 'whether' to trust it?

In this case, asking a computer algebra system to evaluate the indefinite integral

$$\int_0^t \frac{(1-x^4)x^4}{(1+x^2)} dx = \frac{1}{7}t^7 - \frac{2}{3}t^6 + t^5 - \frac{4}{3}t^3 + 4t - 4\arctan(t)$$

and differentiation proves the formula completely – after an appeal to the Fundamental theorem of calculus.

The picture in Figure 4 illustrates Archimedes' inequality in Nathalie Sinclair's *Colour calculator* micro-world in which the digits have been coloured modulo 10. This reveals simple patterns in 22/7, more complex in 223/71 and randomness in π. Many new approaches

Archimedes: 223/71 < π < 22/7

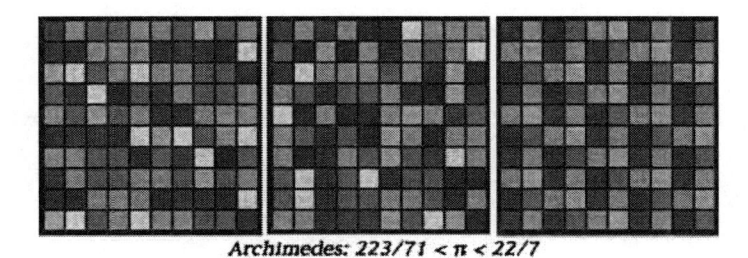

Figure 4. A colour calculator.

to teaching about fractions are made possible by the use of such a visual representation.

In contrast, *Maple* struggles with the following *sophomore's dream*:[19]

$$\int_0^1 \frac{1}{x^x} dx = \sum_{n=1}^{\infty} \frac{1}{n^n},$$

and students asked to confirm this, typically mistake numerical validation for symbolic proof.

Similarly

$$\prod_{n=2}^{\infty} \frac{n^3 - 1}{n^3 + 1} = \frac{2}{3} \tag{2}$$

is rational, while the seemingly simpler ($n = 2$) case

$$\prod_{n=2}^{\infty} \frac{n^2 - 1}{n^2 + 1} = \frac{\pi}{\sinh(\pi)} \tag{3}$$

is irrational, indeed transcendental. Our Inverse Symbolic Calculator can identify the right-hand side of (3) from it numeric value 0.272029054..., and the current versions *Maple* can 'do' both products, but the student learns little or nothing from this unless the software can also recreate the steps of a validation – thereby unpacking the identity. For example, (2) may be rewritten as a lovely telescoping product, and an attempt to evaluate the finite product

$$\prod_{n=2}^{N} \frac{n^2 - 1}{n^2 + 1} \tag{4}$$

leads to a formula involving the *Gamma function*, about which Maple's Help files are quite helpful, and the student can be led to an informative proof on taking the limit in (4) after learning a few basic properties of $\Gamma(x)$. Explicitly, with 'val:=proc(f) f = value(f) end proc;'

THE EXPERIMENTAL MATHEMATICIAN 91

```
> P2:=N->Product((n^2-1)/(n^2+1),n=2..N):

> val(P2(infinity));
                infinity
                ,-------,
                |   |    2
                |   |   n - 1        Pi
                |   |   ------  =  --------
                |   |    2          sinh(Pi)
               n = 2   n + 1

> val(P2(N));
        N
    ,-------,
    |   |    2
    |   |   n - 1     GAMMA(N) GAMMA(N + 2) GAMMA(2 - I) GAMMA(2 + I)
    |   |   ------ = ------------------------------------------------
    |   |    2              2 GAMMA(N + (1 - I)) GAMMA(N + (1 + I))
   n = 2   n + 1
> simplify(%);
        N
    ,-------,
    |   |    2                                   2
    |   |   n - 1            I GAMMA(N)   N (N + 1) Pi
    |   |   ------ = ------------------------------------------------
    |   |    2         GAMMA(N + (1 - I)) GAMMA(N + (1 + I)) sin((2 + I)Pi)
   n = 2   n + 1

> evalc(%);
        N
    ,-------,
    |   |    2                                 2
    |   |   n - 1              GAMMA(N)   N (N + 1) Pi
    |   |   ------ = ------------------------------------------------
    |   |    2         GAMMA(N + (1 - I)) GAMMA(N + (1 + I)) sinh(Pi)
   n = 2   n + 1
```

5.3 *Self-Similarity in Pascal's Triangle*

In any event, in each case so far computing adds reality, making concrete the abstract, and making some hard things simple. This is strikingly the case in *Pascal's Triangle*: www.cecm.sfu.ca/interfaces/ which affords an emphatic example where deep fractal structure is exhibited in the elementary binomial coefficients

$$1, 1, 2, 1, 1, 3, 3, 1, 1, 4, 6, 4, 1, 1, 5, 10, 10, 5, 1$$

92   JONATHAN M. BORWEIN

becomes the parity sequence

$$1,1,0,1,1,1,1,1,1,0,0,0,1,1,1,0,0,1,1$$

and leads to the picture in Figure 5, in which odd elements of the triangle are coloured purple. Thus, as in the $\sqrt{2}$ example notions of *self-similarity* and invariance of scale can be introduced quite early and naturally in the curriculum.

One can also explore what happens if the coefficients are colored modulo three or four – four is nicer. Many other recursive sequences exhibit similar fractal behaviour.[20]

5.4. *Berlinski on Mathematical Experiment*

David Berlinski[21] writes

> "The computer has in turn changed the very nature of mathematical experience, suggesting for the first time that mathematics, like physics, may yet become an empirical discipline, a place where things are discovered because they are seen."

As all sciences rely more on 'dry experiments', via computer simulation, the boundary between physics (e.g., string theory) and mathematics (e.g., by experiment) is delightfully blurred. An early exciting example is provided by gravitational boosting.

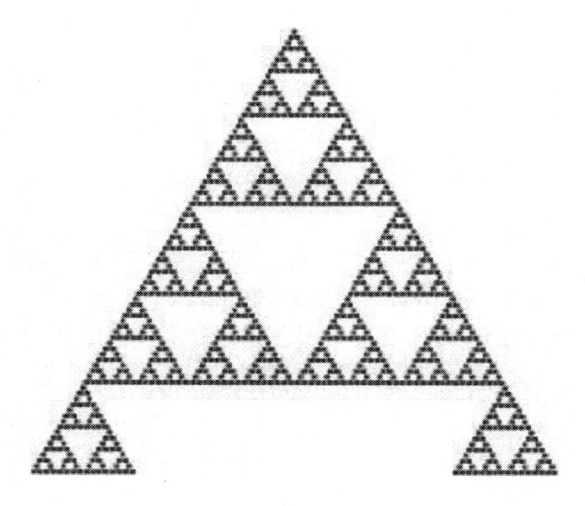

Figure 5. Drawing Pascal's triangle modulo two.

THE EXPERIMENTAL MATHEMATICIAN 93

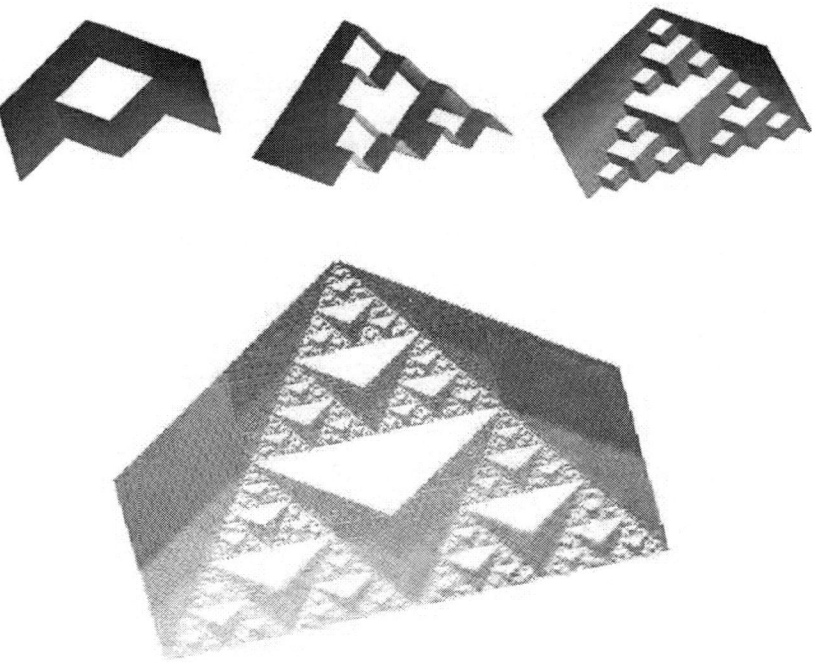

Figure 6. First, second, third and seventh iterates of a Sierpinski triangle.

Gravitational boosting. "The Voyager Neptune Planetary Guide" (JPL Publication 89–24) has an excellent description of Michael Minovitch' computational and unexpected discovery of *gravitational boosting* (otherwise known as slingshot magic) at the Jet Propulsion Laboratory in 1961.

The article starts by quoting Arthur C. Clarke "Any sufficiently advanced technology is indistinguishable from magic." Until Minovitch discovered that the so-called *Hohmann transfer ellipses* were not the minimum energy way of getting to the outer planets, "most planetary mission designers considered the gravity field of a target planet to be somewhat of a nuisance, to be cancelled out, usually by onboard Rocket thrust." For example, without a gravitational boost from the orbits of Saturn, Jupiter and Uranus, the Earth-to-Neptune Voyager mission (achieved in 1989 in little more than a decade) would have taken more than 30 years! We should still be waiting.

5.5. *Making Fractal Postcards*

And yet, as we have seen, not all impressive discoveries require a computer. Elaine Simmt and Brent Davis describe lovely constructions made by repeated regular paper folding and cutting – but no removal of paper – that result in beautiful fractal, self-similar, "pop-up" cards.[22] Nonetheless, in Figure 6, we show various iterates of a pop-up Sierpinski triangle built in software by turning those paper cutting and folding rules into an algorithm. Note the similarity to the triangle in Figure 7. Any regular rule produces a fine card. The pictures should allow the reader to start folding.

Recursive *Maple* code is given below.

```
sierpinski := proc ( n: nonnegint )
          local p1, p2, q1, q2, r1, r2, plotout;
  p1:= [1.,0.,0.]; q1:= [-1.,0.,0.]; r1:= [0.,0.,1.];
  p2:= [1.,1.,0.]; q2:= [-1.,1.,0.]; r2:= [0.,1.,1.];
  plotout:= polys(n, p1, p2, r1, r2, q1, q2);
 return PLOT3D( plotout, SCALING(CONSTRAINED),
  AXESSTYLE(NONE), STYLE(PATCHNOGRID), ORIENTATION(90,45) );
end:

polys:= proc( n::nonnegint, p1, p3, r1, r3, q1, q3 )
          local p2, q2, r2, s1, s2, s3, t1, t2, t3, u2, u3;
  if n=0 then return POLYGONS([p1,p3,r3,r1], [q1,q3,r3,r1]) fi;
  p2:= (p1+p3)/2; q2:= (q1+q3)/2; r2:= (r1+r3)/2;
  s1:= (p1+r1)/2; s2:= (p2+r2)/2; s3:= (p3+r3)/2;
  t1:= (q1+r1)/2; t2:= (q2+r2)/2; t3:= (q3+r3)/2;
  u2:= (p2+q2)/2; u3:= (p3+q3)/2;
 return polys(n-1, p2, p3, s2, s3, u2, u3),
        polys(n-1, s1, s2, r1, r2, t1, t2),
        polys(n-1, u2, u3, t2, t3, q2, q3),
        POLYGONS([p1,p2,s2,s1], [q1,q2,t2,t1]); end:
```

And, as in Figure 7, art can be an additional source of mathematical inspiration.

5.6. *Seeing Patterns in Partitions*

The number of *additive partitions* of n, $p(n)$, is *generated* by

$$1 + \sum_{n \geq 1} p(n) q^n = \frac{1}{\prod_{n \geq 1}(1 - q^n)}. \tag{5}$$

THE EXPERIMENTAL MATHEMATICIAN 95

Figure 7. Self similarity at Chartres.

Thus, $p(5) = 7$ since

$$5 = 4+1 = 3+2 = 3+1+1 = 2+2+1$$
$$= 2+1+1+1 = 1+1+1+1+1.$$

Developing (5) is a nice introduction to enumeration via generating functions of the type discussed in Polya's change example.

Additive partitions are harder to handle than multiplicative factorizations, but again they may be introduced in the elementary school curriculum with questions like: *How many 'trains' of a given length can be built with Cuisenaire rods?*

A more modern computationally driven question is *How hard is p(n) to compute?*

In 1900, it took the father of combinatorics, Major Percy Mac-Mahon (1854–1929), months to compute $p(200)$ using recursions developed from (5). By 2000, *Maple* would produce $p(200)$ in seconds if one simply demands the 200th term of the Taylor series. A few years earlier it required one to be careful to compute the series for $\prod_{n\geq 1}(1-q^n)$ first and then to compute the series for the reciprocal of that series! This seemingly baroque event is occasioned by *Euler's pentagonal number theorem*

$$\prod_{n\geq 1}(1-q^n) = \sum_{n=-\infty}^{\infty}(-1)^n q^{(3n+1)n/2}$$

The reason is that, if one takes the series for (5) directly, the software has to deal with 200 terms on the bottom. But if one takes the series for $\prod_{n\geq 1}(1-q^n)$, the software has only to handle the 23 non-zero terms in series in the pentagonal number theorem. This expost facto algorithmic analysis can be used to facilitate independent student discovery of the pentagonal number theorem, and like results.

If introspection fails, we can find the *pentagonal numbers* occurring above in *Sloane* and Plouffe's on-line 'Encyclopedia of Integer Sequences' www.research.att.com/personal/njas/sequences/eisonline.html.

Ramanujan used MacMahon's table of $p(n)$ to intuit remarkable and deep congruences such as

$$p(5n+4) \equiv 0 \mod 5$$
$$p(7n+5) \equiv 0 \mod 7$$

and

$$p(11n+6) \equiv 0 \mod 11,$$

from relatively limited data like

$$\begin{aligned}P(q) = {} & 1 + q + 2q^2 + 3q^3 + \underline{5}q^{\mathbf{4}} + \overline{7}q^{\mathbf{5}} + 11q^6 + 15q^7 + 22q^8 + \underline{30}q^{\mathbf{9}} \\ & + 42q^{10} + 56q^{11} + \overline{77}q^{\mathbf{12}} + \underline{101}q^{13} + \underline{135}q^{\mathbf{14}} + 176q^{15} + 231q^{16} \\ & + 297q^{17} + 385q^{18} + \overline{490}q^{\mathbf{19}} + 627q^{20}b + 792q^{21} + 1002q^{22} \\ & + \cdots + p(200)q^{200} \ldots \end{aligned} \qquad (6)$$

The exponents and coefficients for the cases $5n + 4$ and $7n + 5$ are highlighted in formula (6). Of course, it is much easier to heuristically confirm than to discover these patterns.

Here we see very fine examples of *Mathematics: the science of patterns* as is the title of Keith Devlin's 1997 book. And much more may similarly be done.

The difficulty of estimating the size of $p(n)$ analytically – so as to avoid enormous or unattainable computational effort – led to some marvellous mathematical advances by researchers including Hardy and Ramanujan, and Rademacher. The corresponding ease of computation may now act as a retardant to mathematical insight. New mathematics is discovered only when prevailing tools run totally out of steam. This raises a caveat against mindless computing: will a student or researcher discover structure when it is easy to compute without needing to think about it? Today, she may thoughtlessly compute $p(500)$ which a generation ago took much, much pain and insight.

Ramanujan typically found results not proofs and sometimes went badly wrong for that reason. So will we all. Thus, we are brought full face to the challenge, such software should be used, but algorithms must be taught and an appropriate appreciation for and facility with proof developed.

For example, even very extended evidence may be misleading. Indeed.

5.7. *Distinguishing Coincidence and Fraud*

Coincidences do occur. The approximations

$$\pi \approx \frac{3}{\sqrt{163}}\log(640320)$$

and

$$\pi \approx \sqrt{2}\frac{9801}{4412}$$

occur for deep number theoretic reasons – the first good to 15 places, the second to eight. By contrast

$$e^{\pi} - \pi = \mathbf{19.999099979}189475768\ldots$$

most probably for no good reason. This seemed more bizarre on an eight digit calculator. Likewise, as spotted by Pierre Lanchon recently, in base-two

$$e = 10.\overline{1011011111 1000010}10100010110001010\ldots$$

$$\pi = 11.001\overline{0001000011111101101010}0100010001\ldots$$

have 19 bits agreeing – with one read right to left.

More extended coincidences are almost always contrived, as in the following due to Kurt Mahler early last century. Below '$[x]$' denotes the integer part of x. Consider:

$$\sum_{n=1}^{\infty} \frac{[n \ \tanh(\pi)]}{10^n} \stackrel{?}{=} \frac{1}{81}$$

is valid to **268** places; while

$$\sum_{n=1}^{\infty} \frac{[n \ \tanh(\frac{\pi}{2})]}{0^n} \stackrel{?}{=} \frac{1}{81}$$

is valid to **12** places. Both are actually transcendental numbers.

Correspondingly, the *simple continued fractions* for $\tanh(\pi)$ and $\tanh(\frac{\pi}{2})$ are respectively,

$$[0, 1, \mathbf{267}, 4, 14, 1, 2, 1, 2, 2, 1, 2, 3, 8, 3, 1, \ldots]$$

and

$$[0, 1, \mathbf{11}, 14, 4, 1, 1, 1, 3, 1, 295, 4, 4, 1, 5, 17, 7, \ldots].$$

This is, as they say, no coincidence! While the reasons (Borwein and Bailey, 2003) are too advanced to explain here, it is easy to conduct experiments to discover what happens when $\tanh(\pi)$ is replaced by another irrational number, say $\log(2)$.

It also affords a great example of fundamental objects that are hard to compute by hand (high precision sums or continued fractions) but easy even on a small computer or calculator. Indeed, I would claim that continued fractions fell out of the undergraduate curriculum precisely because they are too hard to work with by hand. And, of course the main message, is again that computation without insight is mind numbing and destroys learning.

6. COMPUTER DISCOVERY OF BITS OF π

Bailey, P. Borwein and Plouffe (1996) discovered a series for π (and corresponding ones for some other *polylogarithmic constants*) which somewhat disconcertingly allows one to compute hexadecimal digits of π *without* computing prior digits. The algorithm needs very little memory and no multiple precision. The running time grows only slightly faster than linearly in the order of the digit being computed. Until that point it was broadly considered impossible to compute digits of such a number without computing most of the preceding ones.

The key, found as described above, is

$$\pi = \sum_{k=0}^{\infty} \left(\frac{1}{16}\right)^k \left(\frac{4}{8k+1} - \frac{2}{8k+4} - \frac{1}{8k+5} - \frac{1}{8k+6}\right).$$

Knowing an algorithm would follow they spent several months hunting by computer using integer relation methods (Bailey and Borwein, 2000; Borwein and Lisoněk, 2000; Dongarra and Sullivan, 2000) for such a formula. Once found, it is easy to prove in *Mathematica*, in *Maple* or by hand – and provides a very nice calculus exercise. This discovery was a most successful case of **REVERSE MATHEMATICAL ENGINEERING**.

The algorithm is entirely practicable, God reaches her hand deep into π: in September 1997 Fabrice Bellard (INRIA) used a variant of this formula to compute 152 binary digits of π, starting at the *trillionth position* (10^{12}). This took 12 days on 20 work-stations working in parallel over the Internet.

In August 1998 Colin Percival (SFU, age 17) finished a similar naturally or "embarrassingly parallel" computation of the *five trillionth bit* (using 25 machines at about 10 times the speed of Bellard). In *hexadecimal notation* he obtained

$$0\underline{7}E45733CC790B5B5979.$$

The corresponding binary digits of π starting at the 40 trillionth place are

$$\underline{0}0000111110011111.$$

By September 2000, the quadrillionth bit had been found to be '0' (using 250 cpu years on 1734 machines from 56 countries). Starting at the 999, 999, 999, 999, 997th bit of π one has

$$111\underline{0}001100010000101101011000000110.$$

100   JONATHAN M. BORWEIN

Why should we believe this calculation? One good reason is that it was done twice starting at different digits, in which case the algorithm performs entirely different computations. For example, computing 40 hexadecimal digits commencing at the trillionth and trillion-less-tenth place, respectively, should produce 30 shared hex-digits. The probability of those coinciding by chance, at least heuristically, is about

$$\frac{1}{16^{30}} \approx \frac{1}{10^{36.23\ldots}},$$

a stunning small probability. Moreover, since many different machines were engaged no one machine error plays a significant role. I like Hersh – as we shall see later – would be hard pressed to find complex proofs affording such a level of certainty.

In the final mathematical section we attempt to capture all of the opportunities in one more fleshed-out, albeit more advanced, example.

7. A SYMBOLIC-NUMERIC EXAMPLE

I illustrate more elaborately some of the continuing and engaging mathematical challenges with a specific problem, proposed in the *American Mathematical Monthly* (November, 2000), originally discussed in Borwein and Borwein (2001) and Borwein and Bailey (2003).

10832. *Donald E. Knuth, Stanford University, Stanford, CA.* Evaluate

$$\sum_{k=1}^{\infty}\left(\frac{k^k}{k!e^k} - \frac{1}{\sqrt{2\pi k}}\right).$$

1. A very rapid *Maple* computation yielded -0.08406950872765600 ... as the first 16 digits of the sum.
2. The Inverse Symbolic Calculator has a 'smart lookup' feature[23] that replied that this was probably $-(2/3) - \zeta(\frac{1}{2})/\sqrt{2\pi}$.
3. Ample experimental confirmation was provided by checking this to 50 digits. Thus within minutes we *knew* the answer.
4. As to why? A clue was provided by the surprising speed with which *Maple* computed the slowly convergent infinite sum. The package clearly knew something the user did not. Peering under

the covers revealed that it was using the *LambertW* function, W, which is the inverse of $w = z \exp(z)$.[24]

5. The presence of $\zeta(1/2)$ and standard Euler–MacLaurin techniques, using Stirling's formula (as might be anticipated from the question), led to

$$\sum_{k=1}^{\infty}\left(\frac{1}{\sqrt{2\pi k}} - \frac{1}{\sqrt{2}}\frac{(\frac{1}{2})_{k-1}}{(k-1)!}\right) = \frac{\zeta(\frac{1}{2})}{\sqrt{2\pi}}, \tag{7}$$

where the binomial coefficients in (7) are those of $(1/\sqrt{2-2z})$. Now (7) is a formula *Maple* can 'prove'.

6. It remains to show

$$\sum_{k=1}^{\infty}\left(\frac{k^k}{k!e^k} - \frac{1}{\sqrt{2}}\frac{(\frac{1}{2})_{k-1}}{(k-1)!}\right) = -\frac{2}{3}. \tag{8}$$

7. Guided by the presence of W and its series $\sum_{k=1}^{\infty}\frac{(-k)^{k-1}z^k}{k!}$, an appeal to Abel's limit theorem lets one deduce the need to evaluate

$$\lim_{z\to 1}\left(\frac{d}{dz}W(-\frac{z}{e}) + \frac{1}{\sqrt{2-2z}}\right) = \frac{2}{3} \tag{9}$$

Again *Maple* happily does know (9).

Of course this all took a fair amount of human mediation and insight. It will be many years before such computational discovery can be fully automated.

8. PROOF VERSUS TRUTH

By some accounts Colin Percival's web-computation of π^{25} is one of the largest computations ever done. It certainly shows the possibility to use inductive engineering-like methods in mathematics, if one keeps ones eye on the ball. As we saw, to assure accuracy the algorithm could be run twice starting at different points – say starting at 40 trillion minus 10. The overlapping digits will differ if any error has been made. If 20 hex-digits agree we can argue heuristically that the probability of error is roughly 1 part in 10^{25}.

102   JONATHAN M. BORWEIN

While this is not a proof of correctness, it is certainly much less likely to be wrong than any really complicated piece of human mathematics. For example, perhaps 100 people alive can, given enough time, digest *all* of Andrew Wiles' extraordinarily sophisticated proof of *Fermat's Last Theorem* and it relies on a century long program. If there is even a 1% chance that each has overlooked the *same* subtle error[26] – probably in prior work not explicitly in Wiles' corrected version – then, clearly, many computational based ventures are much more secure.

This would seem to be a good place to address another common misconception. No amount of simple-minded case checking constitutes a proof (Figure 8). The 1976–1967 'proof' of the *Four Colour Theorem*[27] was a proof because prior mathematical analysis had reduced the problem to showing that a large but finite number of potentially bad configurations could be ruled out. The proof was viewed as somewhat flawed because the case analysis was inelegant, complicated and originally incomplete. In the last few years, the computation has been redone after a more satisfactory analysis.[28] Of course, Figure 7 is a proof for the USA.

Though many mathematicians still yearn for a simple proof in both cases, there is no particular reason to think that all elegant true conjectures have accessible proofs. Nor indeed given Goedel's or Turing's work need they have proofs at all.

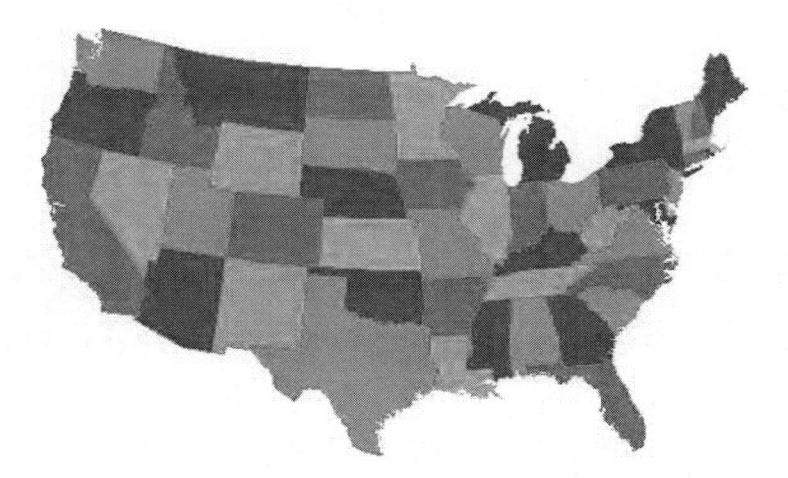

Figure 8. A four colouring of the continental USA.

8.1. *The Kepler Conjecture*

Kepler's conjecture that *The Densest Way to Stack Spheres is in a Pyramid* is the oldest problem in discrete geometry. It is also the most interesting recent example of computer-assisted proof. Published in the elite *Annals of Mathematics* with an "only 99% checked" disclaimer, this has triggered very varied reactions. While the several hundred pages of computer related work is clearly very hard to check, I do not find it credible that all other papers published in the Annals have exceeded such a level of verification.[29]

The proof of the *Kepler Conjecture*, that of the *Four Colour Theorem* and Clement Lams' computer-assisted proof of *The Non-existence of a Projective Plane of Order 10*, raise and answer quite distinct philosophical and mathematical questions – both real and specious. But one thing is certain such proofs will become more and more common.

8.2. *Kuhn and Planck on Paradigm Shifts*

Much of what I have described in detail or in passing involves changing set modes of thinking. Many profound thinkers view such changes as difficult:

> "The issue of paradigm choice can never be unequivocally settled by logic and experiment alone. ··· in these matters neither proof nor error is at issue. The transfer of allegiance from paradigm to paradigm is a conversion experience that cannot be forced." (Thomas Kuhn[30])

and

> "... a new scientific truth does not triumph by convincing its opponents and making them see the light, but rather because its opponents die and a new generation grows up that's familiar with it." (Albert Einstein quoting Max Planck[31])

8.3. *Hersh's Humanist Philosophy*

However hard such paradigm shifts and whatever the outcome of these discourses, mathematics is and will remain a uniquely human undertaking. Indeed Reuben Hersh's arguments for a humanist philosophy of mathematics, as paraphrased below, become more convincing in our setting:

1. *Mathematics is human.* It is part of and fits into human culture. It does not match Frege's concept of an abstract, timeless, tenseless, objective reality.
2. *Mathematical knowledge is fallible.* As in science, mathematics can advance by making mistakes and then correcting or even re-correcting them. The "fallibilism" of mathematics is brilliantly argued in Lakatos' *Proofs and Refutations*.
3. *There are different versions of proof or rigor.* Standards of rigor can vary depending on time, place, and other things. The use of computers in formal proofs, exemplified by the computer-assisted proof of the four color theorem in 1977, is just one example of an emerging nontraditional standard of rigor.
4. *Empirical evidence, numerical experimentation and probabilistic proof all can help us decide what to believe in mathematics.* Aristotelian logic isn't necessarily always the best way of deciding.
5. *Mathematical objects are a special variety of a social-cultural-historical object.* Contrary to the assertions of certain post-modern detractors, mathematics cannot be dismissed as merely a new form of literature or religion. Nevertheless, many mathematical objects can be seen as shared ideas, like Moby Dick in literature, or the Immaculate Conception in religion.[32]

To this I would add that for me mathematics is not ultimately about proof but about secure mathematical knowledge. Georg Friedrich Bernhard Riemann (1826–1866) was one of the most influential thinkers of the past 200 years. Yet he proved very few theorems, and many of the proofs were flawed. But his conceptual contributions, such as through Riemannian geometry and the Riemann zeta function, and to elliptic and Abelian function theory, were epochal. The experimental method is an addition not a substitute for proof, and its careful use is an example of Hersh's 'nontraditional standard of rigor'.

The recognition that 'quasi-intuitive' methods may be used to gain mathematical insight can dramatically assist in the learning and discovery of mathematics. Aesthetic and intuitive impulses are shot through our subject, and honest mathematicians will acknowledge their role. But a student who never masters proof will not be able to profitably take advantage of these tools.

8.4. *A Few Final Observations*

As we have already seen, the stark contrast between the deductive and the inductive has always been exaggerated. Herbert A. Simon, in the final edition of *The Sciences of the Artificial*,[33] wrote:

> "This skyhook-skyscraper construction of science from the roof down to the yet unconstructed foundations was possible because the behaviour of the system at each level depended only on a very approximate, simplified, abstracted characterization at the level beneath.[13] "[34]

> "This is lucky, else the safety of bridges and airplanes might depend on the correctness of the "Eightfold Way" of looking at elementary particles."

It is precisely this *'post hoc ergo propter hoc'* part of theory building that Russell so accurately typifies that makes him an articulate if surprising advocate of my own views.

And finally, I wish to emphasize that good software packages can make very difficult concepts accessible (e.g., *Mathematica, MatLab* and SketchPad) and radically assist mathematical discovery. Nonetheless, introspection is here to stay.

In Kieran Egan's words, "We are Pleistocene People." Our minds can subitize, but were not made for modern mathematics. We need all the help we can get. While proofs are often out of reach to students or indeed lie beyond present mathematics, understanding, even certainty, is not.

Perhaps indeed, "Progress is made 'one funeral at a time'."[35] In any event, as Thomas Wolfe put it "You can't go home again."

ACKNOWLEDGEMENTS

I wish to thank Nathalie Sinclair, Terry Stanway and Rina Zaskis for many pertinent discussions during the creation of this article.

Research supported by NSERC, by the MITACS-Network of Centres of Excellence, and by the Canada Research Chair Program.

NOTES

[1] Henry Briggs is describing his first meeting in 1617 with Napier whom he had travelled from London to Edinburgh to meet. Quoted from H.W. Turnbull's *The Great Mathematicians*, Methuen, 1929.

[2] Chapter in *Making the Connection: Research and Practice in Undergraduate Mathematics*, MAA Notes, 2004 in Press.

[3] Ibid.

[4] CECM now averages well over a million accesses a month, many by humans.

[5] Quoted by Ram Murty in *Mathematical Conversations, Selections from The Mathematical Intelligencer*, compiled by Robin Wilson and Jeremy Gray, Springer-Verlag, New York, 2000.

[6] My own experience is principally at the tertiary level. An excellent middle school illustration is afforded by Nathalie Sinclair. (2001) "The aesthetics is relevant," *for the learning of mathematics*, 21: 25 – 32.

[7] E.g., `www.cecm.sfu.ca/personal/jborwein/talks.html`, `www.cs.dal.ca//`, `/jborwein`, and `personal/loki/Papers/Numbers/`.

[8] In *Mathematical Discovery: On Understanding, Learning and Teaching Problem Solving, 1968*.

[9] Illustration courtesy the Mathematical Association of America.

[10] Ibid.

[11] See Isaac Asimov (1988) book of science and nature quotations. In Isaac Asimov and J.A. Shulman (Eds), New York: Weidenfield and Nicolson, p. 115.

[12] In E. Borel, "Lecons sur la theorie des fonctions," 1928, quoted by George Polya (1981) in *Mathematical discovery: On understanding, learning, and teaching problem solving* (Combined Edition), New York: John Wiley, pp. 2–126.

[13] Others on a short list would include Poincaré and Weil.

[14] I originally typed this by mistake for Methodology.

[15] ISC space limits have changed from 10 Mb being a constraint in 1985 to 10 Gb being 'easily available' today. A version of 'Reveng' is available in current versions of *Maple*. Typing 'identify($\sqrt{2.0} + \sqrt{3.0}$)' will return the symbolic answer $\sqrt{2} + \sqrt{3}$ from the numerical input 3.146264370.

[16] Described as one of the top ten "Algorithm's for the Ages," Random Samples, Science, Feb. 4, 2000, and [10].

[17] From Peter Medawar's wonderful *Advice to a Young Scientist*, Harper (1979).

[18] *MAA Monthly*, November 2000, 241–242.

[19] In that the integrand and the summand agree.

[20] Many examples are given in P. Borwein and L. Jörgenson, "Visible Structures in Number Theory", `www.cecm.sfu.ca/preprints/1998pp.html`.

[21] A quote I agree with from his "A Tour of the Calculus," Pantheon Books, 1995.

[22] Fractal Cards: A Space for Exploration in Geometry and Discrete Mathematics, *Mathematics Teacher* 91 (198), 102–108.

[23] Alternatively, a sufficiently robust integer relation finder could be used.

[24] A search for 'Lambert W function' on MathSciNet provided 9 references – all since 1997 when the function appears named for the first time in *Maple* and *Mathematica*.

[25] Along with *Toy Story 2*.

[26] And they may be psychologically predisposed so to do!

[27] Every planar map can be coloured with four colours so adjoining countries are never the same colour.
[28] This is beautifully described at `www.math.gatech.edu/personal/thomas/FC/fourcolor.html`.
[29] See "In Math, Computers Don't Lie. Or Do They?" *New York Times*, April 6, 2004.
[30] In Ed Regis, *Who got Einstein's Office?* Addison-Wesley, 1986.
[31] From F.G. Major, *The Quantum Beat,* Springer, 1998.
[32] From "Fresh Breezes in the Philosophy of Mathematics," *American Mathematical Monthly*, August–September 1995, 589–594.
[33] MIT Press, 1996, page 16.
[34] Simon quotes Russell at length ...
  [13] "... More than fifty years ago Bertrand Russell made the same point about the architecture of mathematics. See the "Preface" to *Principia Mathematica* "... the chief reason in favour of any theory on the principles of mathematics must always be inductive, i.e., it must lie in the fact that the theory in question allows us to deduce ordinary mathematics. In mathematics, the greatest degree of self-evidence is usually not to be found quite at the beginning, but at some later point; hence the early deductions, until they reach this point, give reason rather for believing the premises because true consequences follow from them, than for believing the consequences because they follow from the premises." Contemporary preferences for deductive formalisms frequently blind us to this important fact, which is no less true today than it was in 1910."
[35] This harsher version of Planck's comment is sometimes attributed to Niels Bohr.
[36] All journal references are available at `www.cecm.sfu.ca/preprints/`.

REFERENCES

Bailey, D.H. and Borwein, J.M. (2000). Experimental mathematics: Recent developments and future outlook. In B. Engquist and W. Schmid (Eds), *Mathematics Unlimited – 2∞1 and Beyond* (Vol. 1, pp. 51–66). Springer-Verlag. [CECM Preprint 99:143].[36]

Borwein, J.M. and Bailey, D.H. (2003). *Mathematics by Experiment: Plausible Reasoning in the 21st Century*. AK Peters Ltd.

Borwein, J.M., Bailey, D.H. and Girgensohn, R. (2004). *Experimentation in Mathematics: Computational Paths to Discovery*. AK Peters Ltd.

Borwein, J. M. (2004). "Aesthetics for the Working Mathematician," in *Beauty and the Mathematical Beast*, in press (2004). [CECM Preprint 01:165].

Borwein, J.M. and Borwein, P.B. (2001). Challenges for mathematical computing. *Computing in Science and Engineering* 3: 48–53. [CECM Preprint 00:160]

Borwein, J.M., Borwein, P.B., Girgensohn, R. and Parnes, S. (1996) Making sense of experimental mathematics. *Mathematical Intelligencer* 18(4): 12–18. [CECM Preprint 95:032]

Borwein, J.M. and Bradley, D.M. (1997). Empirically determined Apéry-like formulae for Zeta($4n+3$). *Experimental Mathematics* 6: 181–194. [CECM 96:069]

Borwein, J. M. and Corless, R. (1999). Emerging tools for experimental mathematics. *American Mathematical Monthly* 106: 889–909. [CECM Peprint 98:110]

108   JONATHAN M. BORWEIN

Borwein, J. M. and Lisonek, P. (2000). Applications of integer relation algorithms. *Discrete Mathematics* (Special issue for FPSAC 1997) 217: 65–82. [CECM Preprint 97:104]

Dongarra, J. and Sullivan, F. (2000). The top 10 algorithms. *Computing in Science and Engineering* 2: 22–23. (See `www.cecm.sfu.ca/personal/jborwein/algorithms.html`)

Canada Research Chair,
Director Dalhousie DRIVE Dalhousie University,
Halifax, NS, Canada
E-mail: jborwein@mail.cs.dal.ca

10. Experimental mathematics: Examples, methods and implications

Discussion

This article continues in the vein of Chapter 7, which was written five years earlier. Notable in this article is the growing presence of pictures. This presence reflects both the increasing ease of producing rich mathematical images and their growing importance in our own process of discovery. This process of discovery is at least as rewarding as its fruits. No one has said this better than has Carl Friedrich Gauss:

> It is not knowledge, but the act of learning, not possession but the act of getting there, which grants the greatest enjoyment. When I have clarified and exhausted a subject, then I turn away from it, in order to go into darkness again; the never-satisfied man is so strange if he has completed a structure, then it is not in order to dwell in it peacefully, but in order to begin another. I imagine the world conqueror must feel thus, who, after one kingdom is scarcely conquered, stretches out his arms for others.[1]

Source

D.H. Bailey and J.M Borwein, "Experimental Mathematics: Examples, Methods and Implications," *Notices Amer. Math. Soc.*, **52** No. 5 (2005), 502–514.

---

[1] Carl Friedrich Gauss, 1777-1855, in an 1808 letter to his friend Farkas Bolyai (the father of Janos Bolyai).

Experimental Mathematics: Examples, Methods and Implications

David H. Bailey and Jonathan M. Borwein

> The object of mathematical rigor is to sanction and legitimize the conquests of intuition, and there was never any other object for it.
> —Jacques Hadamard[1]

> If mathematics describes an objective world just like physics, there is no reason why inductive methods should not be applied in mathematics just the same as in physics.
> —Kurt Gödel[2]

Introduction

Recent years have seen the flowering of "experimental" mathematics, namely the utilization of modern computer technology as an active tool in mathematical research. This development is not limited to a handful of researchers nor to a handful of universities, nor is it limited to one particular field of mathematics. Instead, it involves hundreds of individuals, at many different institutions, who have turned to the remarkable new computational tools now available to assist in their research, whether it be in number theory, algebra, analysis, geometry, or even topology. These tools are being used to work out specific examples, generate plots, perform various algebraic and calculus manipulations, test conjectures, and explore routes to formal proof. Using computer tools to test conjectures is by itself a major timesaver for mathematicians, as it permits them to quickly rule out false notions.

Clearly one of the major factors here is the development of robust symbolic mathematics software. Leading the way are the Maple and Mathematica products, which in the latest editions are far more expansive, robust, and user-friendly than when they first appeared twenty to twenty-five years ago. But numerous other tools, some of which emerged only in the past few years, are also playing key roles. These include: (1) the Magma computational algebra package, developed at the University of Sydney in Australia; (2) Neil Sloane's online integer sequence recognition tool, available at http://www.research.att.com/njas/sequences; (3) the inverse symbolic calculator (an online numeric constant recognition facility), available at http://www.cecm.sfu.ca/projects/ISC; (4) the electronic geometry site at http://www.eg-models.de; and numerous others. See

---

David H. Bailey is at the Lawrence Berkeley National Laboratory, Berkeley, CA 94720. His email address is dhbailey@lbl.gov. This work was supported by the Director, Office of Computational and Technology Research, Division of Mathematical, Information, and Computational Sciences of the U.S. Department of Energy, under contract number DE-AC03-76SF00098.

Jonathan M. Borwein is Canada Research Chair in Collaborative Technology and Professor of Computer Science and of Mathematics at Dalhousie University, Halifax, NS, B3H 2W5, Canada. His email address is jborwein@cs.dal.ca. This work was supported in part by NSERC and the Canada Research Chair Programme.

[1] Quoted at length in E. Borel, Leçons sur la theorie des fonctions, *1928*.

[2] *Kurt Gödel*, Collected Works, Vol. III, *1951*.

http://www.experimentalmath.info for a more complete list, with links to their respective websites.

We must of course also give credit to the computer industry. In 1965 Gordon Moore, before he served as CEO of Intel, observed:

> The complexity for minimum component costs has increased at a rate of roughly a factor of two per year.... Certainly over the short term this rate can be expected to continue, if not to increase. Over the longer term, the rate of increase is a bit more uncertain, although there is no reason to believe it will not remain nearly constant for at least 10 years. [29]

Nearly forty years later, we observe a record of sustained exponential progress that has no peer in the history of technology. Hardware progress alone has transformed mathematical computations that were once impossible into simple operations that can be done on any laptop.

Many papers have now been published in the experimental mathematics arena, and a full-fledged journal, appropriately titled *Experimental Mathematics*, has been in operation for twelve years. Even older is the AMS journal *Mathematics of Computation*, which has been publishing articles in the general area of computational mathematics since 1960 (since 1943 if you count its predecessor). Just as significant are the hundreds of other recent articles that mention computations but which otherwise are considered entirely mainstream work. All of this represents a major shift from when the present authors began their research careers, when the view that "real mathematicians don't compute" was widely held in the field.

In this article, we will summarize some of the discoveries and research results of recent years, by ourselves and by others, together with a brief description of some of the key methods employed. We will then attempt to ascertain at a more fundamental level what these developments mean for the larger world of mathematical research.

Integer Relation Detection

One of the key techniques used in experimental mathematics is integer relation detection, which in effect searches for linear relationships satisfied by a set of numerical values. To be precise, given a real or complex vector $(x_1, x_2, \cdots, x_n)$, an integer relation algorithm is a computational scheme that either finds the n integers (a_i), not all zero, such that $a_1 x_1 + a_2 x_2 + \cdots a_n x_n = 0$ (to within available numerical accuracy) or else establishes that there is no such integer vector within a ball of radius A about the origin, where the metric is the Euclidean norm: $A = (a_1^2 + a_2^2 + \cdots + a_n^2)^{1/2}$. Integer relation computations require very high precision in the input vector x to obtain numerically meaningful results—at least dn-digit precision, where $d = \log_{10} A$. This is the principal reason for the interest in very high-precision arithmetic in experimental mathematics. In one recent integer relation detection computation, 50,000-digit arithmetic was required to obtain the result [9].

At the present time, the best-known integer relation algorithm is the PSLQ algorithm [26] of mathematician-sculptor Helaman Ferguson, who, together with his wife, Claire, received the 2002 Communications Award of the Joint Policy Board for Mathematics (AMS-MAA-SIAM). Simple formulations of the PSLQ algorithm and several variants are given in [10]. The PSLQ algorithm, together with related lattice reduction schemes such as LLL, was recently named one of ten "algorithms of the century" by the publication *Computing in Science and Engineering* [4]. PSLQ or a variant is implemented in current releases of most computer algebra systems.

Arbitrary Digit Calculation Formulas

The best-known application of PSLQ in experimental mathematics is the 1995 discovery, by means of a PSLQ computation, of the "BBP" formula for π:

$$\pi = \sum_{k=0}^{\infty} \frac{1}{16^k} \left(\frac{4}{8k+1} - \frac{2}{8k+4} - \frac{1}{8k+5} - \frac{1}{8k+6} \right). \tag{1}$$

This formula permits one to directly calculate binary or hexadecimal digits beginning at the n-th digit, without needing to calculate any of the first $n - 1$ digits [8], using a simple scheme that requires very little memory and no multiple-precision arithmetic software.

It is easiest to see how this individual digit-calculating scheme works by illustrating it for a similar formula, known at least since Euler, for $\log 2$:

$$\log 2 = \sum_{n=1}^{\infty} \frac{1}{n 2^n}.$$

Note that the binary expansion of $\log 2$ beginning after the first d binary digits is simply $\{2^d \log 2\}$, where by $\{\cdot\}$ we mean fractional part. We can write

$$\{2^d \log 2\} = \left\{ \sum_{n=1}^{\infty} \frac{2^{d-n}}{n} \right\} = \left\{ \sum_{n=1}^{d} \frac{2^{d-n}}{n} \right\} + \left\{ \sum_{n=d+1}^{\infty} \frac{2^{d-n}}{n} \right\}$$
$$= \left\{ \sum_{n=1}^{d} \frac{2^{d-n} \bmod n}{n} \right\} + \left\{ \sum_{n=d+1}^{\infty} \frac{2^{d-n}}{n} \right\}, \tag{2}$$

where we insert "mod n" in the numerator of the first term of (2), since we are interested only in the fractional part after division by n. Now the expression $2^{d-n} \bmod n$ may be evaluated very rapidly by means of the binary algorithm for exponentiation, where each multiplication is reduced modulo n. The entire scheme indicated by formula (2) can be implemented on a computer using ordinary 64-bit or 128-bit arithmetic; high-precision arithmetic software is not required. The resulting floating-point value, when expressed in binary format, gives the first few digits of the binary expansion of $\log 2$ beginning at position $d + 1$. Similar calculations applied to each of the four terms in formula (1) yield a similar result for π. The largest computation of this type to date is binary digits of π beginning at the quadrillionth (10^{15}-th) binary digit, performed by an international network of computers organized by Colin Percival.

Figure 1. Ferguson's "Figure Eight Knot Complement" sculpture.

The BBP formula for π has even found a practical application: it is now employed in the g95 Fortran compiler as part of transcendental function evaluation software.

Since 1995 numerous other formulas of this type have been found and proven using a similar experimental approach. Several examples include:

(3)
$$\pi\sqrt{3} = \frac{9}{32} \sum_{k=0}^{\infty} \frac{1}{64^k} \left(\frac{16}{6k+1} - \frac{8}{6k+2} - \frac{2}{6k+4} - \frac{1}{6k+5} \right),$$

(4)
$$\pi^2 = \frac{1}{8} \sum_{k=0}^{\infty} \frac{1}{64^k} \left[\frac{144}{(6k+1)^2} - \frac{216}{(6k+2)^2} - \frac{72}{(6k+3)^2} - \frac{54}{(6k+4)^2} + \frac{9}{(6k+5)^2} \right],$$

$$\pi^2 = \frac{2}{27} \sum_{k=0}^{\infty} \frac{1}{729^k} \left[\frac{243}{(12k+1)^2} - \frac{405}{(12k+2)^2} - \frac{81}{(12k+4)^2} - \frac{27}{(12k+5)^2} \right.$$

(5)
$$\left. - \frac{72}{(12k+6)^2} - \frac{9}{(12k+7)^2} - \frac{9}{(12k+8)^2} - \frac{5}{(12k+10)^2} + \frac{1}{(12k+11)^2} \right],$$

(6)
$$\sqrt{3} \arctan\left(\frac{\sqrt{3}}{7}\right) = \sum_{k=0}^{\infty} \frac{1}{27^k} \left(\frac{3}{3k+1} + \frac{1}{3k+2} \right),$$

(7)
$$\frac{25}{2} \log \left[\frac{781}{256} \left(\frac{57 - 5\sqrt{5}}{57 + 5\sqrt{5}} \right)^{\sqrt{5}} \right] = \sum_{k=0}^{\infty} \frac{1}{5^{5k}} \left(\frac{5}{5k+2} + \frac{1}{5k+3} \right).$$

Formulas (3) and (4) permit arbitrary-position binary digits to be calculated for $\pi\sqrt{3}$ and π^2. Formulas (5) and (6) permit the same for ternary (base-3) expansions of π^2 and $\sqrt{3}\arctan(\sqrt{3}/7)$. Formula (7) permits the same for the base-5 expansion of the curious constant shown. A compendium of known BBP-type formulas, with references, is available at [5].

One interesting twist here is that the hyperbolic volume of one of Ferguson's sculptures (the "Figure Eight Knot Complement";[3] see Figure 1), which is given by

$$V = 2\sqrt{3} \sum_{n=1}^{\infty} \frac{1}{n\binom{2n}{n}} \sum_{k=n}^{2n-1} \frac{1}{k}$$
$$= 2.029883212819307250042405108549\ldots,$$

has been identified in terms of a BBP-type formula by application of Ferguson's own PSLQ algorithm. In particular, British physicist David Broadhurst found in 1998, using a PSLQ program, that

$$V = \frac{\sqrt{3}}{9} \sum_{n=0}^{\infty} \frac{(-1)^n}{27^n}$$
$$\times \left[\frac{18}{(6n+1)^2} - \frac{18}{(6n+2)^2} - \frac{24}{(6n+3)^2} - \frac{6}{(6n+4)^2} + \frac{2}{(6n+5)^2} \right].$$

This result is proven in [15, Chap. 2, Prob. 34].

Does Pi Have a Nonbinary BBP Formula?

Since the discovery of the BBP formula for π in 1995, numerous researchers have investigated, by means of computational searches, whether there is a similar formula for calculating arbitrary digits of π in other number bases (such as base 10). Alas, these searches have not been fruitful.

Recently, one of the present authors (JMB), together with David Borwein (Jon's father) and William Galway, established that there is no degree-1 BBP-type formula for π for bases other than powers of two (although this does not rule out some other scheme for calculating individual digits). We will sketch this result here. Full details and some related results can be found in [20].

In the following, $\Re(z)$ and $\Im(z)$ denote the real and imaginary parts of z, respectively. The integer $b > 1$ is not a *proper power* if it cannot be written as c^m for any integers c and $m > 1$. We will use the notation $\text{ord}_p(z)$ to denote the p-adic order of the rational $z \in Q$. In particular, $\text{ord}_p(p) = 1$ for prime p, while $\text{ord}_p(q) = 0$ for primes $q \neq p$, and $\text{ord}_p(wz) = \text{ord}_p(w) + \text{ord}_p(z)$. The notation $\nu_b(p)$ will mean the order of the integer b in the multiplicative group of the integers modulo p. We will say that p is a *primitive prime factor* of $b^m - 1$ if m is the least integer such that $p | (b^m - 1)$. Thus p is a primitive prime factor of $b^m - 1$ provided $\nu_b(p) = m$. Given the Gaussian integer $z \in Q[i]$ and the rational prime $p \equiv 1 \pmod 4$, let $\theta_p(z)$ denote $\text{ord}_{\mathfrak{p}}(z) - \text{ord}_{\bar{\mathfrak{p}}}(z)$, where $\mathfrak{p}$ and $\bar{\mathfrak{p}}$ are the two conjugate Gaussian primes dividing p and where we require $0 < \Im(\mathfrak{p}) < \Re(\mathfrak{p})$ to make the definition of θ_p unambiguous. Note that

(8)
$$\theta_p(wz) = \theta_p(w) + \theta_p(z).$$

Given $\kappa \in R$, with $2 \leq b \in Z$ and b not a proper power, we say that κ has a Z-linear or Q-linear

---
[3] *Reproduced by permission of the sculptor.*

Machin-type BBP arctangent formula to the base b if and only if κ can be written as a Z-linear or Q-linear combination (respectively) of generators of the form

$$\arctan\left(\frac{1}{b^m}\right) = \Im\log\left(1 + \frac{i}{b^m}\right)$$
$$(9) \qquad = b^m \sum_{k=0}^{\infty} \frac{(-1)^k}{b^{2mk}(2k+1)}.$$

We shall also use the following result, first proved by Bang in 1886:

Theorem 1. *The only cases where $b^m - 1$ has no primitive prime factor(s) are when $b = 2$, $m = 6$, $b^m - 1 = 3^2 \cdot 7$ or when $b = 2^N - 1, N \in Z$, $m = 2$, $b^m - 1 = 2^{N+1}(2^{N-1} - 1)$.*

We can now state the main result:

Theorem 2. *Given $b > 2$ and not a proper power, there is no Q-linear Machin-type BBP arctangent formula for π.*

Proof: It follows immediately from the definition of a Q-linear Machin-type BBP arctangent formula that any such formula has the form

$$(10) \qquad \pi = \frac{1}{n}\sum_{m=1}^{M} n_m \Im\log(b^m - i),$$

where $n > 0 \in Z$, $n_m \in Z$, and $M \geq 1$, $n_M \neq 0$. This implies that

$$(11) \qquad \prod_{m=1}^{M}(b^m - i)^{n_m} \in e^{ni\pi}Q^\times = Q^\times.$$

For any $b > 2$ and not a proper power, it follows from Bang's Theorem that $b^{4M} - 1$ has a primitive prime factor, say p. Furthermore, p must be odd, since $p = 2$ can only be a *primitive* prime factor of $b^m - 1$ when b is odd and $m = 1$. Since p is a primitive prime factor, it does not divide $b^{2M} - 1$, and so p must divide $b^{2M} + 1 = (b^M + i)(b^M - i)$. We cannot have both $p | b^M + i$ and $p | b^M - i$, since this would give the contradiction that $p|(b^M + i) - (b^M - i) = 2i$. It follows that $p \equiv 1 \pmod 4$ and that p factors as $p = \mathfrak{p}\bar{\mathfrak{p}}$ over $Z[i]$, with exactly one of $\mathfrak{p}, \bar{\mathfrak{p}}$ dividing $b^M - i$. Referring to the definition of θ, we see that we must have $\theta_\mathfrak{p}(b^M - i) \neq 0$. Furthermore, for any $m < M$, neither $\mathfrak{p}$ nor $\bar{\mathfrak{p}}$ can divide $b^m - i$, since this would imply $p | b^{4m} - 1$, $4m < 4M$, contradicting the fact that p is a primitive prime factor of $b^{4M} - 1$. So for $m < M$, we have $\theta_\mathfrak{p}(b^m - i) = 0$. Referring to equation (10) and using equation (8) and the fact that $n_M \neq 0$, we get the contradiction

$$0 \neq n_M \theta_\mathfrak{p}(b^M - i)$$
$$(12) \qquad = \sum_{m=1}^{M} n_m \theta_\mathfrak{p}(b^m - i) = \theta_\mathfrak{p}(Q^\times) = 0.$$

Thus our assumption that there was a b-ary Machin-type BBP arctangent formula for π must be false.

Normality Implications of the BBP Formulas

One interesting (and unanticipated) discovery is that the existence of these computer-discovered BBP-type formulas has implications for the age-old question of normality for several basic mathematical constants, including π and $\log 2$. What's more, this line of research has recently led to a full-fledged proof of normality for an uncountably infinite class of explicit real numbers.

Given a positive integer b, we will define a real number α to be b-*normal* if every m-long string of base-b digits appears in the base-b expansion of α with limiting frequency b^{-m}. In spite of the apparently stringent nature of this requirement, it is well known from measure theory that almost all real numbers are b-normal, for all bases b. Nonetheless, there are very few explicit examples of b-normal numbers, other than the likes of *Champernowne's constant* $0.123456789101112131415\ldots$. In particular, although computations suggest that virtually all of the well-known irrational constants of mathematics (such as $\pi, e, \gamma, \log 2, \sqrt{2}$, etc.) are normal to various number bases, there is not a single proof—not for any of these constants, not for any number base.

Recently one of the present authors (DHB) and Richard Crandall established the following result.

Let $p(x)$ and $q(x)$ be integer-coefficient polynomials, with $\deg p < \deg q$, and $q(x)$ having no zeroes for positive integer arguments. By an *equidistributed* sequence in the unit interval we mean a sequence (x_n) such that for every subinterval (a, b), the fraction $\#[x_n \in (a, b)]/n$ tends to $b - a$ in the limit. The result is as follows:

Theorem 3. *A constant α satisfying the BBP-type formula*

$$\alpha = \sum_{n=1}^{\infty} \frac{p(n)}{b^n q(n)}$$

is b-normal if and only if the associated sequence defined by $x_0 = 0$ and, for $n \geq 1$, $x_n = \{bx_{n-1} + p(n)/q(n)\}$ (where $\{\cdot\}$ denotes fractional part as before), is equidistributed in the unit interval.

For example, $\log 2$ is 2-normal if and only if the simple sequence defined by $x_0 = 0$ and $\{x_n = 2x_{n-1} + 1/n\}$ is equidistributed in the unit interval. For π, the associated sequence is $x_0 = 0$ and

$$x_n = \left\{16x_{n-1} + \frac{120n^2 - 89n + 16}{512n^4 - 1024n^3 + 712n^2 - 206n + 21}\right\}.$$

Full details of this result are given in [11] [15, Section 3.8].

It is difficult to know at the present time whether this result will lead to a full-fledged proof of normality for, say, π or $\log 2$. However, this approach has yielded a solid normality proof for another class of reals: Given $r \in [0, 1)$, let r_n be the n-th binary digit of r. Then for each r in the unit interval, the constant

$$(13) \qquad \alpha_r = \sum_{n=1}^{\infty} \frac{1}{3^n 2^{3^n + r_n}}$$

is 2-normal and transcendental [12]. What's more, it can be shown that whenever $r \neq s$, then $\alpha_r \neq \alpha_s$. Thus (13) defines an uncountably infinite class of distinct 2-normal, transcendental real numbers. A similar conclusion applies when 2 and 3 in (13) are replaced by any pair of relatively prime integers greater than 1.

Here we will sketch a proof of normality for one particular instance of these constants, namely $\alpha_0 = \sum_{n \geq 1} 1/(3^n 2^{3^n})$. Its associated sequence can be seen to be $x_0 = 0$ and $x_n = \{2x_{n-1} + c_n\}$, where $c_n = 1/n$ if n is a power of 3, and zero otherwise. This associated sequence is a very good approximation to the sequence $(\{2^n \alpha_0\})$ of shifted binary fractions of α_0. In fact, $|\{2^n \alpha_0\} - x_n| < 1/(2n)$. The first few terms of the associated sequence are

$0, 0, 0, \frac{1}{3}, \frac{2}{3}, \frac{1}{3}, \frac{2}{3}, \frac{1}{3}, \frac{2}{3}$,

$\frac{4}{9}, \frac{8}{9}, \frac{7}{9}, \frac{5}{9}, \frac{1}{9}, \frac{2}{9}, \frac{4}{9}, \frac{8}{9}, \frac{7}{9}, \frac{5}{9}, \frac{1}{9}, \frac{2}{9}, \frac{4}{9}, \frac{8}{9}, \frac{7}{9}, \frac{5}{9}, \frac{1}{9}, \frac{2}{9}$,

$\frac{13}{27}, \frac{26}{27}, \frac{25}{27}, \frac{23}{27}, \frac{19}{27}, \frac{11}{27}, \frac{22}{27}, \frac{17}{27}, \frac{7}{27}, \frac{14}{27}, \frac{1}{27}, \frac{2}{27}, \frac{4}{27}, \frac{8}{27}, \frac{16}{27}, \frac{5}{27}, \frac{10}{27}, \frac{20}{27},$

$\frac{13}{27}, \frac{26}{27}, \frac{25}{27}, \frac{23}{27}, \frac{19}{27}, \frac{11}{27}, \frac{22}{27}, \frac{17}{27}, \frac{7}{27}, \frac{14}{27}, \frac{1}{27}, \frac{2}{27}, \frac{4}{27}, \frac{8}{27}, \frac{16}{27}, \frac{5}{27}, \frac{10}{27}, \frac{20}{27},$

$\frac{13}{27}, \frac{26}{27}, \frac{25}{27}, \frac{23}{27}, \frac{19}{27}, \frac{11}{27}, \frac{22}{27}, \frac{17}{27}, \frac{7}{27}, \frac{14}{27}, \frac{1}{27}, \frac{2}{27}, \frac{4}{27}, \frac{8}{27}, \frac{16}{27}, \frac{5}{27}, \frac{10}{27}, \frac{20}{27},$

and so forth. The clear pattern is that of triply repeated segments, each of length $2 \cdot 3^m$, where the numerators range over all integers relatively prime to and less than 3^{m+1}.

Note the very even manner in which this sequence fills the unit interval. Given any subinterval (c, d) of the unit interval, it can be seen that this sequence visits this subinterval no more than $3n(d - c) + 3$ times, among the first n elements, provided that $n > 1/(d - c)$. It can then be shown that the sequence $(\{2^j \alpha\})$ visits (c, d) no more than $8n(d - c)$ times, among the first n elements of this sequence, so long as n is at least $1/(d - c)^2$. The 2-normality of α_0 then follows from a result given in [28, p. 77]. Further details on these results are given in [15, Sec. 4.3], [6], [12].

Euler's Multi-Zeta Sums

In April 1993, Enrico Au-Yeung, an undergraduate at the University of Waterloo, brought to the attention of one of us (JMB) the curious result

$$(14) \qquad \sum_{k=1}^{\infty} \left(1 + \frac{1}{2} + \cdots + \frac{1}{k}\right)^2 k^{-2}$$
$$= 4.59987\ldots \approx \frac{17}{4} \zeta(4) = \frac{17\pi^4}{360}$$

where $\zeta(s) = \sum_{n \geq 1} n^{-s}$ is the Riemann zeta function. Au-Yeung had computed the sum in (14) to 500,000 terms, giving an accuracy of five or six decimal digits. Suspecting that his discovery was merely a modest numerical coincidence, Borwein sought to compute the sum to a higher level of precision. Using Fourier analysis and *Parseval's equation*, he wrote

$$(15) \qquad \frac{1}{2\pi} \int_0^\pi (\pi - t)^2 \log^2(2 \sin \frac{t}{2}) dt = \sum_{n=1}^{\infty} \frac{(\sum_{k=1}^n \frac{1}{k})^2}{(n + 1)^2}.$$

The series on the right of (15) permits one to evaluate (14), while the integral on the left can be computed using the numerical quadrature facility of Mathematica or Maple. When he did this, Borwein was surprised to find that the conjectured identity (14) holds to more than 30 digits. We should add here that by good fortune, $17/360 = 0.047222\ldots$ has period one and thus can plausibly be recognized from its first six digits, so that Au-Yeung's numerical discovery was not entirely far-fetched.

Borwein was not aware at the time that (14) follows directly from a 1991 result due to De Doelder and had even arisen in 1952 as a problem in the *American Mathematical Monthly*. What's more, it turns out that Euler considered some related summations. Perhaps it was just as well that Borwein was not aware of these earlier results—and indeed of a large, quite deep and varied literature [21]—because pursuit of this and similar questions had led to a line of research that continues to the present day.

First define the *multi-zeta* constant

$$\zeta(s_1, s_2, \ldots, s_k) := \sum_{n_1 > n_2 > \cdots > n_k > 0} \prod_{j=1}^{k} n_j^{-|s_j|} \sigma_j^{-n_j},$$

where the $s_1, s_2, \ldots, s_k$ are nonzero integers and the $\sigma_j := \operatorname{signum}(s_j)$. Such constants can be considered as generalizations of the Riemann zeta function at integer arguments in higher dimensions.

The analytic evaluation of such sums has relied on fast methods for computing their numerical values. One scheme, based on *Hölder Convolution*, is discussed in [22] and implemented in EZFace+, an online tool available at http://www.cecm.sfu.ca/projects/ezface+. We will illustrate its application to one specific case, namely the analytic identification of the sum

$$(16) \qquad S_{2,3} = \sum_{k=1}^{\infty} \left(1 - \frac{1}{2} + \cdots + (-1)^{k+1} \frac{1}{k}\right)^2 (k + 1)^{-3}.$$

Expanding the squared term in (16), we have

$$(17) \quad \sum_{\substack{0<i,j<k \\ k>0}} \frac{(-1)^{i+j+1}}{ijk^3} = -2\zeta(3,-1,-1) + \zeta(3,2).$$

Evaluating this in EZFace+, we quickly obtain

$S_{2,3} = 0.156166933381176915881035909687988193685776709840303872957529354497075037440295791455205653709358147578\ldots$

Given this numerical value, PSLQ or some other integer-relation-finding tool can be used to see if this constant satisfies a rational linear relation of certain constants. Our experience with these evaluations has suggested that likely terms would include: π^5, $\pi^4 \log(2)$, $\pi^3 \log^2(2)$, $\pi^2 \log^3(2)$, $\pi \log^4(2)$, $\log^5(2)$, $\pi^2 \zeta(3)$, $\pi \log(2)\zeta(3)$, $\log^2(2)\zeta(3)$, $\zeta(5)$, $\text{Li}_5(1/2)$. The result is quickly found to be:

$$S_{2,3} = 4\,\text{Li}_5\left(\frac{1}{2}\right) - \frac{1}{30}\log^5(2) - \frac{17}{32}\zeta(5)$$
$$- \frac{11}{720}\pi^4 \log(2) + \frac{7}{4}\zeta(3)\log^2(2)$$
$$+ \frac{1}{18}\pi^2 \log^3(2) - \frac{1}{8}\pi^2 \zeta(3).$$

This result has been proven in various ways, both analytic and algebraic. Indeed, all evaluations of sums of the form $\zeta(\pm a_1, \pm a_2, \cdots, \pm a_m)$ with weight $w := \sum_k a_m$, for $k < 8$, as in (17) are established.

One general result that is reasonably easily obtained is the following, true for all n:

$$(18) \quad \zeta(\{3\}_n) = \zeta(\{2,1\}_n).$$

On the other hand, a general proof of

$$(19) \quad \zeta(\{2,1\}_n) \stackrel{?}{=} 2^{3n} \zeta(\{-2,1\}_n)$$

remains elusive. There has been abundant evidence amassed to support the conjectured identity (19) since it was discovered experimentally in 1996. The first eighty-five instances of (19) were recently affirmed in calculations by Petr Lisoněk to 1000 decimal place accuracy. Lisonek also checked the case $n = 163$, a calculation that required ten hours run time on a 2004-era computer. The only proof known of (18) is a change of variables in a multiple integral representation that sheds no light on (19) (see [21]).

Evaluation of Integrals

This same general strategy of obtaining a high-precision numerical value, then attempting by means of PSLQ or other numeric-constant recognition facilities to identify the result as an analytic expression, has recently been applied with significant success to the age-old problem of evaluating definite integrals. Obviously Maple and Mathematica have some rather effective integration facilities, not only for obtaining analytic results directly, but also for obtaining high-precision numeric values. However, these products do have limitations, and their numeric integration facilities are typically limited to 100 digits or so, beyond which they tend to require an unreasonable amount of run time.

Fortunately, some new methods for numerical integration have been developed that appear to be effective for a broad range of one-dimensional integrals, typically producing up to 1000 digit accuracy in just a few seconds' (or at most a few minutes') run time on a 2004-era personal computer, and that are also well suited for parallel processing [13], [14], [16, p. 312]. These schemes are based on the *Euler-Maclaurin summation* formula [3, p. 180], which can be stated as follows: Let $m \geq 0$ and $n \geq 1$ be integers, and define $h = (b-a)/n$ and $x_j = a + jh$ for $0 \leq j \leq n$. Further assume that the function $f(x)$ is at least $(2m+2)$-times continuously differentiable on $[a,b]$. Then

$$(20) \quad \int_a^b f(x)\,dx = h\sum_{j=0}^n f(x_j) - \frac{h}{2}(f(a) + f(b))$$
$$- \sum_{i=1}^m \frac{h^{2i} B_{2i}}{(2i)!}\left(f^{(2i-1)}(b) - f^{(2i-1)}(a)\right) - E(h),$$

where B_{2i} denote the Bernoulli numbers, and

$$E(h) = \frac{h^{2m+2}(b-a)B_{2m+2}f^{(2m+2)}(\xi)}{(2m+2)!}$$

for some $\xi \in (a,b)$. In the circumstance where the function $f(x)$ and all of its derivatives are zero at the endpoints a and b (as in a smooth, bell-shaped function), the second and third terms of the Euler-Maclaurin formula (20) are zero, and we conclude that the error $E(h)$ goes to zero more rapidly than any power of h.

This principle is utilized by transforming the integral of some C^∞ function $f(x)$ on the interval $[-1,1]$ to an integral on $(-\infty, \infty)$ using the change of variable $x = g(t)$. Here $g(x)$ is some monotonic, infinitely differentiable function with the property that $g(x) \to 1$ as $x \to \infty$ and $g(x) \to -1$ as $x \to -\infty$, and also with the property that $g'(x)$ and all higher derivatives rapidly approach zero for large positive and negative arguments. In this case we can write, for $h > 0$,

$$\int_{-1}^1 f(x)\,dx = \int_{-\infty}^\infty f(g(t))g'(t)\,dt$$
$$= h\sum_{j=-\infty}^\infty w_j f(x_j) + E(h),$$

where $x_j = g(hj)$ and $w_j = g'(hj)$ are abscissas and weights that can be precomputed. If $g'(t)$ and its derivatives tend to zero sufficiently rapidly for large t, positive and negative, then even in cases where $f(x)$ has a vertical derivative or an integrable singularity at one or both endpoints, the resulting integrand $f(g(t))g'(t)$ is, in many cases, a smooth bell-shaped function for which the Euler-Maclaurin formula applies. In these cases, the error $E(h)$ in this approximation decreases faster than any power of h.

Three suitable g functions are $g_1(t) = \tanh t$, $g_2(t) = \operatorname{erf} t$, and $g_3(t) = \tanh(\pi/2 \cdot \sinh t)$. Among these three, $g_3(t)$ appears to be the most effective for typical experimental math applications. For many integrals, "tanh-sinh" quadrature, as the resulting scheme is known, achieves quadratic convergence: reducing the interval h in half roughly doubles the number of correct digits in the quadrature result. This is another case where we have more heuristic than proven knowledge.

As one example, recently the present authors, together with Greg Fee of Simon Fraser University in Canada, were inspired by a recent problem in the *American Mathematical Monthly* [2]. They found by using a tanh-sinh quadrature program, together with a PSLQ integer relation detection program, that if $C(a)$ is defined by

$$C(a) = \int_0^1 \frac{\arctan(\sqrt{x^2 + a^2})\,dx}{\sqrt{x^2 + a^2}(x^2 + 1)},$$

then

$$C(0) = \pi \log 2/8 + G/2,$$
$$C(1) = \pi/4 - \pi\sqrt{2}/2 + 3\arctan(\sqrt{2})/\sqrt{2},$$
$$C(\sqrt{2}) = 5\pi^2/96.$$

Here $G = \sum_{k \geq 0} (-1)^k/(2k+1)^2$ is *Catalan's constant*—the simplest number whose irrationality is not established but for which abundant numerical evidence exists. These experimental results then led to the following general result, rigorously established, among others:

$$\int_0^\infty \frac{\arctan(\sqrt{x^2 + a^2})\,dx}{\sqrt{x^2 + a^2}(x^2 + 1)}$$
$$= \frac{\pi}{2\sqrt{a^2 - 1}} \left[2\arctan(\sqrt{a^2 - 1}) - \arctan(\sqrt{a^4 - 1}) \right].$$

As a second example, recently the present authors empirically determined that

$$\frac{2}{\sqrt{3}} \int_0^1 \frac{\log^6(x) \arctan[x\sqrt{3}/(x-2)]}{x+1}\,dx = \frac{1}{81648}\,[-229635 L_3(8)$$
$$+ 29852550 L_3(7)\log 3 - 1632960 L_3(6)\pi^2 + 27760320 L_3(5)\zeta(3)$$
$$- 275184 L_3(4)\pi^4 + 36288000 L_3(3)\zeta(5) - 30008 L_3(2)\pi^6$$
$$- 57030120 L_3(1)\zeta(7)\,],$$

where $L_3(s) = \sum_{n=1}^\infty [1/(3n-2)^s - 1/(3n-1)^s]$. Based on these experimental results, general results of this type have been conjectured but not yet rigorously established.

A third example is the following:

$$(21) \quad \frac{24}{7\sqrt{7}} \int_{\pi/3}^{\pi/2} \log \left| \frac{\tan t + \sqrt{7}}{\tan t - \sqrt{7}} \right| dt \stackrel{?}{=} L_{-7}(2)$$

where

$$L_{-7}(s) = \sum_{n=0}^\infty \left[\frac{1}{(7n+1)^s} + \frac{1}{(7n+2)^s} - \frac{1}{(7n+3)^s} \right.$$
$$\left. + \frac{1}{(7n+4)^s} - \frac{1}{(7n+5)^s} - \frac{1}{(7n+6)^s} \right].$$

The "identity" (21) has been verified to over 5000 decimal digit accuracy, but a proof is not yet known. It arises from the volume of an ideal tetrahedron in hyperbolic space, [15, pp. 90-1]. For algebraic topology reasons, it is known that the ratio of the left-hand to the right-hand side of (21) is rational.

A related experimental result, verified to 1000 digit accuracy, is

$$0 \stackrel{?}{=} -2J_2 - 2J_3 - 2J_4 + 2J_{10} + 2J_{11} + 3J_{12} + 3J_{13} + J_{14} - J_{15}$$
$$- J_{16} - J_{17} - J_{18} - J_{19} + J_{20} + J_{21} - J_{22} - J_{23} + 2J_{25},$$

where J_n is the integral in (21), with limits $n\pi/60$ and $(n+1)\pi/60$.

The above examples are ordinary one-dimensional integrals. Two-dimensional integrals are also of interest. Along this line we present a more recreational example discovered experimentally by James Klein—and confirmed by *Monte Carlo* simulation. It is that the expected distance between two random points on different sides of a unit square is

$$\frac{2}{3} \int_0^1 \int_0^1 \sqrt{x^2 + y^2}\,dx\,dy + \frac{1}{3} \int_0^1 \int_0^1 \sqrt{1 + (u-v)^2}\,du\,dv$$
$$= \frac{1}{9}\sqrt{2} + \frac{5}{9} \log(\sqrt{2} + 1) + \frac{2}{9},$$

and the expected distance between two random points on different sides of a unit cube is

$$\frac{4}{5} \int_0^1 \int_0^1 \int_0^1 \int_0^1 \sqrt{x^2 + y^2 + (z-w)^2}\,dw\,dx\,dy\,dz$$
$$+ \frac{1}{5} \int_0^1 \int_0^1 \int_0^1 \int_0^1 \sqrt{1 + (y-u)^2 + (z-w)^2}\,du\,dw\,dy\,dz$$
$$= \frac{4}{75} + \frac{17}{75}\sqrt{2} - \frac{2}{25}\sqrt{3} - \frac{7}{75}\pi$$
$$+ \frac{7}{25} \log\left(1 + \sqrt{2}\right) + \frac{7}{25} \log\left(7 + 4\sqrt{3}\right).$$

See [7] for details and some additional examples. It is not known whether similar closed forms exist for higher-dimensional cubes.

Ramanujan's AGM Continued Fraction

Given $a, b, \eta > 0$, define

$$R_\eta(a,b) = \cfrac{a}{\eta + \cfrac{b^2}{\eta + \cfrac{4a^2}{\eta + \cfrac{9b^2}{\eta + \ddots}}}}.$$

This continued fraction arises in Ramanujan's *Notebooks*. He discovered the beautiful fact that

$$\frac{R_\eta(a,b) + R_\eta(b,a)}{2} = R_\eta\left(\frac{a+b}{2}, \sqrt{ab}\right).$$

The authors wished to record this in [15] and wished to computationally check the identity. A first attempt to numerically compute $R_1(1,1)$ directly failed miserably, and with some effort only three reliable digits were obtained: $0.693\ldots$. With hindsight, the slowest convergence of the fraction occurs in the mathematically simplest case, namely when $a = b$. Indeed $R_1(1,1) = \log 2$, as the first primitive numerics had tantalizingly suggested.

Attempting a direct computation of $R_1(2,2)$ using a depth of 20000 gives us two digits. Thus we must seek more sophisticated methods. From formula (1.11.70) of [16] we see that for $0 < b < a$,

$$\mathcal{R}_1(a,b)$$
(22)
$$= \frac{\pi}{2} \sum_{n \in \mathbb{Z}} \frac{aK(k)}{K^2(k) + a^2 n^2 \pi^2} \operatorname{sech}\left(n\pi \frac{K(k')}{K(k)}\right),$$

where $k = b/a = \theta_2^2/\theta_3^2, k' = \sqrt{1-k^2}$. Here θ_2, θ_3 are Jacobian theta functions and K is a complete elliptic integral of the first kind.

Writing the previous equation as a Riemann sum, we have

(23)
$$\mathcal{R}(a) := \mathcal{R}_1(a,a) = \int_0^\infty \frac{\operatorname{sech}(\pi x/(2a))}{1+x^2} dx$$
$$= 2a \sum_{k=1}^\infty \frac{(-1)^{k+1}}{1+(2k-1)a},$$

where the final equality follows from the Cauchy-Lindelof Theorem. This sum may also be written as $\mathcal{R}(a) = \frac{2a}{1+a} F\left(\frac{1}{2a} + \frac{1}{2}, 1; \frac{1}{2a} + \frac{3}{2}; -1\right)$. The latter form can be used in Maple or Mathematica to determine

$$\mathcal{R}(2) = 0.97499098879872209671990033452 9\ldots$$

This constant, as written, is a bit difficult to recognize, but if one first divides by $\sqrt{2}$, one can obtain, using the *Inverse Symbolic Calculator*, an online tool available at the URL http://www.cecm.sfu.ca/projects/ISC/ISCmain.html, that the quotient is $\pi/2 - \log(1+\sqrt{2})$. Thus we conclude, experimentally, that

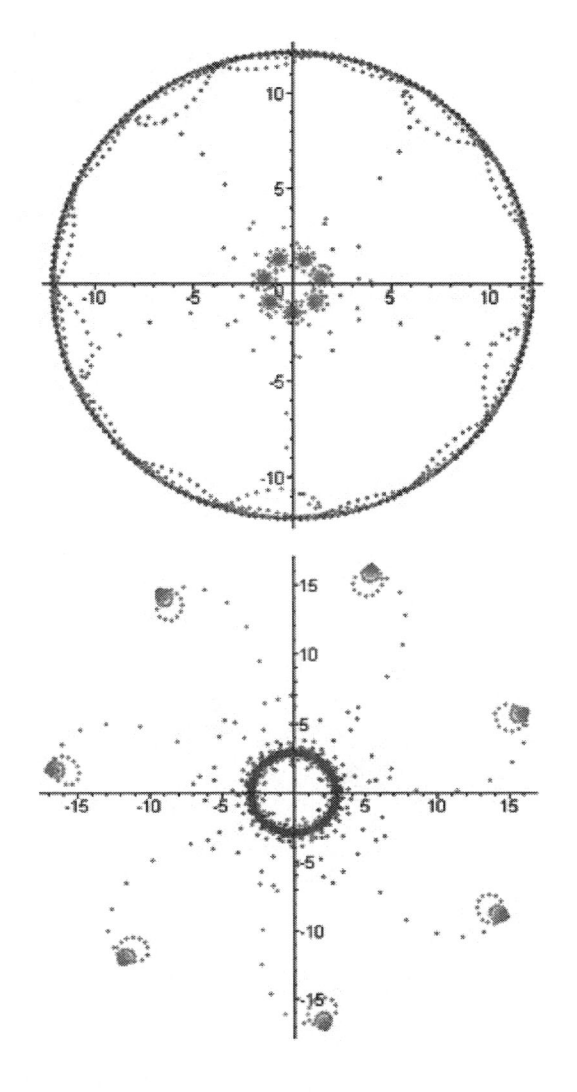

Figure 2. Dynamics and attractors of various iterations.

$$\mathcal{R}(2) = \sqrt{2}[\pi/2 - \log(1+\sqrt{2})].$$

Indeed, it follows (see [19]) that

$$\mathcal{R}(a) = 2\int_0^1 \frac{t^{1/a}}{1+t^2} dt.$$

Note that $\mathcal{R}(1) = \log 2$. No nontrivial closed form is known for $\mathcal{R}(a,b)$ with $a \neq b$, although

$$\mathcal{R}_1\left(\frac{1}{4\pi}\beta\left(\frac{1}{4},\frac{1}{4}\right), \frac{\sqrt{2}}{8\pi}\beta\left(\frac{1}{4},\frac{1}{4}\right)\right) = \frac{1}{2}\sum_{n \in \mathbb{Z}} \frac{\operatorname{sech}(n\pi)}{1+n^2}$$

is close to closed. Here β denotes the classical *Beta function*. It would be pleasant to find a direct proof of (23). Further details are to be found in [19], [17], [16].

Study of these Ramanujan continued fractions has been facilitated by examining the closely related dynamical system $t_0 = 1, t_1 = 1$, and

Chapter 10

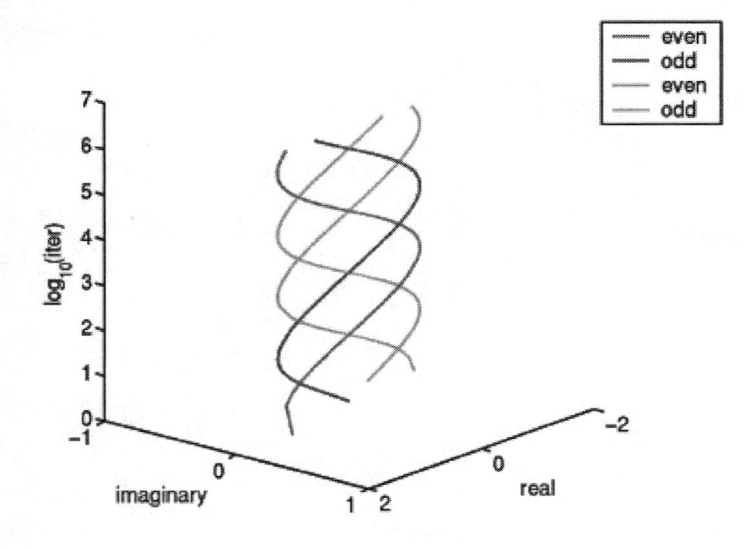

Experimental mathematics: Examples, methods and implications

Figure 3. The subtle fourfold serpent.

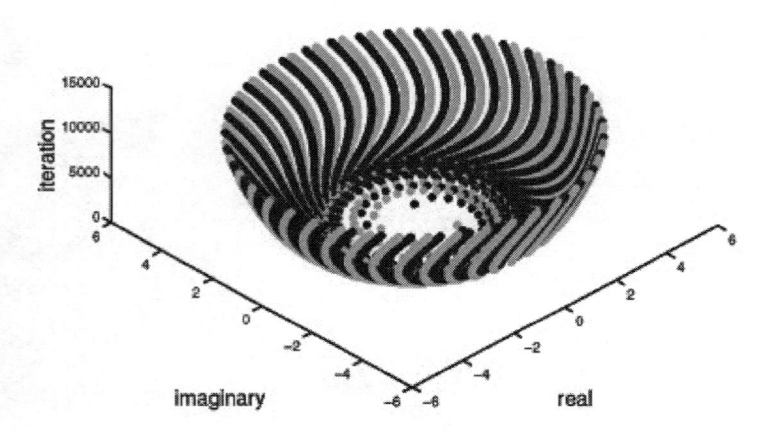

Figure 4. A period three dynamical system (odd and even iterates).

$$(24) \quad t_n := t_n(a,b) = \frac{1}{n} + \omega_{n-1}\left(1 - \frac{1}{n}\right)t_{n-2},$$

where $\omega_n = a^2$ or b^2 (from the Ramanujan continued fraction definition), depending on whether n is even or odd.

If one studies this based only on numerical values, nothing is evident; one only sees that $t_n \to 0$ fairly slowly. However, if we look at this iteration pictorially, we learn significantly more. In particular, if we plot these iterates in the complex plane and then scale by $\sqrt{n}$ and color the iterations blue or red depending on odd or even n, then some remarkable fine structures appear; see Figure 2. With assistance of such plots, the behavior of these iterates (and the Ramanujan continued fractions) is now quite well understood. These studies have ventured into matrix theory, real analysis, and even the theory of martingales from probability theory [19], [17], [18], [23].

There are some exceptional cases. *Jacobsen-Masson theory* [17], [18] shows that the even/odd fractions for $\mathcal{R}_1(i,i)$ behave "chaotically"; neither converge. Indeed, when $a = b = i$, $(t_n(i,i))$ exhibit a fourfold quasi-oscillation, as n runs through values mod 4. Plotted versus n, the (real) sequence $t_n(\mathbf{i})$ exhibits the serpentine oscillation of four separate "necklaces". The detailed asymptotic is

$$t_n(i,i) = \sqrt{\frac{2}{\pi}\cosh\frac{\pi}{2}} \frac{1}{\sqrt{n}}\left(1 + O\left(\frac{1}{n}\right)\right)$$
$$\times \begin{cases} (-1)^{n/2}\cos(\theta - \log(2n)/2) & n \text{ is even} \\ (-1)^{(n+1)/2}\sin(\theta - \log(2n)/2) & n \text{ odd} \end{cases}$$

where $\theta := \arg \Gamma((1+i)/2)$.

Analysis is easy given the following striking hypergeometric parametrization of (24) when $a = b \neq 0$ (see [18]), which was both *experimentally discovered* and is *computer provable*:

$$(25) \quad t_n(a,a) = \frac{1}{2}F_n(a) + \frac{1}{2}F_n(-a),$$

where

$$F_n(a) := -\frac{a^n 2^{1-\omega}}{\omega\,\beta(n+\omega,-\omega)}\, {}_2F_1\left(\omega,\omega;n+1+\omega;\frac{1}{2}\right).$$

Here

$$\beta(n+1+\omega,-\omega) := \frac{\Gamma(n+1)}{\Gamma(n+1+\omega)\Gamma(-\omega)}, \text{ and}$$
$$\omega := \frac{1-1/a}{2}.$$

Indeed, once (25) was discovered by a combination of insight and methodical computer experiment, its proof became highly representative of the changing paradigm: both sides satisfy the same recursion and the same initial conditions. This can be checked in Maple, and if one looks inside the computation, one learns which *confluent hypergeometric identities* are needed for an explicit human proof.

As noted, study of $\mathcal{R}$ devolved to *hard but compelling* conjectures on complex dynamics, with many interesting *proven* and *unproven* generalizations. In [23] consideration is made of continued fractions like

$$S_1(a) = \cfrac{1^2 a_1^2}{1 + \cfrac{2^2 a_2^2}{1 + \cfrac{3^2 a_3^2}{1 + \ddots}}}$$

for *any* sequence $a \equiv (a_n)_{n=1}^\infty$ and convergence properties obtained for deterministic and random sequences (a_n). For the deterministic case the best results obtained are for periodic sequences, satisfying $a_j = a_{j+c}$ for all j and some finite c. The dynamics are considerably more varied, as illustrated in Figure 4.

Coincidence and Fraud

Coincidences do occur, and such examples drive home the need for reasonable caution in this enterprise. For example, the approximations

$$\pi \approx \frac{3}{\sqrt{163}} \log(640320), \quad \pi \approx \sqrt{2}\frac{9801}{4412}$$

occur for deep number theoretic reasons: the first good to fifteen places, the second to eight. By contrast

$$e^\pi - \pi = 19.999099979189475768\ldots,$$

most probably for no good reason. This seemed more bizarre on an eight-digit calculator. Likewise, as spotted by Pierre Lanchon recently,

$$e = \overline{10.1011011111100001}0101000101100\ldots$$

while

$$\pi = 11.0010\overline{0100001111110110101}01000\ldots$$

have 19 bits agreeing in base two—with one reading right to left. More extended coincidences are almost always contrived, as illustrated by the following:

$$\sum_{n=1}^\infty \frac{[n\tanh(\pi/2)]}{10^n} \approx \frac{1}{81}, \quad \sum_{n=1}^\infty \frac{[n\tanh(\pi)]}{10^n} \approx \frac{1}{81}.$$

The first holds to **12** decimal places, while the second holds to **268** places. This phenomenon can be understood by examining the continued fraction expansion of the constants $\tanh(\pi/2)$ and $\tanh(\pi)$: the integer **11** appears as the third entry of the first, while **267** appears as the third entry of the second.

Bill Gosper, commenting on the extraordinary effectiveness of continued-fraction expansions to "see" what is happening in such problems, declared, "It looks like you are cheating God somehow."

A fine illustration is the unremarkable decimal $\alpha = 1.4331274267223117583\ldots$ whose continued fraction begins $[1,2,3,4,5,6,7,8,9\ldots]$ and so most probably is a ratio of Bessel functions. Indeed, $I_0(2)/I_1(2)$ was what generated the decimal. Similarly, π and e are quite different as continued fractions, less so as decimals.

A more sobering example of high-precision "fraud" is the integral

$$(26) \quad \pi_2 := \int_0^\infty \cos(2x) \prod_{n=1}^\infty \cos\left(\frac{x}{n}\right) dx.$$

The computation of a high-precision numerical value for this integral is rather challenging, due in part to the oscillatory behavior of $\prod_{n\geq 1} \cos(x/n)$ (see Figure 2), but mostly due to the difficulty of computing high-precision evaluations of the integrand function. Note that evaluating thousands of terms of the infinite product would produce only

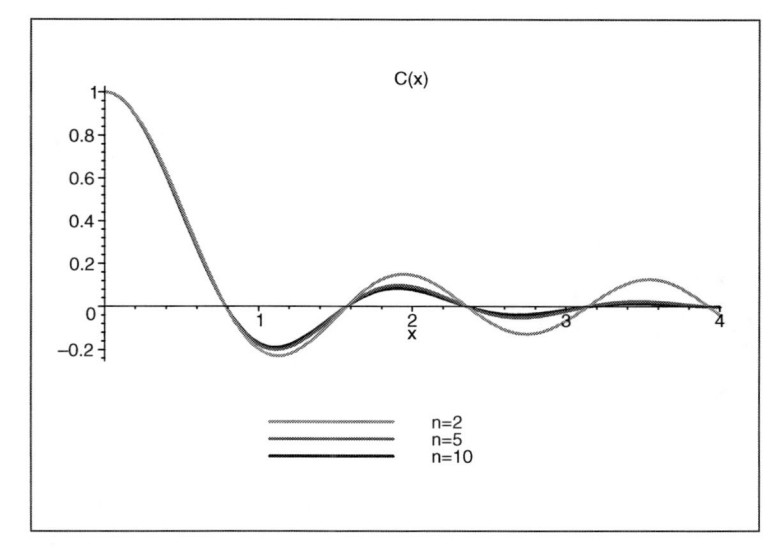

Figure 5. First few terms of $\prod_{n\geq 1} \cos(x/k)$.

a few correct digits. Thus it is necessary to rewrite the integrand function in a form more suitable for computation. This can be done by writing

$$(27) \quad f(x) = \cos(2x) \left[\prod_1^m \cos(x/k)\right] \exp(f_m(x)),$$

where we choose $m > x$, and where

$$(28) \quad f_m(x) = \sum_{k=m+1}^\infty \log \cos\left(\frac{x}{k}\right).$$

The log cos evaluation can be expanded in a Taylor series [1, p. 75], as follows:

$$\log \cos\left(\frac{x}{k}\right) = \sum_{j=1}^\infty \frac{(-1)^j 2^{2j-1}(2^{2j}-1)B_{2j}}{j(2j)!}\left(\frac{x}{k}\right)^{2j},$$

where B_{2j} are *Bernoulli numbers*. Note that since $k > m > x$ in (28), this series converges. We can now write

$$\begin{aligned}
f_m(x) &= \sum_{k=m+1}^\infty \sum_{j=1}^\infty \frac{(-1)^j 2^{2j-1}(2^{2j}-1)B_{2j}}{j(2j)!}\left(\frac{x}{k}\right)^{2j} \\
&= -\sum_{j=1}^\infty \frac{(2^{2j}-1)\zeta(2j)}{j\pi^{2j}} \left[\sum_{k=m+1}^\infty \frac{1}{k^{2j}}\right] x^{2j} \\
&= -\sum_{j=1}^\infty \frac{(2^{2j}-1)\zeta(2j)}{j\pi^{2j}} \left[\zeta(2j) - \sum_{k=1}^m \frac{1}{k^{2j}}\right] x^{2j}.
\end{aligned}$$

This can now be written in a compact form for computation as

$$(29) \quad f_m(x) = -\sum_{j=1}^\infty a_j b_{j,m} x^{2j},$$

where

$$(30) \quad \begin{aligned} a_j &= \frac{(2^{2j}-1)\zeta(2j)}{j\pi^{2j}}, \\ b_{j,m} &= \zeta(2j) - \sum_{k=1}^m 1/k^{2j}. \end{aligned}$$

Figure 6. Advanced Collaborative Environment in Vancouver.

Computation of these b coefficients must be done to a much higher precision than that desired for the quadrature result, since two very nearly equal quantities are subtracted here.

The integral can now be computed using, for example, the tanh-sinh quadrature scheme. The first 60 digits of the result are the following:

0.39269908169872415480782304229099
37860524645434187231595926812....

At first glance, this appears to be $\pi/8$. But a careful comparison with a high-precision value of $\pi/8$, namely

0.39269908169872415480782304229099
37860524646174921888227621868...,

reveals that they are *not* equal: the two values differ by approximately 7.407×10^{-43}. Indeed, these two values are provably distinct. The reason is governed by the fact that $\sum_{n=1}^{55} 1/(2n+1) > 2 > \sum_{n=1}^{54} 1/(2n+1)$. See [16, Chap. 2] for additional details.

A related example is the following. Recall the *sinc* function
$$\mathrm{sinc}(x) := \frac{\sin x}{x}.$$
Consider the seven highly oscillatory integrals below.

$$I_1 := \int_0^\infty \mathrm{sinc}(x)\,dx = \frac{\pi}{2},$$
$$I_2 := \int_0^\infty \mathrm{sinc}(x)\mathrm{sinc}\left(\frac{x}{3}\right)dx = \frac{\pi}{2},$$
$$I_3 := \int_0^\infty \mathrm{sinc}(x)\mathrm{sinc}\left(\frac{x}{3}\right)\mathrm{sinc}\left(\frac{x}{5}\right)dx = \frac{\pi}{2},$$
$$\vdots$$
$$I_6 := \int_0^\infty \mathrm{sinc}(x)\mathrm{sinc}\left(\frac{x}{3}\right)\cdots \mathrm{sinc}\left(\frac{x}{11}\right)dx = \frac{\pi}{2},$$
$$I_7 := \int_0^\infty \mathrm{sinc}(x)\mathrm{sinc}\left(\frac{x}{3}\right)\cdots \mathrm{sinc}\left(\frac{x}{13}\right)dx = \frac{\pi}{2}.$$

However,
$$I_8 := \int_0^\infty \mathrm{sinc}(x)\mathrm{sinc}\left(\frac{x}{3}\right)\cdots \mathrm{sinc}\left(\frac{x}{15}\right)dx$$
$$= \frac{467807924713440738696537864469}{935615849440640907310521750000}\pi$$
$$\approx 0.499999999992646\pi.$$

When this was first found by a researcher using a well-known computer algebra package, both he and the software vendor concluded there was a "bug" in the software. Not so! It is easy to see that the limit of these integrals is $2\pi_1$, where

(31) $\quad \pi_1 := \int_0^\infty \cos(x) \prod_{n=1}^\infty \cos\left(\frac{x}{n}\right) dx.$

This can be seen via *Parseval's theorem*, which links the integral
$$I_N := \int_0^\infty \mathrm{sinc}(a_1 x)\mathrm{sinc}(a_2 x)\cdots \mathrm{sinc}(a_N x)\,dx$$
with the volume of the polyhedron P_N given by
$$P_N := \{x : |\sum_{k=2}^N a_k x_k| \leq a_1, |x_k| \leq 1, 2 \leq k \leq N\},$$
where $x := (x_2, x_3, \cdots, x_N)$. If we let
$$C_N := \{(x_2, x_3, \cdots, x_N) : -1 \leq x_k \leq 1, 2 \leq k \leq N\},$$
then
$$I_N = \frac{\pi}{2a_1}\frac{\mathrm{Vol}(P_N)}{\mathrm{Vol}(C_N)}.$$

Thus, the value drops precisely when the constraint $\sum_{k=2}^N a_k x_k \leq a_1$ becomes *active* and bites the hypercube C_N. That occurs when $\sum_{k=2}^N a_k > a_1$. In the above, $\frac{1}{3} + \frac{1}{5} + \cdots + \frac{1}{13} < 1$, but on addition of the term $\frac{1}{15}$, the sum exceeds 1, the volume drops, and $I_N = \frac{\pi}{2}$ no longer holds. A similar analysis applies to π_2. Moreover, it is fortunate that we began with π_1 or the falsehood of the identity analogous to that displayed above would have been much harder to see.

Further Directions and Implications

In spite of the examples of the previous section, it must be acknowledged that computations can in many cases provide very compelling evidence for mathematical assertions. As a single example, recently Yasumasa Kanada of Japan calculated π to over one trillion decimal digits (and also to over one trillion hexadecimal digits). Given that such computations—which take many hours on large, state-of-the-art supercomputers—are prone to many types of error, including hardware failures, system software problems, and especially programming bugs, how can one be confident in such results?

In Kanada's case, he first used two different arctangent-based formulas to evaluate π to over one trillion hexadecimal digits. Both calculations agreed that the hex expansion beginning at position 1,000,000,000,001 is B4466E8D21 5388C4E014. He then applied a variant of the BBP formula for π, mentioned in Section 3, to calculate these hex digits directly. The result agreed exactly. Needless to say, it is exceedingly unlikely that three different computations, each using a completely distinct computational approach, would all perfectly agree on these digits unless all three are correct.

Another, much more common, example is the usage of probabilistic primality testing schemes. Damgard, Landrock, and Pomerance showed in 1993 that if an integer n has k bits, then the probability that it is prime, provided it passes the most commonly used probabilistic test, is greater than $1 - k^2 4^{2-\sqrt{k}}$, and for certain k is even higher [25]. For instance, if n has 500 bits, then this probability is greater than $1 - 1/4^{28m}$. Thus a 500-bit integer that passes this test even once is prime with prohibitively safe odds: the chance of a false declaration of primality is less than one part in Avogadro's number (6×10^{23}). If it passes the test for four pseudorandomly chosen integers a, then the chance of false declaration of primality is less than one part in a googol (10^{100}). Such probabilities are many orders of magnitude more remote than the chance that an undetected hardware or software error has occurred in the computation. Such methods thus draw into question the distinction between a probabilistic test and a "provable" test.

Another interesting question is whether these experimental methods may be capable of discovering facts that are fundamentally beyond the reach of formal proof methods, which, due to Gödel's result, we know must exist; see also [24].

One interesting example, which has arisen in our work, is the following. We mentioned in Section 3 the fact that the question of the 2-normality of π reduces to the question of whether the chaotic iteration $x_0 = 0$ and

$$x_n = \left\{ 16 x_{n-1} + \frac{120 n^2 - 89 n + 16}{512 n^4 - 1024 n^3 + 712 n^2 - 206 n + 21} \right\},$$

where $\{\cdot\}$ denotes fractional part, are equidistributed in the unit interval.

It turns out that if one defines the sequence $y_n = \lfloor 16 x_n \rfloor$ (in other words, one records which of the 16 subintervals of $(0, 1)$, numbered 0 through 15, x_n lies in), that the sequence (y_n), when interpreted as a hexadecimal string, appears to precisely generate the hexadecimal digit expansion of π. We have checked this to 1,000,000 hex digits and have found no discrepancies. It is known that (y_n) is a very good approximation to the hex digits of π, in the sense that the expected value of the number of errors is finite [15, Section 4.3] [11]. Thus one can argue, by the second Borel-Cantelli lemma, that in a heuristic sense the probability that there

Figure 7. Polyhedra in an immersive environment.

is any error among the remaining digits after the first million is less than 1.465×10^{-8} [15, Section 4.3]. Additional computations could be used to lower this probability even more.

Although few would bet against such odds, these computations do not constitute a rigorous proof that the sequence (y_n) is identical to the hexadecimal expansion of π. Perhaps someday someone will be able to prove this observation rigorously. On the other hand, maybe not—maybe this observation is in some sense an "accident" of mathematics, for which no proof will ever be found. Perhaps numerical validation is all we can ever achieve here.

Conclusion

We are only now beginning to digest some very old ideas:

> Leibniz's idea is very simple and very profound. It's in section VI of the *Discours [de métaphysique]*. It's the observation that the concept of law becomes vacuous if arbitrarily high mathematical complexity is permitted, for then there is always a law. Conversely, if the law has to be extremely complicated, then the data is irregular, lawless, random, unstructured, patternless, and also incompressible and irreducible. A theory has to be simpler than the data that it explains, otherwise it doesn't explain anything. —Gregory Chaitin [24]

Chaitin argues convincingly that there are many mathematical truths which are logically and computationally irreducible—they have *no good reason* in the traditional rationalist sense. This in turn adds force to the desire for evidence even when proof may not be possible. Computer experiments can provide precisely the sort of evidence that is required.

Although computer technology had its roots in mathematics, the field is a relative latecomer to the application of computer technology, compared, say, with physics and chemistry. But now this is changing, as an army of young mathematicians, many of whom have been trained in the usage of sophisticated computer math tools from their high school years, begin their research careers. Further advances in software, including compelling new mathematical visualization environments (see Figures 6 and 7), will have their impact. And the remarkable trend towards greater miniaturization (and corresponding higher power and lower cost) in computer technology, as tracked by Moore's Law, is pretty well assured to continue for at least another ten years, according to Gordon Moore himself and other industry analysts. As Richard Feynman noted back in 1959, "There's plenty of room at the bottom" [27]. It will be interesting to see what the future will bring.

References

[1] MILTON ABRAMOWITZ and IRENE A. STEGUN, *Handbook of Mathematical Functions*, New York, 1970.

[2] ZAFAR AHMED, Definitely an integral, *Amer. Math. Monthly* **109** (2002), 670-1.

[3] KENDALL E. ATKINSON, *An Introduction to Numerical Analysis*, Wiley and Sons, New York, 1989.

[4] DAVID H. BAILEY, Integer relation detection, *Comput. Sci. Engineering* **2** (2000), 24-8.

[5] ———, A compendium of BBP-type formulas for mathematical constants, http://crd.lbl.gov/~dhbailey/dhbpapers/bbp-formulas.pdf (2003).

[6] ———, A hot spot proof of normality for the alpha constants, http://crd.lbl.gov/~dhbailey/dhbpapers/alpha-normal.pdf (2005).

[7] DAVID H. BAILEY, JONATHAN M. BORWEIN, VISHAA KAPOOR, and ERIC WEISSTEIN, Ten problems of experimental mathematics, http://crd.lbl.gov/~dhbailey/dhbpapers/tenproblems.pdf (2004).

[8] DAVID H. BAILEY, PETER B. BORWEIN, and SIMON PLOUFFE, On the rapid computation of various polylogarithmic constants, *Math. of Comp.* **66** (1997), 903-13.

[9] DAVID H. BAILEY and DAVID J. BROADHURST, A seventeenth-order polylogarithm ladder, http://crd.lbl.gov/~dhbailey/dhbpapers/ladder.pdf (1999).

[10] ———, Parallel integer relation detection: Techniques and applications, *Math. of Comp.* **70** (2000), 1719-36.

[11] DAVID H. BAILEY and RICHARD E. CRANDALL, Random generators and normal numbers, *Experiment. Math.* **10** (2001), 175-90.

[12] ———, Random generators and normal numbers, *Experiment. Math.* **11** (2004), 527-46.

[13] DAVID H. BAILEY and XIAOYE S. LI, A comparison of three high-precision quadrature schemes, http://crd.lbl.gov/~dhbailey/dhbpapers/quadrature.pdf (2004).

[14] DAVID H. BAILEY and SINAI ROBINS, Highly parallel, high-precision numerical quadrature, http://crd.lbl.gov/~dhbailey/dhbpapers/quadparallel.pdf (2004).

[15] JONATHAN BORWEIN and DAVID BAILEY, *Mathematics by Experiment*, A K Peters Ltd., Natick, MA, 2004.

[16] JONATHAN BORWEIN, DAVID BAILEY, and ROLAND GIRGENSOHN, *Experimentation in Mathematics: Computational Paths to Discovery*, A K Peters Ltd., Natick, MA, 2004.

[17] JONATHAN BORWEIN and RICHARD CRANDALL, On the Ramanujan AGM fraction. Part II: The complex-parameter case, *Experiment. Math.* **13** (2004), 287-96.

[18] JONATHAN BORWEIN, RICHARD CRANDALL, DAVID BORWEIN, and RAYMOND MAYER, On the dynamics of certain recurrence relations, *Ramanujan J.* (2005).

[19] JONATHAN BORWEIN, RICHARD CRANDALL, and GREG FEE, On the Ramanujan AGM fraction. Part I: The real-parameter case, *Experiment. Math.* **13** (2004), 275-86.

[20] JONATHAN M. BORWEIN, DAVID BORWEIN, and WILLIAM F. GALWAY, Finding and excluding b-ary Machin-type BBP formulae, *Canadian J. Math.* **56** (2004), 897-925.

[21] JONATHAN M. BORWEIN and DAVID M. BRADLEY, On two fundamental identities for Euler sums, http://www.cs.dal.ca/~jborwein/z21.pdf (2005).

[22] JONATHAN M. BORWEIN, DAVID M. BRADLEY, DAVID J. BROADHURST, and PETR LISONĚK, Special values of multiple polylogarithms, *Trans. Amer. Math. Soc.* **353** (2001), 907-41.

[23] JONATHAN M. BORWEIN and D. RUSSELL LUKE, Dynamics of generalizations of the AGM continued fraction of Ramanujan. Part I: Divergence, http://www.cs.dal.ca/~jborwein/BLuke.pdf (2004).

[24] GREGORY CHAITIN, Irreducible complexity in pure mathematics, http://arxiv.org/math.HO/0411091 (2004).

[25] I. DAMGARD, P. LANDROCK, and C. POMERANCE, Average case error estimates for the strong probable prime test, *Math. of Comp.* **61** (1993), 177-94.

[26] HELAMAN R. P. FERGUSON, DAVID H. BAILEY, and STEPHEN ARNO, Analysis of PSLQ, an integer relation finding algorithm, *Math. of Comp.* **68** (1999), 351-69.

[27] RICHARD FEYNMAN, There's plenty of room at the bottom, http://engr.smu.edu/ee/smuphotonics/Nano/FeynmanPlentyofRoom.pdf (1959).

[28] L. KUIPERS and H. NIEDERREITER, *Uniform Distribution of Sequences*, Wiley-Interscience, Boston, 1974.

[29] GORDON E. MOORE, Cramming more components onto integrated circuits, *Electronics* **38** (1965), 114-7.

11. Ten problems in experimental mathematics

Discussion

This article poses and resolves ten computation mathematics problems whose solution requires both numeric and symbolic computation. Such hybrid computation is now more the rule than the exception for the resolution of sophisticated mathematical computation problems. The collection was stimulated by Nick Trefethen's ten problems whose solutions are beautifully described by Folkmar Bornemann, Dirk Laurie, Stan Wagon, and Jörg Waldvogel in *The SIAM 100-Digit Challenge: A Study In High-accuracy Numerical Computing*, SIAM 2004. Here is an excerpt from the challenge review by Jonathan Borwein:

> Lists, challenges and competitions have a long and primarily lustrous history in mathematics. Consider the Hilbert and the Millennium problems. This is the story of a recent highly successful challenge. The book under review also makes it clear that with the continued advance of computing power and accessibility, the view that "real mathematicians don't compute" has little traction, especially for a newer generation of mathematicians who may readily take advantage of the maturation of computational packages such as *Maple*, *Mathematica* and phMATLAB.[1]

Source

D. Bailey, J. Borwein, V. Kapoor and E. Weisstein, "Ten Problems in Experimental Mathematics," *Amer. Math Monthly,* **113**, June-July 2006, 481–509.

---

[1] From an extended review of *The SIAM 100-Digit Challenge* in the *Mathematical Intelligencer*, **27** (4) (2005), 40–48.

Ten Problems in Experimental Mathematics

David H. Bailey, Jonathan M. Borwein, Vishaal Kapoor, and Eric W. Weisstein

1. INTRODUCTION. This article was stimulated by the recent SIAM "100 Digit Challenge" of Nick Trefethen, beautifully described in [12] (see also [13]). Indeed, these ten numeric challenge problems are also listed in [15, pp. 22–26], where they are followed by the ten symbolic/numeric challenge problems that are discussed in this article. Our intent in [15] was to present ten problems that are characteristic of the sorts of problems that commonly arise in "experimental mathematics" [15], [16]. The challenge in each case is to obtain a high precision numeric evaluation of the quantity and then, if possible, to obtain a symbolic answer, ideally one with proof. Our goal in this article is to provide solutions to these ten problems and, at the same time, to present a concise account of how one combines symbolic and numeric computation, which may be termed "hybrid computation," in the process of mathematical discovery.

The passage from object α to answer ω often relies on being able to compute the object to sufficiently high precision, for example, to determine numerically whether α is algebraic or is a rational combination of known constants. While some of this is now automated in mathematical computing software such as *Maple* and *Mathematica*, in most cases intelligence is needed, say in choosing the search space and in deciding the degree of polynomial to hunt for. In a similar sense, using symbolic computing tools such as those incorporated into *Maple* and *Mathematica* often requires significant human interaction to produce material results. Such matters are discussed in greater detail in [15] and [16].

Integer relation detection. Several of these solutions involve the usage of integer relation detection schemes to find experimentally a likely relationship. For a given real vector $(x_1, x_2, \ldots, x_n)$ an integer relation algorithm is a computational scheme that either finds the n-tuple of integers $(a_1, a_2, \ldots, a_n)$, not all zero, such that $a_1 x_1 + a_2 x_2 + \cdots a_n x_n = 0$ or else establishes that there is no such integer vector within a ball of some radius about the origin, where the metric is the Euclidean norm $(a_1^2 + a_2^2 + \cdots + a_n^2)^{1/2}$.

At the present time, the best known integer relation algorithm is the PSLQ algorithm [25] of Helaman Ferguson, who is well known in the community for his mathematical sculptures. Simple formulations of the PSLQ algorithm and several variants are given in [7]. Another widely used integer relation detection scheme involves the Lenstra-Lenstra-Lovasz (LLL) algorithm. The PSLQ algorithm, together with related lattice reduction schemes such as LLL, was recently named one of ten "algorithms of the century" by the publication *Computing in Science and Engineering* [3].

Perhaps the best-known application of PSLQ is the 1995 discovery, by means of a PSLQ computation, of the "BBP" formula for π:

$$\pi = \sum_{k=0}^{\infty} \frac{1}{16^k} \left(\frac{4}{8k+1} - \frac{2}{8k+4} - \frac{1}{8k+5} - \frac{1}{8k+6} \right).$$

This formula permits one to calculate directly binary or hexadecimal digits beginning at the nth digit, without the need to calculate any of the first $n-1$ digits [6]. This result has, in turn, led to more recent results that suggest a possible route to a proof that π and some other mathematical constants are 2-normal (i.e., that every m-long binary string occurs in the binary expansion with limiting frequency b^{-m} [8], [9]). The BBP formula even has some practical applications: it is used, for example, in the g95 compiler for transcendental function evaluations [34].

All integer relation schemes require very high precision arithmetic, both in the input data and in the operation of the algorithms. Simple reckoning shows that if an integer relation solution vector $(a_i, a_2, \ldots, a_n)$ has Euclidean norm 10^d, then the input data must be specified to at least dn digits, lest the true solution be lost in a sea of numerical artifacts. In some cases, including one mentioned at the end of the next section, thousands of digits are required before a solution can be found with these methods. This is the principal reason for the great interest in high-precision numerical evaluations in experimental mathematics research. It is the also the motivation behind this set of ten challenge problems.

2. THE BIFURCATION POINT B_3.

Problem 1. *Compute the value of r for which the chaotic iteration*

$$x_{n+1} = r x_n (1 - x_n),$$

starting with some x_0 in $(0, 1)$, exhibits a bifurcation between four-way periodicity and eight-way periodicity. Extra credit: This constant is an algebraic number of degree not exceeding twenty. Find the minimal polynomial with integer coefficients that it satisfies.

History and context. The chaotic iteration $x_{n+1} = r x_n(1 - x_n)$ has been studied since the early days of chaos theory in the 1950s. It is often called the "logistic iteration," since it mimics the behavior of an ecological population that, if its growth one year outstrips its food supply, often falls back in numbers for the following year, thus continuing to vary in a highly irregular fashion. When r is less than one iterates of the logistic iteration converge to zero. For r in the range $1 < r < B_1 = 3$ iterates converge to some nonzero limit. If $B_1 < r < B_2 = 1 + \sqrt{6} = 3.449489\ldots$, the limiting behavior bifurcates—every other iterate converges to a distinct limit point. For r with $B_2 < r < B_3$ iterates hop between a set of four distinct limit points; when $B_3 < r < B_4$, they select between a set of eight distinct limit points; this pattern repeats until $r > B_\infty = 3.569945672\ldots$, when the iteration is completely chaotic (see Figure 1). The limiting ratio $\lim_n (B_n - B_{n-1})/(B_{n+1} - B_n) = 4.669201\ldots$ is known as *Feigenbaum's delta constant*.

A very readable description of the logistic iteration and its role in modern chaos theory are given in Gleick's book [26]. Indeed, John von Neumann had suggested using the logistic map as a random number generator in the late 1940s. Work by W. Ricker in 1954 and detailed analytic studies of logistic maps beginning in the 1950s with Paul Stein and Stanislaw Ulam showed the existence of complicated properties of this type of map beyond simple oscillatory behavior [35, pp. 918–919].

Solution. We first describe how to obtain a highly accurate numerical value of B_3 using a relatively straightforward search scheme. Other schemes could be used to find B_3; we present this one to underscore the fact that computational results sufficient for the purposes of experimental mathematics can often be obtained without resorting to highly sophisticated techniques.

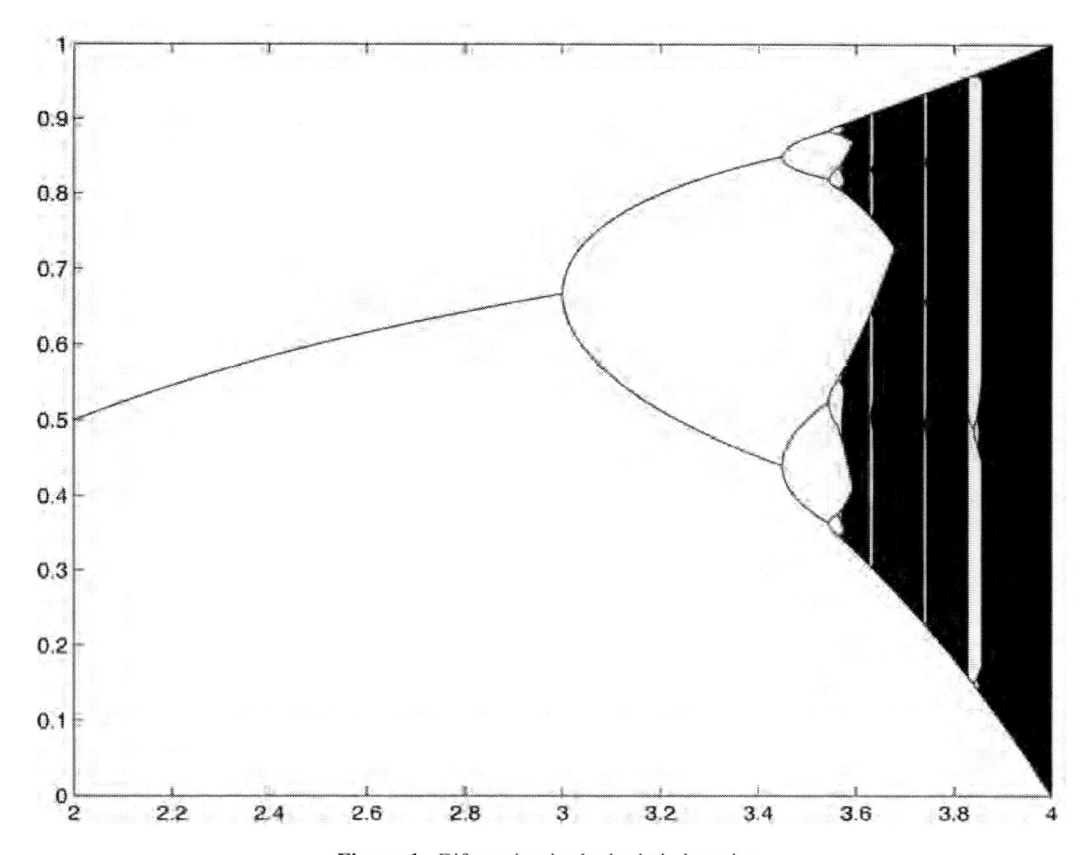

Figure 1. Bifurcation in the logistic iteration.

Let $f_8(r, x)$ be the eight-times iterated evaluation of $rx(1-x)$, and let $g_8(r, x) = f_8(r, x) - x$. Imagine a three-dimensional graph, where r ranges from left to right and x ranges from bottom to top (as in Figure 1), and where $g_8(r, x)$ is plotted in the vertical (out-of-plane) dimension. Given some initial r slightly less than B_3, we compute a "comb" of function values at n evenly spaced x values (with spacing h_x) near the limit of the iteration $x_{n+1} = f_8(r, x_n)$. In our implementation, we use $n = 12$, and we start with $r = 3.544$, $x = 0.364$, $h_r = 10^{-4}$, and $h_x = 5 \times 10^{-4}$. With this construction, the comb has $n/2$ negative function values, followed by $n/2$ positive function values. We then increment r by h_r and reevaluate the "comb," continuing in this fashion until two sign changes are observed among the n function values of the "comb." This means that a bifurcation occurred just prior to the current value of r, so we restore r to its previous value (by subtracting h_r), reduce h_r, say by a factor of four, and also reduce the h_x roughly by a factor of 2.5. We continue in this fashion, moving the value of r and its associated "comb" back and forth near the bifurcation point with progressively smaller intervals h_r. The center of the comb in the x-direction must be adjusted periodically to ensure that $n/2$ negative function values are followed by $n/2$ positive function values, and the spacing parameter h_x must be adjusted as well to ensure that two sign changes are disclosed when this occurs. We quit when the smallest of the n function values is within two or three orders of magnitude of the "epsilon" of the arithmetic (e.g., for 2000-digit working precision, "epsilon" is 10^{-2000}). The final value of r is then the desired value B_3, accurate to within a tolerance given by the final value of r_h. With 2000-digit working precision, our implementation of this scheme finds B_3 to 1330-digit accuracy in about five minutes on a 2004-era computer. The first hundred digits are as follows:

$$B_3 = 3.54409035955192285361596598660480454058309984544457367545781$$
$$25303058429428588630122562585664248917999626\ldots.$$

With even a moderately accurate value of r in hand (at least two hundred digits or so), one can use a PSLQ program (such as the PSLQ programs available at the URL http://crd.lbl.gov/~dhbailey/mpdist) to check whether r is an algebraic constant. This is done by computing the vector $(1, r, r^2, \ldots, r^n)$ for various n, beginning with a small value such as two or three, and then searching for integer relations among these $n + 1$ real numbers. When $n \geq 12$, the relation

$$0 = r^{12} - 12r^{11} + 48r^{10} - 40r^9 - 193r^8 + 392r^7 + 44r^6 + 8r^5 - 977r^4$$
$$- 604r^3 + 2108r^2 + 4913 \qquad (1)$$

can be recovered.

A symbolic solution that explicitly produces the polynomial (1) can be obtained as follows. We seek a sequence $x_1, x_2, \ldots, x_4$ that satisfies the equations

$$x_2 = rx_1(1 - x_1), \quad x_3 = rx_2(1 - x_2), \quad x_4 = rx_3(1 - x_3), \quad x_1 = rx_4(1 - x_4),$$

and

$$1 = \left| \prod_{i=1}^{4} r(1 - 2x_i) \right|.$$

The first four conditions represent a period-4 sequence in the logistic equation $x_{n+1} = rx_n(1 - x_n)$, and the last condition represents the stability of the cycle, which must be 1 or -1 for a bifurcation point (see [33] for details).

First, we deal with the system corresponding to $1 + \prod_{i=1}^{4} r(1 - 2x_i) = 0$. We compute the lexicographic Groebner basis in *Maple*:

```
with(Groebner):
L := [x2 - r*x1*(1-x1),x3 - r*x2*(1-x2),x4 - r*x3*(1-x3),
      x1 - r*x4*(1-x4),r^4*(1-2*x1)*(1-2*x2)*(1-2*x3)*(1-2*x4) + 1];
      gbasis(L,plex(x1,x2,x3,x4,r));
```

After a cup of coffee, we discover the univariate element

$$(r^4 + 1)(r^4 - 8r^3 + 24r^2 - 32r + 17) \times (r^4 - 4r^3 - 4r^2 + 16r + 17)$$
$$\times (r^{12} - 12r^{11} + 48r^{10} - 40r^9 - 193r^8 + 392r^7 + 44r^6 + 8r^5 - 977r^4$$
$$- 604r^3 + 2108r^2 + 4913)$$

in the Groebner basis, in which the monomial ordering is lexicographical with r last.

The first three of these polynomials have no real roots, and the fourth has four real roots. Using trial and error, it is easy to determine that B_3 is the root of the minimal polynomial

$$r^{12} - 12r^{11} + 48r^{10} - 40r^9 - 193r^8 + 392r^7$$
$$+ 44r^6 + 8r^5 - 977r^4 - 604r^3 + 2108r^2 + 4913,$$

which has the numerical value stated earlier. The corresponding *Mathematica* code reads:

```
GroebnerBasis[{x2 - r x1(1 - x1), x3 - r x2(1 - x2),
 x4 - r x3(1 - x3), x1 - r x4(1 - x4),
 r^4(1 - 2x1)(1 - 2x2)(1 - 2x3)(1 - 2x4) + 1},
 r,
{x1, x2, x3, x4}, MonomialOrder -> EliminationOrder]//Timing
```

This requires only 1.2 seconds on a 3 GHz computer. These computations can also be recreated very quickly in *Magma*, an algebraic package available at http://magma.maths.usyd.edu.au/magma:

```
Q := RationalField(); P<x,y,z,w,r> := PolynomialRing(Q,5);
 I:= ideal< P| y - r*x*(1-x), z - r*y*(1-y), w - r*z*(1-z),
     x - r*w*(1-w), r^4*(1-2*x)*(1-2*y)*(1-2*z)*(1-2*w)+1>;
time B := GroebnerBasis(I);
```

This took 0.050 seconds on a 2.4Ghz Pentium 4.

The significantly more challenging problem of computing and analyzing the constant $B_4 = 3.564407266095\ldots$ is discussed in [7]. In this study, conjectural reasoning suggested that B_4 might satisfy a 240-degree polynomial, and, in addition, that $\alpha = -B_4(B_4 - 2)$ might satisfy a 120-degree polynomial. The constant α was then computed to over 10,000-digit accuracy, and an advanced three-level multi-pair PSLQ program was employed, running on a parallel computer system, to find an integer relation for the vector $(1, \alpha, \alpha^2, \ldots, \alpha^{120})$. A numerically significant solution was obtained, with integer coefficients descending monotonically from 257^{30}, which is a 73-digit integer, to the final value, which is one (a striking result that is exceedingly unlikely to be a numerical artifact). This experimentally discovered polynomial was recently confirmed in a large symbolic computation [30].

Additional information on the Logistic Map is available at http://mathworld.wolfram.com/LogisticMap.html.

3. MADELUNG'S CONSTANT.

Problem 2. *Evaluate*

$$\sum_{(m,n,p)\neq 0} \frac{(-1)^{m+n+p}}{\sqrt{m^2 + n^2 + p^2}}, \qquad (2)$$

where convergence means the limit of sums over the integer lattice points enclosed in increasingly large cubes surrounding the origin. Extra credit: Usefully identify this constant.

History and context. Highly conditionally convergent sums like this are very common in physical chemistry, where they are usually written down with no thought of convergence. The sum in question arises as an idealization of the electrochemical stability of NaCl. One computes the total potential at the origin when placing a positive or negative charge at each nonzero point of the cubic lattice [16, chap. 4].

Solution. It is important to realize that this sum must be viewed as the limit of the sum in successively larger cubes. The sum diverges when spheres are used instead. To clarify this consider, for complex s, the series

$$b_2(s) = \sum_{(m,n)\neq 0} \frac{(-1)^{m+n}}{(m^2+n^2)^{s/2}}, \qquad b_3(s) = \sum_{(m,n,p)\neq 0} \frac{(-1)^{m+n+p}}{(m^2+n^2+p^2)^{s/2}}. \qquad (3)$$

These converge in two and three dimensions, respectively, over increasing "cubes," provided that $\operatorname{Re} s > 0$. When $s = 1$, one *may* sum over circles in the plane but not spheres in three-space, and one may *not* sum over diamonds in dimension two. Many chemists do not know that $b_3(1) \neq \sum_n (-1)^n r_3(n)/\sqrt{n}$, a series that arises by summing over increasing spheres but that diverges. Indeed, the number $r_3(n)$ of representations of n as a sum of three squares is quite irregular—no number of the form $8n + 7$ has such a representation—and is not $O(n^{1/2})$. This matter is somewhat neglected in the discussion of Madelung's constant in Julian Havil's deservedly popular recent book *Gamma: Exploring Euler's Constant* [27], which contains a wealth of information related to each of our problems in which Euler had a hand.

Straightforward methods to compute (3) are extremely unproductive. Such techniques produce at most three digits—indeed, the physical model should have a solar-system sized salt crystal to justify ignoring the boundary. Thus, we are led to using more sophisticated methods. We note that

$$b_3(s) = {\sum}' \frac{(-1)^{i+j+k}}{(i^2+j^2+k^2)^{s/2}},$$

where $\sum'$ signifies a sum over $\mathbb{Z}^3 \setminus \{(0,0,0)\}$, and let $M_s(f)$ denote the Mellin transform

$$M_s(f) = \int_0^\infty f(x) x^{s-1}\, dx.$$

The quantity that we wish to compute is $b_3(1)$. It follows by symmetry that

$$\begin{aligned}
b_3(1) &= {\sum}' \frac{(-1)^{i+j+k}(i^2+j^2+k^2)}{(i^2+j^2+k^2)^{3/2}} \\
&= 3 {\sum}' \frac{(-1)^i (i^2)(-1)^{j+k}}{(i^2+j^2+k^2)^{3/2}}.
\end{aligned} \qquad (4)$$

We observe that $M_s(e^{-t}) = \Gamma(s)$, so

$$M_{3/2}\bigl(q^{n^2+j^2+k^2}\bigr) = \Gamma\left(\frac{3}{2}\right)(n^2+j^2+k^2)^{-3/2},$$

where n, j, and k are arbitrary integers and $q = e^{-t}$. Continuing, we rewrite equation (4) as

$$\Gamma\left(\frac{3}{2}\right) b_3(1) = 3 M_{3/2}\left(\sum_{n=-\infty}^\infty (-1)^n n^2 q^{n^2} \theta_4^2(x) \right),$$

where $\theta_4(x) = \sum_{-\infty}^\infty (-1)^n x^{n^2}$ is the usual Jacobi theta-function. Since the *theta transform*—a form of Poisson summation—yields $\theta_4(e^{-\pi/s}) = \sqrt{s}\,\theta_2(e^{-s\pi})$, it follows that

$$\Gamma\left(\frac{3}{2}\right) b_3(1) = 3 \sum_{n=-\infty}^{\infty} n^2 M_{3/2}\left(\sum (-1)^n n^2 q^{n^2} \frac{\pi}{x} \theta_2^2\left(\frac{\pi^2}{x}\right)\right).$$

Also, $\Gamma(3/2) = \sqrt{\pi}/2$, so

$$b_3(1) = 12\sqrt{\pi} \sum_{n=1}^{\infty} (-1)^n n^2 \sum_{(j,k) \text{ odd}} \int_0^{\infty} \left[e^{-n^2 x - (\pi^2/4x)(j^2+k^2)}\right] x^{-1/2} dx.$$

The integral is evaluated in [19, Exercise 4, sec. 2.2] and is $(\pi/n^2)^{1/2} e^{-\pi n \sqrt{j^2+k^2}}$, whence

$$b_3(1) = 48\pi \sum_{k=0}^{\infty} \sum_{j=0}^{\infty} \sum_{n=1}^{\infty} (-1)^n n e^{-\pi n \sqrt{(2j+1)^2 + (2k+1)^2}}.$$

Finally, when $a > 0$,

$$4 \sum_{n=1}^{\infty} (-1)^{n+1} n e^{-an} = \frac{4e^{-a}}{(1+e^{-a})^2} = \text{sech}^2\left(\frac{a}{2}\right),$$

from which we obtain

$$b_3(1) = 12\pi \sum_{\substack{m,n \geq 1 \\ m,n \text{ odd}}} \text{sech}^2\left(\frac{\pi}{2}(m^2 + n^2)^{1/2}\right). \tag{5}$$

Summing over m and n from 1 up to 81 in (5) gives

$$b_3(1) = 1.74756459463318219063621203554439740348516143662474175$$
$$8152825350765040623532761179890758362694607891\ldots.$$

It is possible to accelerate the convergence further still. Details can be found in [19], [16].

There are closed forms for sums with an even number of variables, up to 24 and beyond. For example, $b_2(2s) = -4\alpha(s)\beta(s)$, where

$$\alpha(s) = \sum_{n \geq 0} (-1)^n / (n+1)^s$$

and

$$\beta(s) = \sum_{n \geq 0} (-1)^n / (2n+1)^s.$$

In particular, $b_2(2) = -\pi \log 2$. No such closed form for b_3 is known, while much work has been expended looking for one. The formula for b_2 is due to Lorenz (1879). It was rediscovered by G. H. Hardy and is equivalent to Jacobi's Lambert series formula for $\theta_3^2(q)$:

$$\theta_3^2(q) - 1 = 4 \sum_{n \geq 0} (-1)^n \frac{q^{2n+1}}{1 - q^{2n+1}}.$$

This, in turn, is equivalent to the formula for the number $r_2(n)$ of representations of n as a sum of two squares, counting order and sign,

$$r_2(n) = 4(d_1(n) - d_3(n)),$$

where d_k is the number of divisors of n congruent to k modulo four. The analysis of three squares is notoriously harder.

Additional information on Madelung's constant and lattice sums is available at http://mathworld.wolfram.com/MadelungConstants.html and http://mathworld.wolfram.com/LatticeSum.html.

4. DOUBLE EULER SUMS.

Problem 3. *Evaluate the sum*

$$C = \sum_{k=1}^{\infty} \left(1 - \frac{1}{2} + \cdots + (-1)^{k+1}\frac{1}{k}\right)^2 \frac{1}{(k+1)^3}. \tag{6}$$

Extra credit: Evaluate this constant as a multiterm expression involving well-known mathematical constants. This expression has seven terms and involves π, $\log 2$, $\zeta(3)$, and $\mathrm{Li}_5(1/2)$, where $\mathrm{Li}_n(x) = \sum_{k>0} x^n/n^k$ is the nth polylogarithm. (Hint: The expression is "homogenous," in the sense that each term has the same total "degree." The degrees of π and $\log 2$ are each 1, the degree of $\zeta(3)$ is 3, the degree of $\mathrm{Li}_5(1/2)$ is 5, and the degree of α^n is n times the degree of α.)

History and context. In April 1993, Enrico Au-Yeung, an undergraduate at the University of Waterloo, brought to the attention of one of us (Borwein) the curious result

$$\sum_{k=1}^{\infty} \left(1 + \frac{1}{2} + \cdots + \frac{1}{k}\right)^2 \frac{1}{k^2} = 4.59987\ldots \approx \frac{17}{4}\zeta(4) = \frac{17\pi^4}{360}. \tag{7}$$

The function $\zeta(s)$ in (7) is the classical *Riemann zeta-function*:

$$\zeta(s) = \sum_{n=1}^{\infty} \frac{1}{n^s}.$$

Euler had solved Bernoulli's *Basel problem* when he showed that, for each positive integer n, $\zeta(2n)$ is an explicit rational multiple of π^{2n} [**16**, sec. 3.2].

Au-Yeung had computed the sum in (7) to 500,000 terms, giving an accuracy of five or six decimal digits. Suspecting that his discovery was merely a modest numerical coincidence, Borwein sought to compute the sum to a higher level of precision. Using Fourier analysis and Parseval's equation, he obtained

$$\frac{1}{2\pi} \int_0^{\pi} (\pi - t)^2 \log^2\left(2\sin\frac{t}{2}\right) dt = \sum_{n=1}^{\infty} \frac{\left(\sum_{k=1}^{n} \frac{1}{k}\right)^2}{(n+1)^2}. \tag{8}$$

The idea here is that the series on the right of (8) permits one to evaluate (7), while the integral on the left can be computed using the numerical quadrature facility of *Mathematica* or *Maple*. When he did this, Borwein was surprised to find that the conjectured identity holds to more than thirty digits. We should add here that, by good fortune, $17/360 = 0.047222\ldots$ has period one and thus can plausibly be recognized from its first six digits, so that Au-Yeung's numerical discovery was not entirely farfetched.

Solution. We define the multivariate zeta-function by

$$\zeta(s_1, s_2, \ldots, s_k) = \sum_{n_1 > n_2 > \cdots > n_k > 0} \prod_{j=1}^{k} n_j^{-|s_j|} \sigma_j^{-n_j},$$

where the $s_1, s_2, \ldots, s_k$ are nonzero integers and $\sigma_j = \text{signum}(s_j)$. A fast method for computing such sums based on Hölder convolution is discussed in [20] and implemented in the EZFace+ interface, which is available as an online tool at the URL http://www.cecm.sfu.ca/projects/ezface+. Expanding the squared term in (6), we have

$$C = \sum_{0 < i, j < k} \frac{(-1)^{i+j}}{ijk^3} = 2\zeta(3, -1, -1) + \zeta(3, 2). \tag{9}$$

Evaluating this in EZFace+ we quickly obtain

$C = 0.15616693338117691588103590968798819368577670984030387295752935449707503744029579145520565370935814757 8\ldots.$

Given this numerical value, PSLQ or some other integer-relation-finding tool can be used to see if this constant satisfies a rational linear relation with the following constants (as suggested in the hint): π^5, $\pi^4 \log(2)$, $\pi^3 \log^2(2)$, $\pi^2 \log^3(2)$, $\pi \log^4(2)$, $\log^5(2)$, $\pi^2 \zeta(3)$, $\pi \log(2)\zeta(3)$, $\log^2(2)\zeta(3)$, $\zeta(5)$, $\text{Li}_5(1/2)$. The result is quickly found to be

$$C = 4\text{Li}_5\left(\frac{1}{2}\right) - \frac{1}{30}\log^5(2) - \frac{17}{32}\zeta(5) - \frac{11}{720}\pi^4 \log(2) + \frac{7}{4}\zeta(3)\log^2(2)$$
$$+ \frac{1}{18}\pi^2 \log^3(2) - \frac{1}{8}\pi^2 \zeta(3).$$

This result has been proved in various ways, both analytic and algebraic. Indeed, all evaluations of sums of the form $\zeta(\pm a_1, \pm a_2, \ldots, \pm a_m)$ with *weight* $w = \sum_k a_k$ ($w < 8$), as in (9), have been established.

Further history and context. What Borwein did not know at the time was that Au-Yeung's suspected identity follows directly from a related result proved by De Doelder in 1991. In fact, it had cropped up even earlier as a problem in this MONTHLY, but the story goes back further still. Some historical research showed that Euler considered these summations. In response to a letter from Goldbach, he examined sums that are equivalent to

$$\sum_{k=1}^{\infty} \left(1 + \frac{1}{2^m} + \cdots + \frac{1}{k^m}\right) \frac{1}{(k+1)^n}. \tag{10}$$

The great Swiss mathematician was able to give explicit values for certain of these sums in terms of the Riemann zeta-function.

Starting from where we left off in the previous section provides some insight into evaluating related sums. Recall that the Taylor expansion of

$$f(x) = -\frac{1}{2}\log(1-x)\log(1+x)$$

takes the form

$$f(x) = \sum_{k=1}^{\infty}\left(1 - \frac{1}{2} + \frac{1}{3} - \cdots + \frac{1}{2k-1}\right)\frac{x^{2k}}{2k}.$$

Applying Parseval's identity to $f(e^{it})$, we have an effective way of computing

$$\sum_{k=1}^{\infty}\frac{\left(1 - \frac{1}{2} + \frac{1}{3} - \cdots + \frac{1}{2k-1}\right)^2}{(2k)^2}$$

in terms of an integral that can be rapidly evaluated in *Maple* or *Mathematica*.

Alternatively, we may compute

$$\sum_{k=1}^{\infty}\frac{\left(1 + \frac{1}{2} + \frac{1}{3} + \cdots + \frac{1}{k}\right)^2}{k^2}.$$

The Fourier expansions of $(\pi - t)/2$ and $-\log|2\sin(t/2)|$ are

$$\sum_{n=1}^{\infty}\frac{\sin(nt)}{n} = \frac{\pi - t}{2} \qquad (0 < t < 2\pi)$$

and

$$\sum_{n=1}^{\infty}\frac{\cos(nt)}{n} = -\log|2\sin(t/2)| \qquad (0 < t < 2\pi), \tag{11}$$

respectively. Multiplying these together, simplifying, and doing a partial fraction decomposition gives

$$-\log|2\sin(t/2)| \cdot \frac{\pi - t}{2} = \sum_{n=1}^{\infty}\frac{1}{n}\sum_{k=1}^{n-1}\frac{1}{k}\sin(nt)$$

on $(0, 2\pi)$. Applying Parseval's identity results in

$$\frac{1}{4\pi}\int_0^{2\pi}(\pi - t)^2 \log^2(2\sin(t/2))\,dt = \sum_{n=1}^{\infty}\frac{\left(1 + \frac{1}{2} + \frac{1}{3} + \cdots + \frac{1}{n}\right)^2}{(n+1)^2}.$$

The integral may be computed numerically in *Maple* or *Mathematica*, delivering an approximation to the sum.

The *Clausen functions* defined by

$$\text{Cl}_2(\theta) = \sum_{n=1}^{\infty}\frac{\sin(n\theta)}{n^2}, \qquad \text{Cl}_3(\theta) = \sum_{n=1}^{\infty}\frac{\cos(n\theta)}{n^3}, \qquad \text{Cl}_4(\theta) = \sum_{n=1}^{\infty}\frac{\sin(n\theta)}{n^4}, \cdots$$

arise as repeated antiderivatives of (11). They are useful throughout harmonic analysis and elsewhere. For example, with $\alpha = 2\arctan\sqrt{7}$, one discovers with the aid of

PSLQ that

$$6\text{Cl}_2(\alpha) - 6\text{Cl}_2(2\alpha) + 2\text{Cl}_2(3\alpha) \stackrel{?}{=} 7\text{Cl}_2\left(\frac{2\pi}{7}\right) + 7\text{Cl}_2\left(\frac{4\pi}{7}\right) - 7\text{Cl}_2\left(\frac{6\pi}{7}\right) \quad (12)$$

(here the question mark is used because no proof is yet known) or, in what can be shown to be equivalent, that

$$\frac{24}{7\sqrt{7}} \int_{\pi/3}^{\pi/2} \log\left(\left|\frac{\tan(t) + \sqrt{7}}{\tan(t) - \sqrt{7}}\right|\right) dt \stackrel{?}{=} L_{-7}(2) = 1.151925470\ldots. \quad (13)$$

This arises from the volume of an ideal tetrahedron in hyperbolic space [15, pp. 90–91]. (Here $L_{-7}(s) = \sum_{n>0} \chi_{-7}(n) n^{-s}$ is the primitive L-series modulo seven, whose character pattern is $1, 1, -1, 1, -1, -1, 0$, which is given by

$$\chi_{-7}(k) = 2(\sin(k\tau) + \sin(2k\tau) - \sin(3k\tau))/\sqrt{7}$$

with $\tau = 2\pi/7$.)

Although (13) has been checked to twenty thousand decimal digits, by using a numerical integration scheme we shall describe in section 8, and although it is known for K-theoretic reasons that the ratio of the left- and right-hand sides of (12) is rational [14], to the best of our knowledge there is no proof of either (12) or (13). We might add that recently two additional conjectured identities related to (13) have been discovered by PSLQ computations. Let I_n be the definite integral of (13), except with limits $n\pi/24$ and $(n+1)\pi/24$. Then

$$-2I_2 - 2I_3 - 2I_4 - 2I_5 + I_8 + I_9 - I_{10} - I_{11} \stackrel{?}{=} 0,$$
$$I_2 + 3I_3 + 3I_4 + 3I_5 + 2I_6 + 2I_7 - 3I_8 - I_9 \stackrel{?}{=} 0. \quad (14)$$

Readers who attempt to calculate numerical values for either the integral in (13) or the integral I_9 in (14) should note that the integrand has a nasty singularity at $t = \arctan\sqrt{7}$.

In retrospect, perhaps it was for the better that Borwein had not known of De Doelder's and Euler's results, because Au-Yeung's intriguing numerical discovery launched a fruitful line of research by a number of researchers that has continued until the present day. Sums of this general form are known nowadays as "Euler sums" or "Euler-Zagier sums." Euler sums can be studied through a profusion of methods: combinatorial, analytic, and algebraic. The reader is referred to [16, chap. 3] for an overview of Euler sums and their applications. We take up the story again in Problem 9.

Additional information on Euler sums is available at http://mathworld.wolfram.com/EulerSum.html.

5. KHINTCHINE'S CONSTANT.

Problem 4. *Evaluate*

$$K_0 = \prod_{k=1}^{\infty}\left[1 + \frac{1}{k(k+2)}\right]^{\log_2 k} = \prod_{k=1}^{\infty} k^{[\log_2(1+1/k(k+2))]}. \quad (15)$$

Extra credit: Evaluate this constant in terms of a less-well-known mathematical constant.

History and context. Given some particular continued fraction expansion $\alpha = [a_0, a_1, \ldots]$, consider forming the limit

$$K_0(\alpha) = \lim_{n \to \infty} (a_0 a_1 \cdots a_n)^{1/n}.$$

Based on the *Gauss-Kuzmin distribution*, which establishes that the digit distribution of a random continued fraction satisfies $\text{Prob}\{a_k = n\} = \log_2(1 + 1/k(k+2))$, Khintchine showed that the limit exists for almost all continued fractions and is a certain constant, which we now denote K_0. This circle of ideas is accessibly developed in [27]. As such a constant has an interesting interpretation, computation seems like the next step.

Taking logarithms of both sides of (15) and simplifying, we have

$$\log 2 \cdot \log K_0 = \sum_{n=1}^{\infty} \log n \cdot \log\left(1 + \frac{1}{n(n+2)}\right).$$

Such a series converges extremely slowly. Computing the sum of the first 10000 terms gives only two digits of $\log 2 \cdot \log K_0$. Thus, direct computation again proves to be quite difficult.

Solution. Rewriting $\log n$ as the telescoping sum

$$\log n = (\log n - \log(n-1)) + \cdots + (\log 2 - \log 1) = \sum_{k=2}^{n} \log \frac{k}{k-1},$$

we see that

$$\log 2 \cdot \log K_0 = \sum_{n=2}^{\infty} \sum_{k=2}^{n} \log \frac{k}{k-1} \cdot \log \frac{(n+1)^2}{n(n+2)}.$$

We interchange the order of summation to obtain

$$\log 2 \cdot \log K_0 = \sum_{k=2}^{\infty} \sum_{n=k}^{\infty} \log \frac{(n+1)^2}{n(n+2)} \log \frac{k}{k-1}. \tag{16}$$

But

$$\sum_{n=k}^{\infty} \log \frac{(n+1)^2}{n(n+2)} = \log \frac{k+1}{k} = \log\left(1 + \frac{1}{k}\right),$$

so (16) transforms into

$$\log 2 \cdot \log K_0 = -\sum_{k=2}^{\infty} \log\left(1 - \frac{1}{k}\right) \log\left(1 + \frac{1}{k}\right). \tag{17}$$

The Maclaurin series for $-\log(1-x)\log(1+x)$ is

$$\sum_{k=1}^{\infty} \left(1 - \frac{1}{2} + \frac{1}{3} - \cdots + \frac{1}{2k-1}\right) \frac{x^{2k}}{k}.$$

This allows us to rewrite $\log 2 \cdot \log K_0$ as

$$\log 2 \cdot \log K_0 = \sum_{k=1}^{\infty}\left(1 - \frac{1}{2} + \frac{1}{3} - \cdots + \frac{1}{2k-1}\right)\frac{1}{k}\sum_{n=2}^{\infty} n^{-2k}$$

$$= \sum_{k=1}^{\infty}\left(1 - \frac{1}{2} + \frac{1}{3} - \cdots + \frac{1}{2k-1}\right)\frac{1}{k}(\zeta(2k) - 1).$$

Appealing to either *Maple* or *Mathematica*, we can easily compute this sum. Taking the first 161 terms, we obtain one hundred digits of K_0:

$K_0 = 2.68545200106530644530971483548179569382038229399446295$
$30511523455572188595371520028011411749318477709\ldots.$

However, faster convergence is possible, and the constant has now been computed to more than seven thousand places. Moreover, the harmonic and other averages are similarly treated. It appears to satisfy its own predicted behavior (for details, see [5], [32]). Correspondingly, using 10^8 terms one can obtain the approximation $K_0(\pi) \approx 2.675\ldots$. Note however that $K_0(e) = \infty = \lim_{n\to\infty} \sqrt[3n]{(2n)!}$, since e is a member of the measure zero set of exceptions not having $K_0(\alpha) = K_0$, as a result of the non-Gauss-Kuzmin distribution of terms in the continued fraction $e = [2, 1, 2, 1, 1, 4, 1, 1, 6, \ldots]$.

We emphasize that while it is known that almost all numbers α have limits $K_0(\alpha)$ that equal K_0, this has not been exhibited for any explicit number α, excluding artificial examples constructed using their continued fractions [5].

6. RAMANUJAN'S AGM CONTINUED FRACTION.

Problem 5. *For positive real numbers a, b, and η define $R_\eta(a, b)$ by*

$$R_\eta(a, b) = \cfrac{a}{\eta + \cfrac{b^2}{\eta + \cfrac{4a^2}{\eta + \cfrac{9b^2}{\eta + \ddots}}}}.$$

Calculate $R_1(2, 2)$. Extra credit: Evaluate this constant as a two-term expression involving a well-known mathematical constant.

History and context. This continued fraction arises in Ramanujan's *Notebooks*. He discovered the beautiful fact that

$$\frac{R_\eta(a, b) + R_\eta(b, a)}{2} = R_\eta\left(\frac{a+b}{2}, \sqrt{ab}\right).$$

The authors wished to record this in [15] and to check the identity computationally. A first attempt to find $R_1(1, 1)$ by direct numerical computation failed miserably, and with some effort only three reliable digits were obtained: $0.693\ldots$. With hindsight, it was realized that the slowest convergence of the fraction occurs in the mathematically simplest case, namely, when $a = b$. Indeed, $R_1(1, 1) = \log 2$, as the first primitive numerics had tantalizingly suggested.

Solution. Attempting a direct computation of $R_1(2, 2)$ using a depth of twenty thousand gives only two digits. Thus we must seek more sophisticated methods. From [**16**, (1.11.70)] we learn that when $0 < b < a$,

$$\mathcal{R}_1(a, b) = \frac{\pi}{2} \sum_{n \in \mathbb{Z}} \frac{aK(k)}{K^2(k) + a^2 n^2 \pi^2} \operatorname{sech}\left(n\pi \frac{K(k')}{K(k)}\right), \qquad (18)$$

where $k = b/a = \theta_2^2/\theta_3^2$ and $k' = \sqrt{1-k^2}$. Here θ_2 and θ_3 are Jacobian theta-functions, and K is a complete elliptic integral of the first kind.

Writing (18) as a Riemann sum, we find that

$$\begin{aligned}\mathcal{R}(a) = \mathcal{R}_1(a, a) &= \int_0^\infty \frac{\operatorname{sech}(\pi x/(2a))}{1 + x^2}\, dx \\ &= 2a \sum_{k=1}^\infty \frac{(-1)^{k+1}}{1 + (2k-1)a},\end{aligned} \qquad (19)$$

where the final equality follows from the Cauchy-Lindelöf theorem. This sum can also be written as

$$\mathcal{R}(a) = \frac{2a}{1+a}\, {}_2F_1\left(\frac{1}{2a} + \frac{1}{2}, 1; \frac{1}{2a} + \frac{3}{2}; -1\right),$$

where ${}_2F_1(\cdot)$ denotes the hypergeometric function [**1**, p. 556]. The latter form is what we use in *Maple* or *Mathematica* to determine

$\mathcal{R}(2) = 0.97499098879872209671990033452921084400592021999471060574526825128587738745570859435232532091 1129362\ldots.$

This constant, as written, is a bit difficult to recognize, but if one first divides by $\sqrt{2}$ and exploits the *Inverse Symbolic Calculator*, an online tool available at the URL http://www.cecm.sfu.ca/projects/ISC/ISCmain.html, it becomes apparent that the quotient is $\pi/2 - \log(1 + \sqrt{2})$. Thus we conclude, experimentally, that

$$\mathcal{R}(2) = \sqrt{2}[\pi/2 - \log(1 + \sqrt{2})].$$

Indeed, it follows (see [**18**]) that

$$\mathcal{R}(a) = 2 \int_0^1 \frac{t^{1/a}}{1+t^2}\, dt.$$

Note that $\mathcal{R}(1) = \log 2$. No nontrivial closed-form expression is known for $\mathcal{R}(a, b)$ when $a \neq b$, although

$$\mathcal{R}_1\left(\frac{1}{4\pi}\beta\left(\frac{1}{4}, \frac{1}{4}\right), \frac{\sqrt{2}}{8\pi}\beta\left(\frac{1}{4}, \frac{1}{4}\right)\right) = \frac{1}{2} \sum_{n \in \mathbb{Z}} \frac{\operatorname{sech}(n\pi)}{1 + n^2}$$

is almost closed. It would be pleasant to find a direct proof of (19). Further details are to be found in [**18**], [**17**], and [**16**].

7. EXPECTED DISTANCE ON A UNIT SQUARE.

Problem 6. *Calculate the expected distance E_2 between two random points on different sides of the unit square:*

$$E_2 = \frac{2}{3} \int_0^1 \int_0^1 \sqrt{x^2 + y^2}\, dx\, dy + \frac{1}{3} \int_0^1 \int_0^1 \sqrt{1 + (u-v)^2}\, du\, dv. \qquad (20)$$

Extra credit: Express this constant as a three-term expression involving algebraic constants and an evaluation of the natural logarithm with an algebraic argument.

History and context. This evaluation and the next were discovered, in slightly more complicated form, by James D. Klein [16, p. 66]. He computed the numerical integral and compared it with the result of a Monte Carlo simulation. Indeed, a straightforward approach to a quick numerical value for an arbitrary iterated integral is to use a Monte-Carlo simulation, which entails approximating the integral by a sum of function values taken at pseudo-randomly generated points within the region. It is important to use a good pseudo-random number generator for this purpose. We tried doing a Monte Carlo evaluation for this problem, using a pseudo-random number generator based on the recently discovered class of provably normal numbers [9], [15, pp. 169–70]. The results we obtained for the two integrals in question, with 10^8 pseudo-random pairs, are $0.765203\ldots$ and $1.076643\ldots$, respectively, yielding an expected distance of $0.869017\ldots$. Unfortunately, none of these three values immediately suggests a closed form, and they are not sufficiently accurate (because of statistical limitations) to be suitable for PSLQ or other constant recognition tools. More digits are needed.

Solution. It is possible to calculate high-precision numerical values for these two integrals using a two-dimensional quadrature (numerical integration) program. In our program, we employed a two-dimensional version of the "tanh-sinh" quadrature algorithm, which we will discuss in more detail in Problem 8. Two-dimensional quadrature is usually much more expensive than one-dimensional quadrature, at a given precision level, because many more function evaluations must be performed. Often a highly parallel computer system must be used to obtain a high-precision result in reasonable run time [11]. Nonetheless, in this case we were able to evaluate the first of the two integrals to 108-digit accuracy in twenty-one minutes run time on a 2004-era computer, and the second to 118-digit accuracy in just twenty seconds. The first is more difficult due to nondifferentiability of the integrand at the origin.

Indeed, in this case both *Maple* and *Mathematica* are able to evaluate each of these integrals numerically, as is, to over one hundred decimal digit accuracy in just a few minutes of run time, either by evaluating the inner integral symbolically and the outer integral numerically or else by performing full two-dimensional numerical quadrature. *Maple*, *Mathematica*, and the two-dimensional quadrature program all agreed on the following numerical value for the expected distance:

$$\alpha = 0.86900905527453446388497059434540662485671927963168056$$
$$9660350864584179822174693053113213554875435754\ldots.$$

Using PSLQ, with the basis elements α, $\sqrt{2}$, $\log(\sqrt{2}+1)$, and 1, we obtain

$$\alpha = \frac{1}{9}\sqrt{2} + \frac{5}{9}\log(\sqrt{2}+1) + \frac{2}{9}. \qquad (21)$$

An alternate solution is to attempt to evaluate the integrals symbolically! In fact, in this case Version 5.1 of *Mathematica* can do both the integrals "out of the box," whereas in the first case *Maple* appears to need coaxing, for instance, by converting to polar coordinates:

$$2\int_0^{\pi/4}\int_0^{\sec\theta} r^2\,dr\,d\theta = \frac{2}{3}\int_0^{\pi/4}\sec^3\theta\,d\theta = \frac{1}{3}\sqrt{2} - \frac{1}{6}\log(2) + \frac{1}{3}\log(2+\sqrt{2}),$$

since the radius for a given θ is $1/\cos\theta$. As for the second integral, *Maple* and *Mathematica* both give

$$-\frac{1}{3}\sqrt{2} - \frac{1}{2}\log(\sqrt{2}-1) + \frac{1}{2}\log(1+\sqrt{2}) + \frac{2}{3}.$$

To obtain the second integral analytically, write it as $2\int_0^1\int_0^u \sqrt{1+(u-v)^2}\,dv\,du$. Now change variables (set $t = u - v$) to obtain $1/2\int_0^1\{u\sqrt{1+u^2} + \operatorname{arcsinh} u\}\,du$. Thus, the expected distance is

$$\frac{1}{9}\sqrt{2} - \frac{1}{9}\log(2) + \frac{2}{9}\log(2+\sqrt{2}) - \frac{1}{6}\log(\sqrt{2}-1) + \frac{1}{6}\log(1+\sqrt{2}) + \frac{2}{9},$$

which can be simplified to the formula (21).

Additional information on the problem is available at http://mathworld.wolfram.com/SquareLinePicking.html.

8. EXPECTED DISTANCE ON A UNIT CUBE.

Problem 7. *Calculate the expected distance between two random points on different faces of the unit cube. (Hint: This can be expressed in terms of integrals as*

$$E_3 := \frac{4}{5}\int_0^1\int_0^1\int_0^1\int_0^1 \sqrt{x^2+y^2+(z-w)^2}\,dw\,dx\,dy\,dz$$

$$+ \frac{1}{5}\int_0^1\int_0^1\int_0^1\int_0^1 \sqrt{1+(y-u)^2+(z-w)^2}\,du\,dw\,dy\,dz.)$$

Extra credit: Express this constant as a six-term expression involving algebraic constants and two evaluations of the natural logarithm with algebraic arguments.

History and context. As we noted earlier, this evaluation was discovered, in essentially the same form, by Klein [16, p. 66]. As with Problem 6, a Monte Carlo integration scheme can be used to obtain quick approximations to the integrals. The values we obtained were $0.870792\ldots$ and $1.148859\ldots$, respectively, yielding an expected distance of $0.926406\ldots$. Once again, however, these numerical values do not immediately suggest a closed-form evaluation, yet the accuracy is too low to apply PSLQ or other constant recognition schemes. What's more, in this case, unlike Problem 6, neither *Maple* nor *Mathematica* are able to evaluate these four-fold integrals directly—though *Mathematica* comes close. As in most cases "help" is needed, in the form of mathematical manipulation to render these integrals in a form where mathematical computing software can evaluate them—numerically or symbolically.

Solution. Let $_2F_1(\cdot)$ again denote the hypergeometric function [1, p. 556]. One can show that the first integral evaluates to

$$\frac{\sqrt{2\pi}}{5} \sum_{n=2}^{\infty} \frac{{}_2F_1(1/2, -n+2; 3/2; 1/2)}{(2n+1)\Gamma(n+2)\Gamma(5/2-n)} + \frac{4}{15}\sqrt{2} + \frac{2}{5}\log\left(\sqrt{2}+1\right) - \frac{1}{75}\pi$$

and the second generalized hypergeometric function formally evaluates to

$$\frac{\sqrt{\pi}}{10} \sum_{n=0}^{\infty} \frac{{}_4F_3(1, 1/2, -1/2-n, -n-1; 2, 1/2-n, 3/2; -1)}{(2n+1)\Gamma(n+2)\Gamma(3/2-n)}$$

$$-\frac{2}{25} + \frac{\sqrt{2}}{50} + \frac{1}{10}\log\left(\sqrt{2}+1\right).$$

(Although the second diverges as a Riemann sum, both *Maple* and *Mathematica* can handle it, with some human help, producing numerical values of the corresponding Borel sum.) Both expressions are consequences of the binomial theorem, modulo an initial integration with respect to z in the first case. These expansions allow one to compute the expectation to high precision numerically and to express both of the individual integrals in terms of the same set of constants. The numerical value of the desired expectation is

0.92639005517404672921816358654777901444496019010733504673252192127
14185045940366838293134730753499682l2....

An integer relation search in the span of $\{1, \pi, \sqrt{2}, \sqrt{3}, \log(1+\sqrt{2}), \log(2+\sqrt{3})\}$ produces

$$\frac{4}{75} + \frac{17}{75}\sqrt{2} - \frac{2}{25}\sqrt{3} - \frac{7}{75}\pi + \frac{7}{25}\log\left(1+\sqrt{2}\right) + \frac{7}{25}\log\left(7+4\sqrt{3}\right).$$

With substantial effort we were able to nurse the symbolic integral out of *Maple*. We started, as in the previous problem, by integrating with respect to w over $[0, z]$, doubling, and continuing in this fashion until we reduced the problem to showing that

$$E_3 = -\int_0^1 \left(2x^3 + 6x^2 + 3\right) \ln\left(\sqrt{2+x^2} - 1\right) dx$$

$$+ \int_0^1 3\frac{-(x^2+1)\ln\left(\sqrt{2+x^2}-1\right) + \ln\left(\sqrt{2}-1\right)}{x^2(x^2+1)} dx$$

$$= -\frac{5}{3}\pi + \frac{7}{6}\sqrt{2} + \frac{7}{2}\ln\left(1+\sqrt{2}\right) - \frac{3}{2}\ln(2) + \ln\left(1+\sqrt{3}\right) + \frac{37}{24}$$

$$+ \frac{3}{4}\ln\left(1+\sqrt{2}\right)\pi,$$

which we leave to the reader to establish.

Mathematica was more helpful: consider

```
4/5 Integrate[Sqrt[x^2 + y^2 + (z - w)^2], {x, 0, 1}, {y, 0, 1},
  {w, 0, 1}, {z, 0, 1}]// Timing
  {52.483021*Second, (168*Sqrt[2] - 24*Sqrt[3] - 44*Pi + 72*ArcSinh[1] +
  162*ArcSinh[1/Sqrt[2]] + 24*Log[2] - 240*Log[-1 + Sqrt[3]] +
  192*Log[1 + Sqrt[3]] + 20*Log[26 + 15*Sqrt[3]] + 3*Log[70226 +
  40545*Sqrt[3]])/900}
```

This form is what the shipping version of *Mathematica* 5.1 returns on a 3.0 GHz Pentium 4. It evaluates the first integral directly, while the second one can be done with a little help. The combined outcomes can then be simplified symbolically to the result shown.

There is also an ingenious method due to Michael Trott using a Laplace transform to reduce the four-dimensional integrals to integrals over one-dimensional integrands. It proceeds by eliminating the square roots (which cause most of the difficulty in symbolic evaluation of the multiple integrals) at the expense of introducing one additional (but "easy") integral. The original problem can then be written in terms of the *single* integral

$$\int_0^\infty \left[-\frac{14}{25} e^{-z^2} \sqrt{\pi} \operatorname{erf}^2(z) + \frac{28 e^{-2z^2} \operatorname{erf}(z)}{25z} + \frac{7 e^{-z^2} \operatorname{erf}(z)}{25z} - \frac{12 e^{-3z^2}}{25\sqrt{\pi}} \right.$$
$$\left. + \frac{68 e^{-2z^2}}{75\sqrt{\pi}} + \frac{8 e^{-z^2}}{75\sqrt{\pi}} \right] dz,$$

which can be evaluated directly in *Mathematica* to produce the symbolic expression for E_3.

Nonetheless, we must emphasize (i) that one needs to proceed with confidence, since such symbolic computations can take several minutes, and (ii) that phrases like "*Maple* can not" or "*Mathematica* can" are release-specific and may also depend on the skill of the human user to make use of expert knowledge in mathematics, symbolic computation, or both, in order to produce a form of the problem that is most amenable to computation in a given software system. This explains our desire to illustrate various solution paths here and elsewhere.

Additional information on this problem is available at `http://mathworld.wolfram.com/CubeLinePicking.html`. For more information about the Laplace transform trick applied to the related problem of expected distance in a unit hypercube, see `http://mathworld.wolfram.com/HypercubeLinePicking.html`.

9. AN INFINITE COSINE PRODUCT.

Problem 8. *Calculate*

$$\pi_2 = \int_0^\infty \cos(2x) \prod_{n=1}^\infty \cos\left(\frac{x}{n}\right) dx.$$

History and context. The challenge of showing that $\pi_2 < \pi/8$ was posed by Bernard Mares, Jr., along with the problem of demonstrating that

$$\pi_1 = \int_0^\infty \prod_{n=1}^\infty \cos\left(\frac{x}{n}\right) dx < \frac{\pi}{4}.$$

This is indeed true, although the error is remarkably small, as we shall see.

Solution. The computation of a high-precision numerical value for this integral is rather challenging, owing in part to the oscillatory behavior of $\prod_{n\geq 1} \cos(x/n)$ (see Figure 2) but mostly because of the difficulty of computing high-precision evaluations of the integrand. Note that evaluating thousands of terms of the infinite product would

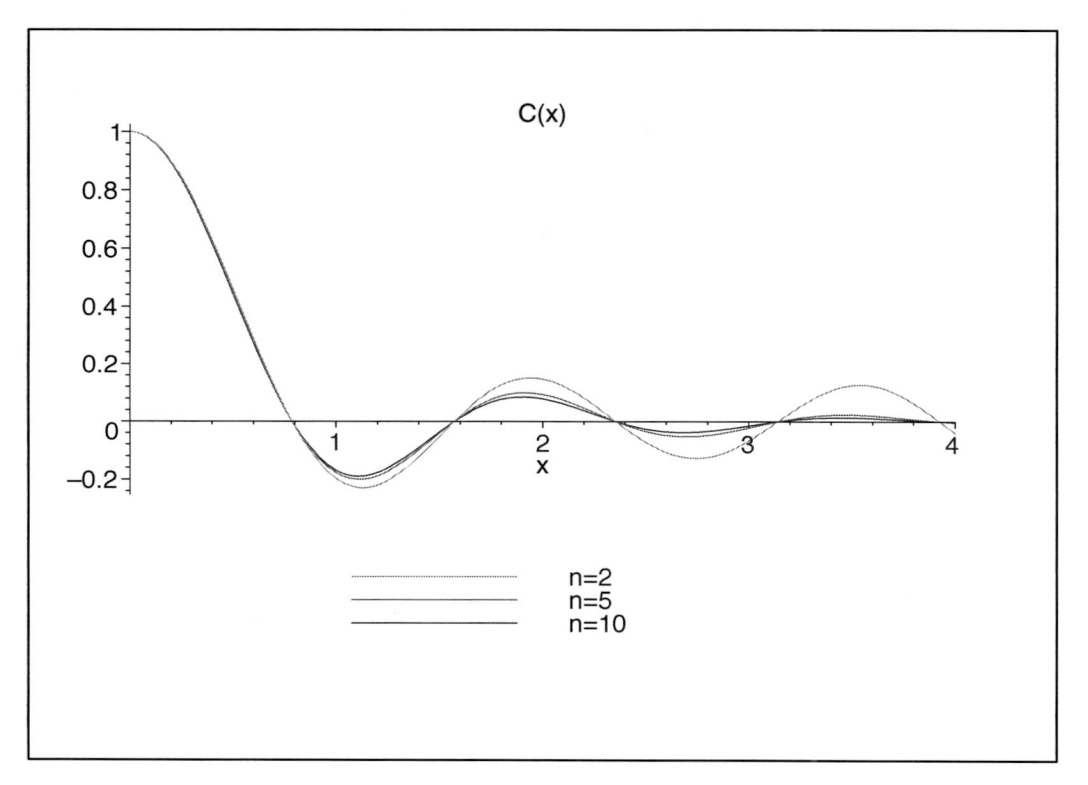

Figure 2. Approximations to $\prod_{n\geq 1} \cos(x/n)$.

produce only a few correct digits. Thus it is necessary to rewrite the integrand in a form more suitable for computation.

Let $f(x)$ signify the integrand. We can express $f(x)$ as

$$f(x) = \cos(2x) \left[\prod_{k=1}^{m} \cos\left(\frac{x}{k}\right) \right] \exp(f_m(x)), \tag{22}$$

where we choose m greater than x and where

$$f_m(x) = \sum_{k=m+1}^{\infty} \log \cos\left(\frac{x}{k}\right). \tag{23}$$

The kth summand can be expanded in a Taylor series [1, p. 75], as follows:

$$\log \cos\left(\frac{x}{k}\right) = \sum_{j=1}^{\infty} \frac{(-1)^j 2^{2j-1}(2^{2j}-1)B_{2j}}{j(2j)!} \left(\frac{x}{k}\right)^{2j},$$

in which B_{2j} are Bernoulli numbers. Observe that since $k > m > x$ in (23), this series converges. We can then write

$$f_m(x) = \sum_{k=m+1}^{\infty} \sum_{j=1}^{\infty} \frac{(-1)^j 2^{2j-1}(2^{2j}-1)B_{2j}}{j(2j)!} \left(\frac{x}{k}\right)^{2j}. \tag{24}$$

After applying the identity [1, p. 807]

$$B_{2j} = \frac{(-1)^{j+1} 2(2j)! \zeta(2j)}{(2\pi)^{2j}}$$

and interchanging the sums, we obtain

$$f_m(x) = -\sum_{j=1}^{\infty} \frac{(2^{2j} - 1)\zeta(2j)}{j\pi^{2j}} \left[\sum_{k=m+1}^{\infty} \frac{1}{k^{2j}}\right] x^{2j}.$$

Note that the inner sum can also be written in terms of the zeta-function, as follows:

$$f_m(x) = -\sum_{j=1}^{\infty} \frac{(2^{2j} - 1)\zeta(2j)}{j\pi^{2j}} \left[\zeta(2j) - \sum_{k=1}^{m} \frac{1}{k^{2j}}\right] x^{2j}.$$

This can now be reduced to a compact form for purposes of computation as

$$f_m(x) = -\sum_{j=1}^{\infty} a_j b_{j,m} x^{2j}, \qquad (25)$$

where

$$a_j = \frac{(2^{2j} - 1)\zeta(2j)}{j\pi^{2j}}, \qquad (26)$$

$$b_{j,m} = \zeta(2j) - \sum_{k=1}^{m} 1/k^{2j}. \qquad (27)$$

We remark that $\zeta(2j)$, a_j, and $b_{j,m}$ can all be precomputed, say for j up to some specified limit and for a variety of m. In our program, which computes this integral to 120-digit accuracy, we precompute $b_{j,m}$ for $m = 1, 2, 4, 8, 16, \ldots, 256$ and for j up to 300. During the quadrature computation, the function evaluation program picks m to be the first power of two greater than the argument x, and then applies formulas (22) and (25). It is not necessary to compute $f(x)$ for x larger than 200, since for these large arguments $|f(x)| < 10^{-120}$ and thus may be presumed to be zero.

The computation of values of the Riemann zeta-function can be done using a simple algorithm due to Peter Borwein [21] or, since what we really require is the entire set of values $\{\zeta(2j) : 1 \leq j \leq n\}$ for some n, by a convolution scheme described in [5]. It is important to note that the computation of both the zeta values and the $b_{j,m}$ must be done with a much higher working precision (in our program, we use 1600-digit precision) than the 120-digit precision required for the quadrature results, since the two terms being subtracted in formula (27) are very nearly equal. These values need to be calculated to a *relative* precision of 120 digits.

With this evaluation scheme for $f(x)$ in hand, the integral (8) can be computed using, for instance, the tanh-sinh quadrature algorithm, which can be implemented fairly easily on a personal computer or workstation and is also well suited to highly parallel processing [10], [11], [16, p. 312]. This algorithm approximates an integral $f(x)$ on $[-1, 1]$ by transforming it to an integral on $(-\infty, \infty)$ via the change of variable $x = g(t)$, where $g(t) = \tanh(\pi/2 \cdot \sinh t)$:

$$\int_{-1}^{1} f(x)\,dx = \int_{-\infty}^{\infty} f(g(t))g'(t)\,dt = h \sum_{j=-\infty}^{\infty} w_j f(x_j) + E(h). \qquad (28)$$

Here $x_j = g(hj)$ and $w_j = g'(hj)$ are abscissas and weights for the tanh-sinh quadrature scheme (which can be precomputed), and $E(h)$ is the error in this approximation.

The function $g'(t) = \pi/2 \cdot \cosh t \cdot \mathrm{sech}^2(\pi/2 \cdot \sinh t)$ and its derivatives tend to zero very rapidly for large $|t|$. Thus, even if the function $f(t)$ has an infinite derivative, a blow-up discontinuity, or oscillatory behavior at an endpoint, the product function $f(g(t))g'(t)$ is in many cases quite well behaved, going rapidly to zero (together with all of its derivatives) for large $|t|$. In such cases, the Euler-Maclaurin summation formula [2, p. 180] can be invoked to conclude that the error $E(h)$ in the approximation (28) decreases very rapidly—faster than any power of h. In many applications, the tanh-sinh algorithm achieves quadratic convergence (i.e., reducing the size h of the interval in half produces twice as many correct digits in the result).

The tanh-sinh quadrature algorithm is designed for a finite integration interval. In this problem, where the interval of integration is $[0, \infty)$, it is necessary to convert the integral to a problem on a finite interval. This can be done with the simple substitution $s = 1/(x+1)$, which yields an integral from 0 to 1.

In spite of the substantial computation required to construct the zeta- and b-arrays, as well as the abscissas x_j and weights w_j needed for tanh-sinh quadrature, the entire calculation requires only about one minute on a 2004-era computer, using the ARPREC arbitrary precision software package available at http://crd.lbl.gov/~dhbailey/mpdist. The first hundred digits of the result are the following:

0.39269908169872415480783042290993786052464543418723159592681228516209324713993854617901651274745536677....

A *Mathematica* program capable of producing 100 digits of this constant is available on Michael Trott's website: http://www.mathematicaguidebooks.org/downloads/N_2_01_Evaluated.nb.

Using the Inverse Symbolic Calculator, for instance, one finds that this constant is likely to be $\pi/8$. But a careful comparison with a high-precision value of $\pi/8$, namely,

0.392699081698724154807830422909937860524646174921888227621868074038477050785776124828504353167764633497...,

reveals that they are *not* equal—the two values differ by approximately 7.407×10^{-43}. Indeed, these two values are provably distinct. This follows from the fact that

$$\sum_{n=1}^{55} 1/(2n+1) > 2 > \sum_{n=1}^{54} 1/(2n+1).$$

See [16, chap. 2] for additional details. We do not know a concise closed-form expression for this constant.

Further history and context. Recall the *sinc* function

$$\mathrm{sinc}\, x = \frac{\sin x}{x},$$

and consider, the seven highly oscillatory integrals:

$$I_1 = \int_0^\infty \text{sinc}\, x\, dx = \frac{\pi}{2},$$

$$I_2 = \int_0^\infty \text{sinc}\, x\, \text{sinc}\left(\frac{x}{3}\right) dx = \frac{\pi}{2},$$

$$I_3 = \int_0^\infty \text{sinc}\, x\, \text{sinc}\left(\frac{x}{3}\right) \text{sinc}\left(\frac{x}{5}\right) dx = \frac{\pi}{2},$$

$$\vdots$$

$$I_6 = \int_0^\infty \text{sinc}\, x\, \text{sinc}\left(\frac{x}{3}\right) \cdots \text{sinc}\left(\frac{x}{11}\right) dx = \frac{\pi}{2},$$

$$I_7 = \int_0^\infty \text{sinc}\, x\, \text{sinc}\left(\frac{x}{3}\right) \cdots \text{sinc}\left(\frac{x}{13}\right) dx = \frac{\pi}{2}.$$

It comes as something of a surprise, therefore, that

$$I_8 = \int_0^\infty \text{sinc}\, x\, \text{sinc}\left(\frac{x}{3}\right) \cdots \text{sinc}\left(\frac{x}{15}\right) dx$$

$$= \frac{467807924713440738696537864469}{935615849440640907310521750000}\pi \approx 0.4999999999992646\pi.$$

When this was first discovered by a researcher, using a well-known computer algebra package, both he and the software vendor concluded there was a "bug" in the software. Not so! It is fairly easy to see that the limit of the sequence of such integrals is $2\pi_1$. Our analysis, via Parseval's theorem, links the integral

$$I_N = \int_0^\infty \text{sinc}(a_1 x)\, \text{sinc}(a_2 x) \cdots \text{sinc}(a_N x)\, dx$$

with the volume of the polyhedron P_N described by

$$P_N = \left\{ x : \left|\sum_{k=2}^N a_k x_k\right| \leq a_1, |x_k| \leq 1, 2 \leq k \leq N \right\}$$

for $x = (x_2, x_3, \ldots, x_N)$. If we let

$$C_N = \{(x_2, x_3, \ldots, x_N) : -1 \leq x_k \leq 1, 2 \leq k \leq N\},$$

then

$$I_N = \frac{\pi}{2a_1} \frac{\text{Vol}(P_N)}{\text{Vol}(C_N)}.$$

Thus, the value drops precisely when the constraint $\sum_{k=2}^N a_k x_k \leq a_1$ becomes *active* and bites the hypercube C_N. That occurs when $\sum_{k=2}^N a_k > a_1$. In the foregoing,

$$\frac{1}{3} + \frac{1}{5} + \cdots + \frac{1}{13} < 1,$$

but on addition of the term $1/15$, the sum exceeds 1, the volume drops, and $I_N = \pi/2$ no longer holds. A similar analysis applies to π_2. Moreover, it is fortunate that we began with π_1 or the falsehood of $\pi_2 = 1/8$ would have been much harder to see.

Additional information on this problem is available at `http://mathworld.wolfram.com/InfiniteCosineProductIntegral.html` and `http://mathworld.wolfram.com/BorweinIntegrals.html`.

10. A MULTIVARIATE ZETA-FUNCTION.

Problem 9. *Calculate*

$$\sum_{i>j>k>l>0} \frac{1}{i^3 j k^3 l}.$$

Extra credit: Express this constant as a single-term expression involving a well-known mathematical constant.

History and context. We resume the discussion from Problem 3. In the notation introduced there, we ask for the value of $\zeta(3, 1, 3, 1)$. The study of such sums in two variables, as we noted, originated with Euler. These investigations were apparently due to a serendipitous mistake. Goldbach wrote to Euler [15, pp. 99–100]:

> When I recently considered further the indicated sums of the last two series in my previous letter, I realized immediately that the same series arose due to a mere writing error, from which indeed the saying goes, "Had one not erred, one would have achieved less [*Si non errasset, fecerat ille minus*]."

Euler's *reduction formula* is

$$\zeta(s, 1) = \frac{s}{2} \zeta(s+1) - \frac{1}{2} \sum_{k=1}^{s-2} \zeta(k+1)\zeta(s+1-k),$$

which *reduces* the given double Euler sums to a sum of products of classical ζ-values. Euler also noted the first *reflection formulas*

$$\zeta(a, b) + \zeta(b, a) = \zeta(a)\zeta(b) - \zeta(a+b),$$

certainly valid when $a > 1$ and $b > 1$. This is an easy algebraic consequence of adding the double sums. Another marvelous fact is the *sum formula*

$$\sum_{\Sigma a_i = n, a_i \geq 0} \zeta(a_1 + 2, a_2 + 1, \ldots, a_r + 1) = \zeta(n + r + 1) \qquad (29)$$

for nonnegative integers n and r. This, as David Bradley observes, is equivalent to the generating function identity

$$\sum_{n>0} \frac{1}{n^r(n-x)} = \sum_{k_1 > k_2 > \cdots > k_r > 0} \prod_{j=1}^{r} \frac{1}{k_j - x}.$$

The first three nontrivial cases of (29) are $\zeta(3) = \zeta(2, 1)$, $\zeta(4) = \zeta(3, 1) + \zeta(2, 2)$, and $\zeta(2, 1, 1) = \zeta(4)$.

Solution. We notice that such a function is a generalization of the zeta-function. Similar to the definition in section 4, we define

$$\zeta(s_1, s_2, \ldots, s_k; x) = \sum_{n_1 > n_2 > \cdots > n_k > 0} \frac{x_1^n}{n_1^{s_1} n_2^{s_2} \cdots n_r^{s_r}}, \tag{30}$$

for $s_1, s_2, \ldots, s_k$ nonnegative integers. We see that we are asked to compute the value $\zeta(3, 1, 3, 1; 1)$. Such a sum can be evaluated directly using the EZFace+ interface at http://www.cecm.sfu.ca/projects/ezface+, which employs the Hölder convolution, giving us the numerical value

$$0.00522956956353096010093065228389923158989042078463463552254744897214886954466015007497545432485610401627\ldots. \tag{31}$$

Alternatively, we may proceed using differential equations. It is fairly easy to see [**16**, sec. 3.7] that

$$\frac{d}{dx}\zeta(n_1, n_2, \ldots, n_r; x) = \frac{1}{x}\zeta(n_1 - 1, n_2, \ldots, n_r; x) \quad (n_1 > 1), \tag{32}$$

$$\frac{d}{dx}\zeta(n_1, n_2, \ldots, n_r; x) = \frac{1}{1-x}\zeta(n_2, \ldots, n_r; x) \quad (n_1 = 1), \tag{33}$$

with initial conditions $\zeta(n_1; 0) = \zeta(n_1, n_2; 0) = \cdots = \zeta(n_1, \ldots, n_r; 0) = 0$ and $\zeta(\cdot; x) \equiv 1$. Solving

```
> dsys1 =
> diff(y3131(x),x) = y2131(x)/x,
> diff(y2131(x),x) = y1131(x)/x,
> diff(y1131(x),x) = 1/(1-x)*y131(x),
> diff(y131(x),x) = 1/(1-x)*y31(x),
> diff(y31(x),x) = y21(x)/x,
> diff(y21(x),x) = y11(x)/x,
> diff(y11(x),x) = y1(x)/(1-x),
> diff(y1(x),x) = 1/(1-x);
> init1 = y3131(0) = 0,y2131(0) = 0, y1131(0) = 0,
>                  y131(0) = 0,y31(0) = 0,y21(0) = 0,y11(0) = 0,y1(0) = 0;
```

in *Maple*, we obtain $0.0052295695635180396128305365196667669502942$ (this is valid to thirteen decimal places). Maple's `identify` command is unable to identify portions of *this* number, and the inverse symbolic calculator does not return a result. It should be mentioned that both *Maple* and the ISC identified the constant $\zeta(3, 1)$ (see the remark under the "history and context" heading). From the hint for this question, we know this is a single-term expression. Suspecting a form similar to $\zeta(3, 1)$, we search for a constants c and d such that $\zeta(3, 1, 3, 1) = c\pi^d$. This leads to $c = 1/81440 = 2/10!$ and $d = 8$.

Further history and context. We start with the simpler value, $\zeta(3, 1)$. Notice that

$$-\log(1 - x) = x + \frac{1}{2}x^2 + \frac{1}{3}x^3 + \cdots,$$

so

$$f(x) = -\log(1-x)/(1-x) = x + \left(1 + \frac{1}{2}\right)x^2 + \left(1 + \frac{1}{2} + \frac{1}{3}\right)x^3 + \cdots$$
$$= \sum_{n \geq m > 0} \frac{x^n}{m}.$$

As noted in the section on double Euler sums,

$$\frac{(-1)^{m+1}}{\Gamma(m)} \int_0^1 x^n \log^{m-1} x \, dx = \frac{1}{(n+1)^m},$$

so integrating f using this transform for $m = 3$, we obtain

$$\zeta(3, 1) = \frac{1}{2} \int_0^1 f(x) \log^2 x \, dx$$
$$= 0.2705808084277845478790000924\ldots.$$

The corresponding generating function is

$$\sum_{n \geq 0} \zeta(\{3, 1\}_n) x^{4n} = \frac{\cosh(\pi x) - \cos(\pi x)}{\pi^2 x^2},$$

equivalent to Zagier's conjectured identity

$$\zeta(\{3, 1\}_n) = \frac{2\pi^{4n}}{(4n+2)!}.$$

Here $\{3, 1\}_n$ denotes n-fold concatenation of $\{3, 1\}$.

The proof of this identity (see [16, p. 160]) derives from a remarkable factorization of the generating function in terms of hypergeometric functions:

$$\sum_{n \geq 0} \zeta(\{3, 1\}_n) x^{4n} = {}_2F_1\left(x\frac{(1+i)}{2}, -x\frac{(1+i)}{2}; 1; 1\right)$$
$$\times {}_2F_1\left(x\frac{(1-i)}{2}, -x\frac{(1-i)}{2}; 1; 1\right).$$

Finally, it can be shown in various ways that

$$\zeta(\{3\}_n) = \zeta(\{2, 1\}_n)$$

for all n, while a proof of the numerically-confirmed conjecture

$$\zeta(\{2, 1\}_n) \stackrel{?}{=} 2^{3n} \zeta(\{-2, 1\}_n) \tag{34}$$

remains elusive. Only the first case of (34), namely,

$$\sum_{n=1}^{\infty} \frac{1}{n^2} \sum_{m=1}^{n-1} \frac{1}{m} = 8 \sum_{n=1}^{\infty} \frac{(-1)^n}{n^2} \sum_{m=1}^{n-1} \frac{1}{m} \quad (= \zeta(3))$$

has a self-contained proof [16]. Indeed, the only other established case is

$$\sum_{n=1}^{\infty}\frac{1}{n^2}\sum_{m=1}^{n-1}\frac{1}{m}\sum_{p=1}^{m-1}\frac{1}{p^2}\sum_{q=1}^{p-1}\frac{1}{q} = 64\sum_{n=1}^{\infty}\frac{(-1)^n}{n^2}\sum_{m=1}^{n-1}\frac{1}{m}\sum_{p=1}^{m-1}\frac{(-1)^p}{p^2}\sum_{q=1}^{p-1}\frac{1}{q} \quad (=\zeta(3,3)).$$

This is an outcome of a complete set of equations for multivariate zeta-functions of depth four.

There has been abundant evidence amassed to support identity (34) since it was found in 1996. For example, very recently Petr Lisonek checked the first eighty-five cases to one thousand places in about forty-one hours with only the *expected roundoff error*. And he checked $n = 163$ in ten hours. This is the *only* identification of its type of an Euler sum with a distinct multivariate zeta-function.

11. A WATSON INTEGRAL.

Problem 10. *Evaluate*

$$W = \frac{1}{\pi^3}\int_0^\pi\int_0^\pi\int_0^\pi \frac{1}{3 - \cos x - \cos y - \cos z}\,dx\,dy\,dz. \tag{35}$$

History and context. The integral arises in Gaussian and spherical models of ferromagnetism and in the theory of random walks. It leads to one of the most impressive closed-form evaluations of an equivalent multiple integral due to G. N. Watson:

$$\begin{aligned}\widehat{W} &= \int_{-\pi}^\pi\int_{-\pi}^\pi\int_{-\pi}^\pi \frac{1}{3 - \cos x - \cos y - \cos z}\,dx\,dy\,dz \\ &= \frac{1}{96}(\sqrt{3}-1)\Gamma^2\left(\frac{1}{24}\right)\Gamma^2\left(\frac{11}{24}\right) \\ &= 4\pi\left(18 + 12\sqrt{2} - 10\sqrt{3} - 7\sqrt{6}\right)K^2(k_6),\end{aligned} \tag{36}$$

where $k_6 = (2 - \sqrt{3})(\sqrt{3} - \sqrt{2})$ is the sixth singular value. The most self-contained derivation of this very subtle result is due to Joyce and Zucker in [28] and [29], where more background can also be found.

Solution. In [31], it is shown that a simplification can be obtained by applying the formula

$$\frac{1}{\lambda} = \int_0^\infty e^{-\lambda t}\,dt \quad (\mathrm{Re}\,\lambda > 0) \tag{37}$$

to W_3. The three-dimensional integral is then reducible to a single integral by using the identity

$$\frac{1}{\pi}\int_0^\infty \exp(t\cos\theta)d\theta = I_0(t), \tag{38}$$

in which $I_0(t)$ is the modified Bessel function of the first kind. It follows from this that $W = \int_0^\infty \exp(-3t)I_0^3(t)dt$. This integral can be evaluated to one hundred digits in *Maple*, giving

$$W_3 = 0.50546201971732600605200405322714025998512901481742089$$
$$21889934878860287734511738168005372470698960380\ldots \qquad (39)$$

Finally, an integer relation hunt to express $\log W$ in terms of $\log \pi$, $\log 2$, $\log \Gamma(k/24)$, and $\log(\sqrt{3} - 1)$ will produce (36).

We may also write W_3 as a product solely of values of the gamma function. This is what our *Mathematician's ToolKit* returned:

```
0 = -1.* log[w3] + -1.* log[gamma[1/24]] + 4.*log[gamma[3/24]] +
-8.*log[gamma[5/24]] + 1.* log[gamma[7/24]] + 14.*log[gamma[9/24]] +
-6.*log[gamma[11/24]] + -9.*log[gamma[13/24]] + 18.*log[gamma[15/24]] +
-2.*log[gamma[17/24]] + -7.*log[gamma[19/24]]
```

Proving this is achieved by comparing the result with (36) and establishing the implicit gamma representation of $(\sqrt{3} - 1)^2/96$.

Similar searches suggest there is no similar four-dimensional closed form—the relevant Bessel integral is $W_4 = \int_0^\infty \exp(-4t) I_0^4(t)\, dt$. (N.B. $\int_0^\infty \exp(-2t) I_0^2(t)\, dt = \infty$.) In this case it is necessary to compute $\exp(-t) I_0(t)$ carefully, using a combination of the formula

$$\exp(-t) I_0(t) = \exp(-t) \sum_{n=0}^\infty \frac{t^{2n}}{2^{2n}(n!)^2}$$

for t up to roughly $1.2 \cdot d$, where d is the number of significant digits desired for the result, and

$$\exp(-t) I_0(t) \approx \frac{1}{\sqrt{2\pi t}} \sum_{n=0}^N \frac{\prod_{k=1}^n (2k-1)^2}{(8t)^n n!}$$

for large t, where the upper limit N of the summation is chosen to be the first index n such that the summand is less than 10^{-d} (since this is an asymptotic expansion, taking more terms than N may increase, not decrease the error). We have implemented this as 'besselexp' in our *Mathematician's ToolKit*, available at `http://crd.lbl.gov/~dhbailey/mpdist`. Using this software, which includes a PSLQ facility, we found that W_4 is not expressible as a product of powers of $\Gamma(k/120)$ ($0 < k < 120$) with coefficients having fewer than 80 digits. This result does not, of course, rule out the possibility of a larger relation, but it does cast some doubt, in an experimental sense, that such a relation exists—enough to stop looking.

Additional information on this problem is available at `http://mathworld.wolfram.com/WatsonsTripleIntegrals.html`.

12. CONCLUSION. While all the problems described herein were studied with a great deal of experimental computation, clean proofs are known for the final results given (except for Problem 7), and in most cases a lot more has by now been proved. Nonetheless, in each case the underlying object suggests plausible generalizations that are still open.

The "hybrid computations" involved in these solutions are quite typical of modern experimental mathematics. Numerical computations by themselves produce no insight, and symbolic computations frequently fail to produce full-fledged, closed-form solutions. But when used together, with significant human interaction, they are often successful in discovering new facts of mathematics and in suggesting routes to formal proof.

ACKNOWLEDGMENTS. Bailey's work was supported by the Director, Office of Computational and Technology Research, Division of Mathematical, Information, and Computational Sciences of the U.S. Department of Energy, under contract number DE-AC02-05CH11231; also by the NSF, under Grant DMS-0342255. Borwein's work was supported in part by NSERC and the Canada Research Chair Programme. Kapoor's work was performed as part of an independent study project at the University of British Columbia. Weisstein's work on *MathWorld* is supported by Wolfram Research as a free service to the world mathematics and Internet communities.

REFERENCES

1. M. Abramowitz and I. A. Stegun, *Handbook of Mathematical Functions*, Dover, New York, 1970.
2. K. E. Atkinson, *An Introduction to Numerical Analysis*, John Wiley & Sons, New York, 1989.
3. D. H. Bailey, Integer relation detection, *Comp. Sci. & Eng.* **2** (2000) 24–28.
4. D. H. Bailey and J. M. Borwein, Sample problems of experimental mathematics (2003), available at http://www.experimentalmath.info/expmath-probs.pdf.
5. D. H. Bailey, J. M. Borwein, and R. E. Crandall, On the Khintchine constant, *Math. Comp.* **66** (1997) 417–431.
6. D. H. Bailey, P. B. Borwein, and S. Plouffe, On the rapid computation of various polylogarithmic constants, *Math. Comp.* **66** (1997) 903–913.
7. D. H. Bailey and D. J. Broadhurst, Parallel integer relation detection: Techniques and applications, *Math. Comp.* **70** (2000) 1719–1736.
8. D. H. Bailey and R. E. Crandall, On the random character of fundamental constant expansions, *Experiment. Math.* **10** (2001) 175–190.
9. ———, Random generators and normal numbers, *Experiment. Math.* **11** (2002) 527–546.
10. D. H. Bailey, X. S. Li, and K. Jeyabalan, A comparison of three high-precision quadrature schemes, *Experiment. Math.* **14** (2005) 317–329.
11. D. H. Bailey and J. M. Borwein, Highly parallel, high-precision numerical integration (2004), available at http://crd.lbl.gov/~dhbailey/dhbpapers/quadparallel.pdf.
12. F. Bornemann, D. Laurie, S. Wagon, and J. Waldvogel, *The SIAM 100 Digit Challenge: A Study in High-Accuracy Numerical Computing*, SIAM, Philadelphia, 2004.
13. J. M. Borwein, "The 100 digit challenge: An extended review," *Math Intelligencer* **27** (2005) 40–48.
14. J. M. Borwein and D. J. Broadhurst, Determinations of rational Dedekind-zeta invariants of hyperbolic manifolds and Feynman knots and links (1998), available at http://arxiv.org/abs/hep-th/9811173.
15. J. M. Borwein and D. H. Bailey, *Mathematics by Experiment*, A K Peters, Natick, MA, 2004.
16. J. M. Borwein, D. H. Bailey, and R. Girgensohn, *Experimentation in Mathematics: Computational Paths to Discovery*, A K Peters, Natick, MA, 2004.
17. J. Borwein and R. Crandall, On the Ramanujan AGM fraction. Part II: The complex-parameter case, *Experiment. Math.* **13** (2004) 287–295.
18. J. Borwein, R. Crandall, and G. Fee, On the Ramanujan AGM fraction. Part I: The real-parameter case, *Experiment. Math.* **13** (2004) 275–285.
19. J. M. Borwein and P. B. Borwein, *Pi and the AGM*, Canadian Mathematical Society Series of Monographs and Advanced Texts, no. 4, John Wiley & Sons, New York, 1998.
20. J. M. Borwein, D. M. Bradley, D. J. Broadhurst, and P. Lisoněk, Special values of multiple polylogarithms, *Trans. Amer. Math. Soc.* **353** (2001) 907–941.
21. P. Borwein, An efficient algorithm for the Riemann zeta function (1995), available at http://www.cecm.sfu.ca/personal/pborwein/PAPERS/P155.pdf.
22. R. E. Crandall, *Topics in Advanced Scientific Computation*, Springer-Verlag, New York, 1996.
23. ———, New representations for the Madelung constant, *Experiment. Math.* **8** (1999) 367–379.
24. R. E. Crandall and J. P. Buhler, Elementary function expansions for Madelung constants, *J. Phys. A* **20** (1987) 5497–5510.
25. H. R. P. Ferguson, D. H. Bailey, and S. Arno, Analysis of PSLQ, an integer relation finding algorithm, *Mathematics of Computation* **68** (1999) 351–369.
26. J. Gleick, *Chaos: The Making of a New Science*, Penguin Books, New York, 1987.
27. J. Havel, *Gamma: Exploring Euler's Constant*, Princeton University Press, Princeton, 2003.

28. G. S. Joyce and I. J. Zucker, Evaluation of the Watson integral and associated logarithmic integral for the d-dimensional hypercubic lattice, *J. Phys. A* **34** (2001) 7349–7354.
29. ———, On the evaluation of generalized Watson integrals, *Proc. Amer. Math. Soc.* **133** (2005) 71–81.
30. I. Kotsireas and K. Karamanos, Exact computation of the bifurcation point B_4 of the logistic map and the Bailey-Broadhurst conjectures, *Intl. J. Bifurcation and Chaos* **14** (2004) 2417–2423.
31. A. A. Maradudin, E. W. Montroll, G. H. Weiss, R. Herman, and H. W. Milnes, *Green's Functions for Monatomic Simple Cubic Lattices*, Academie Royale de Belgique, Memoires XIV **7** (1960).
32. D. Shanks and J. W. Wrench Jr., Khintchine's constant, this MONTHLY **66** (1959) 276–279.
33. S. H. Strogatz, *Nonlinear Dynamics and Chaos: With Applications to Physics, Biology, Chemistry and Engineering*, Perseus Book Group, Philadelphia, 2001.
34. A. Vaught, personal communication (2004).
35. S. Wolfram, *A New Kind of Science*, Wolfram Media, Champaign, IL, 2002.

DAVID H. BAILEY received his B.S. at Brigham Young University and received his Ph.D. (1976) from Stanford University. He worked for the Department of Defense and for SRI International, before spending fourteen years at NASA's Ames Research Center in California. Since 1997 he has been the chief technologist of the Computational Research Department at the Lawrence Berkeley National Laboratory. In 1993, he was a corecipient of the Chauvenet Prize from the MAA and in that year was also awarded the Sidney Fernbach Award from the IEEE Computer Society. His research spans computational mathematics and high-performance computing. He is the author (with Jonathan Borwein and, for volume two, Roland Girgensohn) of two recent books on experimental mathematics.
Lawrence Berkeley National Laboratory, Berkeley, CA 94720
dhbailey@lbl.gov

JONATHAN M. BORWEIN received his DPhil from Oxford (1974) as a Rhodes Scholar. He taught at Dalhousie, Carnegie-Mellon, and Waterloo, before becoming the Shrum Professor of Science and a Canada Research Chair in Information Technology at Simon Fraser University. At SFU he was founding director of the Centre for Experimental and Constructive Mathematics. In 2004, he rejoined Dalhousie University in the faculty of computer science. Jonathan has received several awards, including the 1993 Chauvenet Prize of the MAA, Fellowship in the Royal Society of Canada, and Fellowship in the American Association for the Advancement of Science. His research spans computational number theory and optimization theory, as well as numerous topics in computer science. He is the author of ten books and is cofounder of Math Resources, Inc., an educational software firm.
Faculty of Computer Science, Dalhousie University, Halifax, NS, B3H 2W5
jmborwein@cs.dal.ca

VISHAAL KAPOOR graduated from Simon Fraser University with a B.S. degree in mathematics and is currently completing his M.S. at the University of British Columbia under the supervision of Greg Martin.
Department of Mathematics, University of BC, Vancouver, BC, V6T 1Z2
vkapoor@math.ubc.ca

ERIC W. WEISSTEIN graduated from Cornell University, with a B.A. degree in physics, and from the California Institute of Technology (M.S., 1993; Ph.D., 1996) with degrees in planetary astronomy. Upon completion of his doctoral thesis, Weisstein became a research scientist in the Department of Astronomy at the University of Virginia in Charlottesville. Since 1999 he has been a member of the Scientific Information group at Wolfram Research, where he holds the official title "Encyclopedist." Eric is best known as the author of the *MathWorld* website (http://mathword.wolfram.com), an online compendium of mathematical knowledge, as well as the author of the *CRC Concise Encyclopedia of Mathematics*.
Wolfram Research Inc., Champaign, IL 61820
eww@wolfram.com

12. Implications of experimental mathematics for the philosophy of mathematics

Discussion

One of most damaging myths in mathematics is the notion that as Phillip Davis puts it: "Well, in principle, you could understand all the talks." A humanist philosophy of mathematics, as advocated by Davis and Reuben Hersh among others in, for example, *The Mathematical Experience* has the most to offer in addressing such misconceptions. Davis writes:

> Once the opening ceremonies were over, the real meat of the Congress was then served up in the form of about 1400 individual talks and posters. I estimated that with luck I might be able to comprehend 2% of them. For two successive weeks in the halls of a single University, ICM'98 perpetuated the myth of the unity of mathematics; which myth is supposedly validated by the repetition of that most weaselly of rhetorical phrases: "Well, in principle, you could understand all the talks."[1]

Source

J.M. Borwein, "Implications of Experimental Mathematics for the Philosophy of Mathematics," Chapter 2, pp. 33–61, and Cover Image in *Proof and Other Dilemmas: Mathematics and Philosophy*, Bonnie Gold and Roger Simons Eds, MAA Spectrum Series. 2008.

---

[1] Describing the 1998 Berlin International Congress of Mathematicians in the October 1998 *SIAM News*.

Implications of Experimental Mathematics for the Philosophy of Mathematics[1]

Jonathan Borwein, FRSC[2]

Christopher Koch [35] accurately captures a great scientific distaste for philosophizing:

> *"Whether we scientists are inspired, bored, or infuriated by philosophy, all our theorizing and experimentation depends on particular philosophical background assumptions. This hidden influence is an acute embarrassment to many researchers, and it is therefore not often acknowledged."* (Christopher Koch, 2004)

That acknowledged, I am of the opinion that mathematical philosophy matters more now than it has in nearly a century. The power of modern computers matched with that of modern mathematical software and the sophistication of current mathematics is changing the way we do mathematics.

In my view it is now both necessary and possible to admit quasi-empirical inductive methods fully into mathematical argument. In doing so carefully we will enrich mathematics and yet preserve the mathematical literature's deserved reputation for reliability—even as the methods and criteria change. What do I mean by reliability? Well, research mathematicians still consult Euler or Riemann to be informed, anatomists only consult Harvey[3] for historical reasons. Mathematicians happily quote old papers as core steps of arguments, physical scientists expect to have to confirm results with another experiment.

1 Mathematical Knowledge as I View It

Somewhat unusually, I can exactly place the day at registration that I became a mathematician and I recall the reason why. I was about to deposit my punch cards in the 'honours history bin'. I remember thinking

> *"If I do study history, in ten years I shall have forgotten how to use the calculus properly. If I take mathematics, I shall still be able to read competently about the War of 1812 or the Papal schism."* (Jonathan Borwein, 1968)

The inescapable reality of objective mathematical knowledge is still with me. Nonetheless, my view then of the edifice I was entering is not that close to my view of the one I inhabit forty years later.

---

[1] The companion web site is at **www.experimentalmath.info**

[2] Canada Research Chair, Faculty of Computer Science, 6050 University Ave, Dalhousie University, Nova Scotia, B3H 1W5 Canada. E-mail: `jborwein@cs.dal.ca`

[3] William Harvey published the first accurate description of circulation, "An Anatomical Study of the Motion of the Heart and of the Blood in Animals," in 1628.

I also know when I became a computer-assisted fallibilist. Reading Imre Lakatos' *Proofs and Refutations*, [38], a few years later while a very new faculty member, I was suddenly absolved from the grave sin of error, as I began to understand that missteps, mistakes and errors are the grist of all creative work.[4] The book, his doctorate posthumously published in 1976, is a student conversation about the Euler characteristic. The students are of various philosophical stripes and the discourse benefits from his early work on Hegel with the Stalinist Lukács in Hungary and from later study with Karl Popper at the London School of Economics. I had been prepared for this dispensation by the opportunity to learn a variety of subjects from Michael Dummett. Dummett was at that time completing his study rehabilitating Frege's status, [23].

A decade later the appearance of the first 'portable' computers happily coincided with my desire to decode Srinivasa Ramanujan's (1887–1920) cryptic assertions about theta functions and elliptic integrals, [13]. I realized that by coding his formulae and my own in the *APL* programming language[5], I was able to rapidly confirm and refute identities and conjectures and to travel much more rapidly and fearlessly down potential blind alleys. I had become a computer-assisted fallibilist; at first somewhat falteringly but twenty years have certainly honed my abilities.

Today, while I appreciate fine proofs and aim to produce them when possible, I no longer view proof as the royal road to secure mathematical knowledge.

2 Introduction

I first discuss my views, and those of others, on the nature of mathematics, and then illustrate these views in a variety of mathematical contexts. A considerably more detailed treatment of many of these topics is to be found in my book with Dave Bailey entitled *Mathematics by Experiment: Plausible Reasoning in the 21st Century*—especially in Chapters One, Two and Seven, [9]. Additionally, [2] contains several pertinent case studies as well as a version of this current chapter.

Kurt Gödel may well have overturned the mathematical apple cart entirely deductively, but nonetheless he could hold quite different ideas about legitimate forms of mathematical reasoning, [28]:

> *"If mathematics describes an objective world just like physics, there is no reason why inductive methods should not be applied in mathematics just the same as in physics."* (Kurt Gödel[6], 1951)

---

[4]Gila Hanna [30] takes a more critical view placing more emphasis on the role of proof and certainty in mathematics; I do not disagree, so much as I place more value on the role of computer-assisted refutation. Also 'certainty' usually arrives late in the development of a proof.

[5]Known as a 'write only' very high level language, APL was a fine tool; albeit with a steep learning curve whose code is almost impossible to read later.

[6]Taken from a previously unpublished work, [28].

While we mathematicians have often separated ourselves from the sciences, they have tended to be more ecumenical. For example, a recent review of *Models. The Third Dimension of Science*, [17], chose a mathematical plaster model of a Clebsch diagonal surface as its only illustration. Similarly, authors seeking examples of the aesthetic in science often choose iconic mathematics formulae such as $E = MC^2$.

Let me begin by fixing a few concepts before starting work in earnest. Above all, I hope to persuade you of the power of mathematical experimentation—it is also fun—and that the traditional accounting of mathematical learning and research is largely an ahistorical caricature. I recall three terms.

mathematics, n. *a group of related subjects, including algebra, geometry, trigonometry and calculus, concerned with the study of number, quantity, shape, and space, and their inter-relationships, applications, generalizations and abstractions.*

This definition–taken from my Collins Dictionary [6]—makes no immediate mention of proof, nor of the means of reasoning to be allowed. The Webster's Dictionary [54] contrasts:

induction, n. *any form of reasoning in which the conclusion, though supported by the premises, does not follow from them necessarily.*; and

deduction, n. *a process of reasoning in which a conclusion follows necessarily from the premises presented, so that the conclusion cannot be false if the premises are true.*
b. a conclusion reached by this process.

Like Gödel, I suggest that both should be entertained in mathematics. This is certainly compatible with the general view of mathematicians that in some sense "mathematical stuff is out there" to be discovered. In this paper, I shall talk broadly about experimental and heuristic mathematics, giving accessible, primarily visual and symbolic, examples.

3 Philosophy of Experimental Mathematics

"The computer has in turn changed the very nature of mathematical experience, suggesting for the first time that mathematics, like physics, may yet become an empirical discipline, a place where things are discovered because they are seen."
(David Berlinski, [4])

The shift from *typographic* to *digital culture* is vexing for mathematicians. For example, there is still no truly satisfactory way of displaying mathematics on the web–and certainly not of asking mathematical questions. Also, we respect *authority*, [29], but value *authorship* deeply—however much the two values are in conflict, [16]. For example, the more I recast someone else's ideas in my own words, the more I enhance my authorship while undermining the original authority of the notions. Medieval scribes had the opposite concern and so took care to attribute their ideas to such as Aristotle or Plato.

And we care more about the *reliability* of our literature than does any other science, Indeed I would argue that we have over-subscribed to this notion and often pay lip-service not real attention to our older literature. How often does one see original sources sprinkled like holy water in papers that make no real use of them–the references offering a false sense of scholarship?

The traditional central role of proof in mathematics is arguably and perhaps appropriately under siege. Via examples, I intend to pose and answer various questions. I shall conclude with a variety of quotations from our progenitors and even contemporaries:

My Questions. What constitutes secure mathematical knowledge? When is computation convincing? Are humans less fallible? What tools are available? What methodologies? What of the 'law of the small numbers'? Who cares for certainty? What is the role of proof? How is mathematics actually done? How should it be? I mean these questions both about the apprehension (discovery) and the establishment (proving) of mathematics. This is presumably more controversial in the formal proof phase.

My Answers. To misquote D'Arcy Thompson (1860–1948) 'form follows function', [52]: rigour (proof) follows reason (discovery); indeed, excessive focus on rigour has driven us away from our wellsprings. Many good ideas are wrong. Not all truths are provable, and not all provable truths are worth proving. Gödel's incompleteness results certainly showed us the first two of these assertions while the third is the bane of editors who are frequently presented with correct but unexceptional and unmotivated generalizations of results in the literature. Moreover, near certainty is often as good as it gets—intellectual context (community) matters. Recent complex human proofs are often very long, extraordinarily subtle and fraught with error—consider, Fermat's last theorem, the Poincaré conjecture, the classification of finite simple groups, presumably any proof of the Riemann hypothesis, [25]. So while we mathematicians publicly talk of certainty we really settle for security.

In all these settings, modern computational tools dramatically change the nature and scale of available evidence. Given an interesting identity buried in a long and complicated paper on an unfamiliar subject, which would give you more confidence in its correctness: staring at the proof, or confirming computationally that it is correct to 10,000 decimal places?

Here is such a formula, [3, p. 20]:

$$\frac{24}{7\sqrt{7}} \int_{\pi/3}^{\pi/2} \log \left| \frac{\tan t + \sqrt{7}}{\tan t - \sqrt{7}} \right| dt \stackrel{?}{=} L_{-7}(2) = \sum_{n=0}^{\infty} \left[\frac{1}{(7n+1)^2} + \frac{1}{(7n+2)^2} - \frac{1}{(7n+3)^2} + \frac{1}{(7n+4)^2} - \frac{1}{(7n+5)^2} - \frac{1}{(7n+6)^2} \right]. \tag{1}$$

This identity links a volume (the integral) to an arithmetic quantity (the sum). It arose out of some studies in quantum field theory, in analysis of the volumes of ideal tetrahedra in hyperbolic space. The question mark is used because, while no hint of a path to a formal proof is yet known, it has been verified numerically to 20,000 digit precision–using 45 minutes on 1024 processors at Virginia Tech.

A more inductive approach can have significant benefits. For example, as there is still some doubt about the proof of the classification of finite simple groups it is important to ask whether the result is true but the proof flawed, or rather if there is still perhaps an 'ogre' sporadic group even larger than the 'monster'? What heuristic, probabilistic or computational tools can increase our confidence that the ogre does or does not exist? Likewise, there are experts who still believe the *Riemann hypothesis*[7] (RH) may be false and that the billions of zeroes found so far are much too small to be representative.[8] In any event, our understanding of the complexity of various crypto-systems relies on (RH) and we should like secure knowledge that any counter-example is enormous.

Peter Medawar (1915–87)—a Nobel prize winning oncologist and a great expositor of science—writing in *Advice to a Young Scientist*, [44], identifies four forms of scientific experiment:

1. The Kantian experiment: generating "the classical non-Euclidean geometries (hyperbolic, elliptic) by replacing Euclid's axiom of parallels (or something equivalent to it) with alternative forms." All mathematicians perform such experiments while the majority of computer explorations are of the following Baconian form.

2. The Baconian experiment is a contrived as opposed to a natural happening, it "is the consequence of 'trying things out' or even of merely messing about." Baconian experiments are the explorations of a happy if disorganized beachcomber and carry little predictive power.

3. Aristotelian demonstrations: "apply electrodes to a frog's sciatic nerve, and lo, the leg kicks; always precede the presentation of the dog's dinner with the ringing of a bell, and lo, the bell alone will soon make the dog dribble." Arguably our 'Corollaries' and 'Examples' are Aristotelian, they reinforce but do not predict. Medawar then says the most important form of experiment is:

4. The Galilean experiment is "a critical experiment – one that discriminates between possibilities and, in doing so, either gives us confidence in the view we are taking or makes us think it in need of correction." The Galilean the only form of experiment which stands to make Experimental Mathematics a serious enterprise. Performing careful, replicable Galilean experiments requires work and care.

Reuben Hersh's arguments for a humanist philosophy of mathematics, especially [31, pp. 590–591] and [32, p. 22], as paraphrased below, become even more convincing in our highly computational setting.

1. Mathematics is human. *It is part of and fits into human culture. It does not match Frege's concept of an abstract, timeless, tenseless, objective reality.*[9]

2. Mathematical knowledge is fallible. *As in science, mathematics can advance by making mistakes and then correcting or even re-correcting them. The "fallibilism" of mathematics is brilliantly argued in Lakatos'* Proofs and Refutations.

3. There are different versions of proof or rigor. *Standards of rigor can vary depending on time, place, and other things. The use of computers in formal proofs, exemplified by*

---

[7] All non-trivial zeroes—not negative even integers—of the zeta function lie on the line with real part 1/2.
[8] See [45] and various of Andrew Odlyzko's unpublished but widely circulated works.
[9] That Frege's view of mathematics is wrong, for Hersh as for me, does not diminish its historical importance.

the computer-assisted proof of the four color theorem in 1977,[10] *is just one example of an emerging nontraditional standard of rigor.*

4. Empirical evidence, numerical experimentation and probabilistic proof all can help us decide what to believe in mathematics. *Aristotelian logic isn't necessarily always the best way of deciding.*

5. Mathematical objects are a special variety of a social-cultural-historical object. *Contrary to the assertions of certain post-modern detractors, mathematics cannot be dismissed as merely a new form of literature or religion. Nevertheless, many mathematical objects can be seen as shared ideas, like Moby Dick in literature, or the Immaculate Conception in religion.*

I entirely subscribe to points 2., 3., 4., and with certain caveats about objective knowledge[11] to points 1. and 5. In any event mathematics is and will remain a uniquely human undertaking.

This version of humanism sits *fairly* comfortably along-side current versions of **social-constructivism** as described next.

> *"The social constructivist thesis is that mathematics is a social construction, a cultural product, fallible like any other branch of knowledge."* (Paul Ernest, [26, §3])

But only if I qualify this with *"Yes, but much-much less fallible than most branches of knowledge."* Associated most notably with the writings of Paul Ernest—an English Mathematician and Professor in the Philosophy of Mathematics Education who in [27] traces the intellectual pedigree for his thesis, a pedigree that encompasses the writings of Wittgenstein, Lakatos, Davis, and Hersh among others—social constructivism seeks to define mathematical knowledge and epistemology through the social structure and interactions of the mathematical community and society as a whole.

This interaction often takes place over very long periods. Many of the ideas our students—and some colleagues—take for granted took a great deal of time to gel. The Greeks suspected the impossibility of the three *classical construction problems* [12] and the irrationality of the golden mean was well known to the Pythagoreans.

While concerns about potential and completed infinities are very old, until the advent of the calculus with Newton and Leibniz and the need to handle fluxions or infinitesimals, the level of need for rigour remained modest. Certainly Euclid is in its geometric domain generally a model of rigour, while also Archimedes' numerical analysis was not equalled until the 19th century.

---

[10] Especially, since a new implementation by Seymour, Robertson and Thomas in 1997 which has produced a simpler, clearer and less troubling implementation.

[11] While it is not Hersh's intention, a superficial reading of point 5. hints at a cultural relativism to which I certainly do not subscribe.

[12] Trisection, circle squaring and cube doubling were taken by the educated to be impossible in antiquity. Already in 414 BCE, in his play *The Birds*, Aristophanes uses 'circle-squarers' as a term for those who attempt the impossible. Similarly, the French Academy stopped accepting claimed proofs a full two centuries before the 19th century achieved proofs of their impossibility.

The need for rigour arrived in full force in the time of Cauchy and Fourier. The treacherous countably infinite processes of analysis and the limitations of formal manipulation came to the fore. It is difficult with a modern sensibility to understand how Cauchy's proof of the continuity of pointwise-limits could coexist in texts for a generation with clear counter-examples originating in Fourier's theory of heat.[13]

By the end of the 19th century Frege's (1848-1925) attempt to base mathematics in a linguistically based *logicism* had foundered on Russell and other's discoveries of the paradoxes of naive set theory. Within thirty five years Gödel—and then Turing's more algorithmic treatment[14]—had similarly damaged both Russell and Whitehead's and Hilbert's programs.

Throughout the twentieth century, bolstered by the armor of abstraction, the great ship Mathematics has sailed on largely unperturbed. During the last decade of the 19th and first few decades of the 20th century the following main streams of philosophy emerged explicitly within mathematics to replace logicism, but primarily as the domain of philosophers and logicians.

- *Platonism.* Everyman's idealist philosophy—stuff exists and we must find it. Despite being the oldest mathematical philosophy, Platonism—still predominant among working mathematicians—was only christened in 1934 by Paul Bernays.[15]

- *Formalism.* Associated mostly with Hilbert—it asserts that mathematics is invented and is best viewed as formal symbolic games without intrinsic meaning.

- *Intuitionism.* Invented by Brouwer and championed by Heyting, intuitionism asks for inarguable monadic components that can be fully analyzed and has many variants; this has interesting overlaps with recent work in cognitive psychology such as Lakoff and Nunez' work, [39], on 'embodied cognition'.[16]

- *Constructivism.* Originating with Markoff and especially Kronecker (1823–1891), and refined by Bishop it finds fault with significant parts of classical mathematics. Its 'I'm from Missouri, tell me how big it is' sensibility is not to be confused with Paul Ernest's 'social constructivism', [27].

The last two philosophies deny the principle of the *excluded middle*, "A or not A", and resonate with computer science—as does some of formalism. It is hard after all to run a deterministic program which does not know which disjunctive logic-gate to follow.

---

[13] Cauchy's proof appeared in his 1821 text on analysis. While counterexamples were pointed out almost immediately, Stokes and Seidel were still refining the missing uniformity conditions in the late 1840s.

[14] The modern treatment of incompleteness leans heavily on Turing's analysis of the *Halting problem* for so-called Turing machines.

[15] See Karlis Podnieks, "Platonism, Intuition and the Nature on Mathematics", available at http://www.ltn.lv/ podnieks/gt1.html

[16] The cognate views of Henri Poincaré (1854–1912), [47, p. 23] on the role of the *subliminal* are reflected in *"The mathematical facts that are worthy of study are those that, by their analogy with other facts are susceptible of leading us to knowledge of a mathematical law, in the same way that physical facts lead us to a physical law."* He also wrote *"It is by logic we prove, it is by intuition that we invent,"* [48].

By contrast the battle between a Platonic idealism (a 'deductive absolutism') and various forms of 'fallibilism'(a quasi-empirical 'relativism') plays out across all four, but fallibilism perhaps lives most easily within a restrained version of intuitionism which looks for 'intuitive arguments' and is willing to accept that 'a proof is what convinces'. As Lakatos shows, an argument that was convincing a hundred years ago may well now be viewed as inadequate. And one today trusted may be challenged in the next century.

As we illustrate in the next section or two, it is only perhaps in the last twenty five years, with the emergence of powerful mathematical platforms, that any approach other than a largely undigested Platonism and a reliance on proof and abstraction has had the tools[17] to give it traction with working mathematicians.

In this light, Hales' proof of Kepler's conjecture that *the densest way to stack spheres is in a pyramid* resolves the oldest problem in discrete geometry. It also supplies the most interesting recent example of intensively computer-assisted proof, and after five years with the review process was published in the *Annals of Mathematics*—with an "only 99% checked" disclaimer.

This process has triggered very varied reactions [34] and has provoked Thomas Hales to attempt a formal computational proof which he expects to complete by 2011, [25]. Famous earlier examples of fundamentally computer-assisted proof include the *Four color theorem* and proof of the *Non-existence of a projective plane of order 10*. The three raise and answer quite distinct questions about computer-assisted proof—both real and specious. For example, there were real concerns about the completeness of the search in the 1976 proof of the Four color theorem but there should be none about the 1997 reworking by Seymour, Robertson and Thomas.[18] Correspondingly, Lam deservedly won the 1992 *Lester R. Ford award* for his compelling explanation of why to trust his computer when it announced there was no plane of order ten, [40]. Finally, while it is reasonable to be concerned about the certainty of Hales' conclusion, was it really the *Annal's* purpose to suggest all other articles have been more than 99% certified?

To make the case as to how far mathematical computation has come we trace the changes over the past half century. The 1949 computation of π to 2,037 places suggested by von Neumann, took 70 hours. A billion digits may now be computed in much less time on a laptop. Strikingly, it would have taken roughly 100,000 ENIAC's to store the Smithsonian's picture—as is possible thanks to *40 years of Moore's law* in action ....[19]
This is an astounding record of sustained exponential progress without peer in the history of technology. Additionally, mathematical tools are now being implemented on parallel platforms, providing *much* greater power to the research mathematician. Amassing huge amounts of processing power will not alone solve many mathematical problems. There are very few mathematical 'Grand-challenge problems', [12] where, as in the physical sciences, a few more orders of computational power will resolve a problem.

---

[17]That is, to broadly implement Hersh's central points (2.-4.).

[18]See http://www.math.gatech.edu/ thomas/FC/fourcolor.html.

[19]**Moore's Law** is now taken to be the assertion that *semiconductor technology approximately doubles in capacity and performance roughly every 18 to 24 months.*

For example, an order of magnitude improvement in computational power currently translates into one more day of accurate weather forecasting, while it is now common for biomedical researchers to design experiments today whose outcome is predicated on 'peta-scale' computation being available by say 2010, [51]. There is, however, much more value in *very rapid 'Aha's'* as can be obtained through "micro-parallelism"; that is, where we benefit by being able to compute many simultaneous answers on a neurologically-rapid scale and so can hold many parts of a problem in our mind at one time.

To sum up, in light of the discussion and terms above, I now describe myself a a social-constructivist, and as a computer-assisted fallibilist with constructivist leanings. I believe that more-and-more the interesting parts of mathematics will be less-and-less susceptible to classical deductive analysis and that Hersh's 'non-traditional standard of rigor' must come to the fore.

4 Our Experimental Mathodology

Despite Picasso's complaint that "computers are useless, they only give answers," the main goal of computation in pure mathematics is arguably to yield *insight*. This demands speed or, equivalently, substantial *micro-parallelism* to provide answers on a cognitively relevant scale; so that we may ask and answer more questions while they remain in our consciousness. This is relevant for rapid verification; for validation; for *proofs* and *especially for refutations* which includes what Lakatos calls "monster barring", [38]. Most of this goes on in the daily small-scale accretive level of mathematical discovery but insight is gained even in cases like the proof of the Four color theorem or the Non-existence of a plane of order ten. Such insight is not found in the case-enumeration of the proof, but rather in the algorithmic reasons for believing that one has at hand a tractable unavoidable set of configurations or another effective algorithmic strategy. For instance, Lam [40] ran his algorithms on known cases in various subtle ways, and also explained why built-in redundancy made the probability of machine-generated error negligible. More generally, the act of programming—if well performed—always leads to more insight about the structure of the problem.

In this setting it is enough to equate *parallelism* with access to requisite *more* space and speed of computation. Also, we should be willing to consider all computations as 'exact' which provide truly reliable answers.[20] This now usually requires a careful *hybrid* of symbolic and numeric methods, such as achieved by *Maple*'s liaison with the *Numerical Algorithms Group* (NAG) Library[21], see [5, 8]. There are now excellent tools for such purposes throughout analysis, algebra, geometry and topology, see [9, 10, 5, 12, 15].

Along the way questions required by—or just made natural by—computing start to force out older questions and possibilities in the way beautifully described a century ago by Dewey regarding evolution.

---

[20]If careful interval analysis can certify that a number known to be integer is larger that 2.5 and less than 3.5, this constitutes an exact computational proof that it is 3.

[21]See http://www.nag.co.uk/.

> "Old ideas give way slowly; for they are more than abstract logical forms and categories. They are habits, predispositions, deeply engrained attitudes of aversion and preference. Moreover, the conviction persists—though history shows it to be a hallucination—that all the questions that the human mind has asked are questions that can be answered in terms of the alternatives that the questions themselves present. But in fact intellectual progress usually occurs through sheer abandonment of questions together with both of the alternatives they assume; an abandonment that results from their decreasing vitality and a change of urgent interest. We do not solve them: we get over them. Old questions are solved by disappearing, evaporating, while new questions corresponding to the changed attitude of endeavor and preference take their place. Doubtless the greatest dissolvent in contemporary thought of old questions, the greatest precipitant of new methods, new intentions, new problems, is the one effected by the scientific revolution that found its climax in the 'Origin of Species.' " (John Dewey, [20])

Lest one think this a feature of the humanities and the human sciences, consider the artisanal chemical processes that have been lost as they were replaced by cheaper industrial versions. And mathematics is far from immune. Felix Klein, quoted at length in the introduction to [11], laments that "now the younger generation hardly knows abelian functions." He goes on to explain that:

> "In mathematics as in the other sciences, the same processes can be observed again and again. First, new questions arise, for internal or external reasons, and draw researchers away from the old questions. And the old questions, just because they have been worked on so much, need ever more comprehensive study for their mastery. This is unpleasant, and so one is glad to turn to problems that have been less developed and therefore require less foreknowledge—even if it is only a matter of axiomatics, or set theory, or some such thing." (Felix Klein, [33, p. 294])

Freeman Dyson has likewise gracefully described how taste changes:

> "I see some parallels between the shifts of fashion in mathematics and in music. In music, the popular new styles of jazz and rock became fashionable a little earlier than the new mathematical styles of chaos and complexity theory. Jazz and rock were long despised by classical musicians, but have emerged as art-forms more accessible than classical music to a wide section of the public. Jazz and rock are no longer to be despised as passing fads. Neither are chaos and complexity theory. But still, classical music and classical mathematics are not dead. Mozart lives, and so does Euler. When the wheel of fashion turns once more, quantum mechanics and hard analysis will once again be in style." (Freeman Dyson, [24])

For example recursively defined objects were once anathema—Ramanujan worked very hard to replace lovely iterations by sometimes-obscure closed-form approximations. Additionally, what is "easy" changes: high performance computing and networking are blurring, merging disciplines and collaborators. This is democratizing mathematics but further challenging authentication—consider how easy it is to find information on *Wikipedia*[22] and how hard it is to validate it.

Moving towards a well articulated Experimental *Mathodology*—both in theory and practice—will take much effort. The need is premised on the assertions that intuition is acquired—we can and must better mesh computation and mathematics, and that visualization is of growing importance—in many settings even three is a lot of dimensions.

"Monster-barring" (Lakatos's term, [38], for refining hypotheses to rule out nasty counter-examples[23]) and "caging" (Nathalie Sinclair tells me this is my own term for imposing needed restrictions in a conjecture) are often easy to enhance computationally, as for example with randomized checks of equations, linear algebra, and primality or graphic checks of equalities, inequalities, areas, etc. Moreover, our mathodology fits well with the kind of pedagogy espoused at a more elementary level (and without the computer) by John Mason in [43].

4.1 Eight Roles for Computation

I next recapitulate eight roles for computation that Bailey and I discuss in our two recent books [9, 10]:

#1. **Gaining insight and intuition or just knowledge.** Working algorithmically with mathematical objects almost inevitably adds insight to the processes one is studying. At some point even just the careful aggregation of data leads to better understanding.

#2. **Discovering new facts, patterns and relationships.** The number of *additive partitions* of a positive integer n, $p(n)$, is *generated* by

$$P(q) := 1 + \sum_{n \geq 1} p(n) q^n = \frac{1}{\prod_{n=1}^{\infty}(1 - q^n)}. \tag{2}$$

Thus, $p(5) = 7$ since

$$5 = 4 + 1 = 3 + 2 = 3 + 1 + 1 = 2 + 2 + 1 = 2 + 1 + 1 + 1 = 1 + 1 + 1 + 1 + 1.$$

Developing (2) is a fine introduction to enumeration via *generating functions*. Additive partitions are harder to handle than multiplicative factorizations, but they are very interesting, [10, Chapter 4]. Ramanujan used Major MacMahon's table of $p(n)$ to intuit remarkable deep congruences such as

$$p(\mathbf{5n+4}) \equiv 0 \mod \mathbf{5}, \quad p(\mathbf{7n+5}) \equiv 0 \mod \mathbf{7}, \quad p(\mathbf{11n+6}) \equiv 0 \mod \mathbf{11},$$

---

[22] *Wikipedia* is an open source project at http://en.wikipedia.org/wiki/Main_Page; "wiki-wiki" is Hawaiian for "quickly".

[23] Is, for example, a polyhedron always convex? Is a curve intended to be simple? Is a topology assumed Hausdorff, a group commutative?

from relatively limited data like

$$\begin{aligned}P(q) &= 1+q+2\,q^2+3\,q^3+\underline{5}\,q^4+\overline{7}\,q^{\mathbf{5}}+11\,q^6+15\,q^7\\ &+\ 22\,q^8+\underline{30}\,q^9+42\,q^{10}+56\,q^{11}+\overline{77}\,q^{\mathbf{12}}+101\,q^{13}+\underline{135}\,q^{14}\\ &+\ 176\,q^{15}+231\,q^{16}+297\,q^{17}+385\,q^{18}+\overline{490}\,q^{19}\\ &+\ 627\,q^{20}b+792\,q^{21}+1002\,q^{22}+\cdots+p(200)q^{200}+\cdots\end{aligned} \quad (3)$$

Cases $5n+4$ and $7n+5$ are flagged in (3). Of course, it is markedly easier to (heuristically) confirm than find these fine examples of *Mathematics: the science of patterns*.[24] The study of such congruences—much assisted by symbolic computation—is very active today.

#3. **Graphing to expose mathematical facts, structures or principles.** Consider Nick Trefethen's fourth challenge problem as described in [5, 8]. It requires one to find ten good digits of:

4. What is the global minimum of the function

$$\exp(\sin(50x))+\sin(60e^y)+\sin(70\sin x)+\sin(\sin(80y))-\sin(10(x+y))+(x^2+y^2)/4?$$

As a foretaste of future graphic tools, one can solve this problem graphically and interactively using current *adaptive 3-D plotting* routines which can catch all the bumps. This does admittedly rely on trusting a good deal of software.

#4. **Rigourously testing and especially falsifying conjectures.** I hew to the Popperian scientific view that we primarily falsify; but that as we perform more and more testing experiments without such falsification we draw closer to firm belief in the truth of a conjecture such as: *the polynomial $P(n) = n^2 - n + p$ has prime values for all $n = 0, 1, \ldots, p-2$, exactly for* Euler's lucky prime numbers, *that is, p= 2, 3, 5, 11, 17, and 41.*[25]

#5. **Exploring a possible result to see if it *merits* formal proof.** A conventional deductive approach to a hard multi-step problem really requires establishing all the subordinate lemmas and propositions needed along the way—especially if they are highly technical and un-intuitive. Now some may be independently interesting or useful, but many are only worth proving if the entire expedition pans out. Computational experimental mathematics provides tools to survey the landscape with little risk of error: only if the view from the summit is worthwhile, does one lay out the route carefully. I discuss this further at the end of the next Section.

#6. **Suggesting approaches for formal proof.** The proof of the *cubic theta function identity* discussed on [10, pp. 210] shows how a fully intelligible human proof can be obtained entirely by careful symbolic computation.

---

[24] The title of Keith Devlin's 1996 book, [21].
[25] See [55] for the answer.

#7. Computing replacing lengthy hand derivations. Who would wish to verify the following prime factorization by hand?

$$6422607578676942838792549775208734746307$$
$$= (2140992015395526641)(1963506722254397)(1527791).$$

Surely, what we value is understanding the underlying algorithm, not the human work?

#8. Confirming analytically derived results. This is a wonderful and frequently accessible way of confirming results. Even if the result itself is not computationally checkable, there is often an accessible corollary. An assertion about bounded operators on Hilbert space may have a useful consequence for three-by-three matrices. It is also an excellent way to error correct, or to check calculus examples before giving a class.

5 Finding Things versus Proving Things

I now illuminate these eight roles with eight mathematical examples. At the end of each I note some of the roles illustrated.

1. **Pictorial comparison** of $y - y^2$ and $y^2 - y^4$ to $-y^2 \ln(y)$, when y lies in the unit interval, is a much more rapid way to divine which function is larger than by using traditional analytic methods.

   Figure 1 below shows that it is clear in the latter case the functions cross, and so it is futile to try to prove one majorizes the other. In the first case, evidence is provided to motivate attempting a proof and often the picture serves to guide such a proof—by showing monotonicity or convexity or some other salient property. ∎

   This certainly illustrates roles #3 and #4, and perhaps role #5.

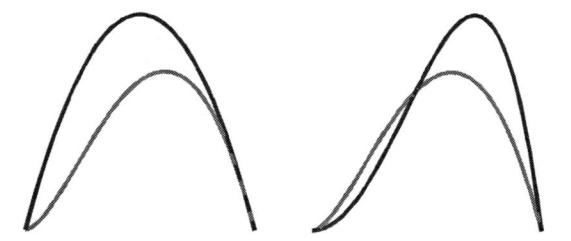

   Figure 1. (Ex. 1.): Graphical comparison of $-x^2 \ln(x)$ (lower local maximum in both graphs) with $x - x^2$ (left graph) and $x^2 - x^4$ (right graph)

2. **A proof and a disproof.** Any modern computer algebra can tell one that

$$0 < \int_0^1 \frac{(1-x)^4 x^4}{1+x^2} \, dx = \frac{22}{7} - \pi, \tag{4}$$

since the integral may be interpreted as the area under a positive curve. We are however no wiser as to why! If however we ask the same system to compute the indefinite integral, we are likely to be told that

$$\int_0^t \cdot = \frac{1}{7} t^7 - \frac{2}{3} t^6 + t^5 - \frac{4}{3} t^3 + 4\,t - 4\arctan(t).$$

Then (4) is now rigourously established by differentiation and an appeal to the Fundamental theorem of calculus. ∎

This illustrates roles #1 and #6. It also falsifies the bad conjecture that $\pi = 22/7$ and so illustrates #4 again. Finally, the computer's proof is easier (#7) and very nice, though probably it is not the one we would have developed by ourselves. The fact that 22/7 is a continued fraction approximation to π has led to many hunts for generalizations of (4), see [10, Chapter 1]. None so far are entirely successful.

3. **A computer discovery and a 'proof' of the series for** $\arcsin^2(x)$. We compute a few coefficients and observe that there is a regular power of 4 in the numerator, and integers in the denominator; or equivalently we look at $\arcsin(x/2)^2$. The generating function package 'gfun' in *Maple*, then predicts a recursion, r, for the denominators and solves it, as R.

```
>with(gfun):
>s:=[seq(1/coeff(series(arcsin(x/2)^2,x,25),x,2*n),n=1..6)]:
>R:=unapply(rsolve(op(1, listtorec(s,r(m))),r(m)),m);[seq(R(m),m=0..8)];
```

yields, $s := [4, 48, 360, 2240, 12600, 66528]$,

$$R := m \mapsto 8 \frac{4^m\,\Gamma(3/2 + m)(m+1)}{\pi^{1/2}\Gamma(1+m)},$$

where Γ is the Gamma function, and then returns the sequence of values

$$[4, 48, 360, 2240, 12600, 66528, 336336, 1647360, 7876440].$$

We may now use Sloane's *Online Encyclopedia of Integer Sequences*[26] to reveal that the coefficients are $R(n) = 2n^2 \binom{2n}{n}$. More precisely, sequence A002544 identifies $R(n+1)/4 = \binom{2n+1}{n}(n+1)^2$.

```
> [seq(2*n^2*binomial(2*n,n),n=1..8)];
```

confirms this with

$$[4, 48, 360, 2240, 12600, 66528, 336336, 1647360].$$

Next we write

---

[26] At www.research.att.com/~njas/sequences/index.html

```
> S:=Sum((2*x)^(2*n)/(2*n^2*binomial(2*n,n)),n=1..infinity):S=values(S);
```

which returns

$$\frac{1}{2}\sum_{n=1}^{\infty}\frac{(2\,x)^{2n}}{n^2\binom{2n}{n}} = \arcsin^2(x).$$

That is, we have discovered—and proven if we trust or verify *Maple*'s summation algorithm—the desired Maclaurin series.

As prefigured by Ramanujan, it transpires that there is a beautiful closed form for $\arcsin^{2m}(x)$ for all $m = 1, 2, \ldots$. In [14] there is a discussion of the use of *integer relation methods*, [9, Chapter 6], to find this closed form and associated proofs are presented. ∎

Here we see an admixture of all of the roles save #3, but above all #2 and #5.

4. **Discovery without proof.** Donald Knuth[27] asked for a closed form evaluation of:

$$\sum_{k=1}^{\infty}\left\{\frac{k^k}{k!\,e^k} - \frac{1}{\sqrt{2\pi k}}\right\} = -0.084069508727655\ldots. \qquad (5)$$

Since about 2000 CE it has been easy to compute 20—or 200—digits of this sum in *Maple* or *Mathematica*; and then to use the 'smart lookup' facility in the *Inverse Symbolic Calculator*(ISC). The ISC at http://oldweb.cecm.sfu.ca/projects/ISC uses a variety of search algorithms and heuristics to predict what a number might actually be. Similar ideas are now implemented as 'identify' in *Maple* and (for algebraic numbers only) as 'Recognize' in *Mathematica*, and are described in [8, 9, 15, 1]. In this case it *rapidly* returns

$$0.084069508727655 \approx \frac{2}{3} + \frac{\zeta(1/2)}{\sqrt{2\pi}}.$$

We thus have a prediction which *Maple* 9.5 on a 2004 laptop *confirms* to 100 places in under 6 seconds and to 500 in 40 seconds. Arguably we are done. After all we were asked to *evaluate* the series and we now know a closed-form answer.

Notice also that the 'divergent' $\zeta(1/2)$ term is formally to be expected in that while $\sum_{n=1}^{\infty} 1/n^{1/2} = \infty$, the *analytic continuation* of $\zeta(s) := \sum_{n=1}^{\infty} 1/n^s$ for $s > 1$ evaluated at $1/2$ does occur! ∎

We have discovered and tested the result and in so doing gained insight and knowledge while illustrating roles #1, #2 and #4. Moreover, as described in [10, pp. 15], one can also be led by the computer to a very satisfactory computer-assisted but also very human proof,

---

[27]Posed as an MAA Problem [36].

thus illustrating role #6. Indeed, the first hint is that the computer algebra system returned the value in (5) very quickly even though the series is very slowly convergent. This suggests the program is doing something intelligent—and it is! Such a use of computing is termed "instrumental" in that the computer is fundamental to the process, see [41].

5. **A striking conjecture with no known proof strategy** (as of spring 2007) given in [10, p. 162] is: for $n = 1, 2, 3 \cdots$

$$8^n \zeta\left(\{\overline{2}, 1\}_n\right) \stackrel{?}{=} \zeta\left(\{2, 1\}_n\right). \tag{6}$$

Explicitly, the first two cases are

$$8 \sum_{n>m>0} \frac{(-1)^n}{n^2 m} = \sum_{n>0} \frac{1}{n^3} \quad \text{and} \quad 64 \sum_{n>m>o>p>0} \frac{(-1)^{n+o}}{n^2 m\, o^2 p} = \sum_{n>m>0} \frac{1}{n^3 m^3}.$$

The notation should now be clear—we use the 'overbar' to denote an alternation. Such alternating sums are called *multi-zeta values* (MZV) and positive ones are called *Euler sums* after Euler who first studied them seriously. They arise naturally in a variety of modern fields from combinatorics to mathematical physics and knot theory.

There is abundant evidence amassed since 'identity' (6) was found in 1996. For example, very recently Petr Lisonek checked the first 85 cases to 1000 places in about 41 HP hours with only the *predicted round-off error*. And the case $n = 163$ was checked in about ten hours. These objects are very hard to compute naively and require substantial computation as a precursor to their analysis.

Formula (6) is the *only* identification of its type of an Euler sum with a distinct MZV and we have no idea why it is true. Any similar MZV proof has been both highly non-trivial and illuminating. To illustrate how far we are from proof: can just the case $n = 2$ be proven *symbolically* as has been the case for $n = 1$? ∎

Chapter 12 229

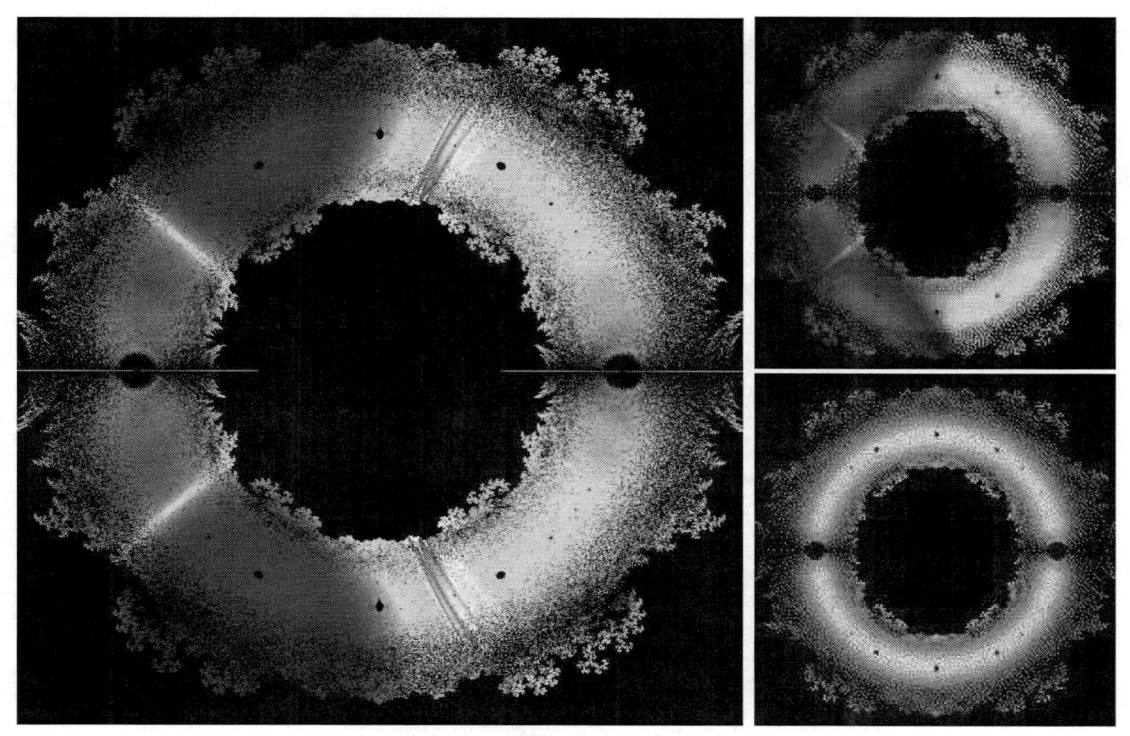

Figure 2. (Ex. 6.): "The price of metaphor is eternal vigilance."
(Arturo Rosenblueth & Norbert Wiener, [42])

This identity was discovered by the British quantum field theorist David Broadhurst and me during a large hunt for such objects in the mid-nineties. In this process we discovered and proved many lovely results (see [9, Chapter 2] and [10, Chapter 4]), thereby illustrating #1,#2, #4, #5 and #7. In the case of 'identity' (6) we have failed with #6, but we have ruled out many sterile approaches. It is one of many examples where we can now have (near) certainty without proof. Another was shown in equation (1) above.

6. **What you draw *is* what you see.** *Roots of polynomials with coefficients 1 or -1 up to degree 18.*

   As the quote suggests, pictures are highly metaphorical. The shading in Figure 2 is determined by a normalized sensitivity of the coefficients of the polynomials to slight variations around the values of the zeros with red indicating low sensitivity and violet indicating high sensitivity.[28] It is hard to see how the structure revealed in the pictures above[29] would be seen other than through graphically data-mining. Note the different shapes—now proven—of the holes around the various roots of unity.

   The striations are unexplained but all re-computations expose them! And the fractal structure is provably there. Nonetheless different ways of measuring the stability of

---

[28] Colour versions may be seen at http://oldweb.cecm.sfu.ca/personal/loki/Projects/Roots/Book/.
[29] We plot all complex zeroes of polynomials with only -1 and 1 as coefficients up to a given degree. As the degree increases some of the holes fill in—at different rates.

17

the calculations reveal somewhat different features. This is very much analogous to a chemist discovering an unexplained but robust spectral line. ∎

This certainly illustrates #2 and #7, but also #1 and #3.

7. **Visual Dynamics.** In recent continued fraction work, Crandall and I needed to study the *dynamical system* $t_0 := t_1 := 1$:

$$t_n := \frac{1}{n} t_{n-1} + \omega_{n-1}\left(1 - \frac{1}{n}\right) t_{n-2},$$

where $\omega_n = a^2, b^2$ for n even, odd respectively, are two unit vectors. Think of this as a **black box** which we wish to examine scientifically. Numerically, all one *sees* is $t_n \to 0$ slowly. Pictorially, in Figure 3, we *learn* significantly more.[30] If the iterates are plotted with colour changing after every few hundred iterates,[31] it is clear that they spiral roman-candle like in to the origin:

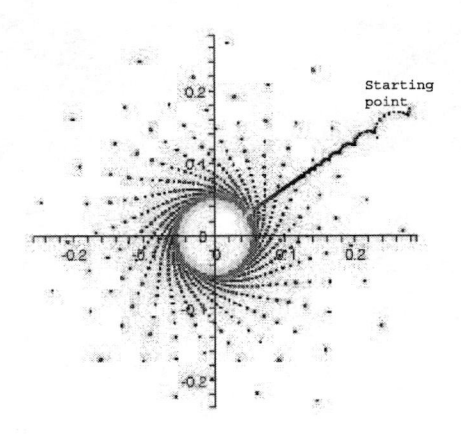

Figure 3. (Ex. 7.): "Visual convergence in the complex plane"

Scaling by $\sqrt{n}$, and distinguishing even and odd iterates, *fine structure* appear in Figure 4. We now observe, predict and validate that the outcomes depend on whether or not one or both of a and b are roots of unity (that is, rational multiples of π). Input a p-th root of unity and out come p spirals, input a non-root of unity and we see a circle. ∎

This forceably illustrates role #2 but also roles #1, #3, #4. It took my coauthors and me, over a year and 100 pages to convert this intuition into a rigorous formal proof, [3]. Indeed, the results are technical and delicate enough that I have more faith in the facts than in the finished argument. In this sentiment, I am not entirely alone.

---

[30]... "Then felt I like a watcher of the skies, when a new planet swims into his ken." From John Keats (1795-1821) poem *On first looking into Chapman's Homer*.

[31]A colour version may be seen on the cover of [2].

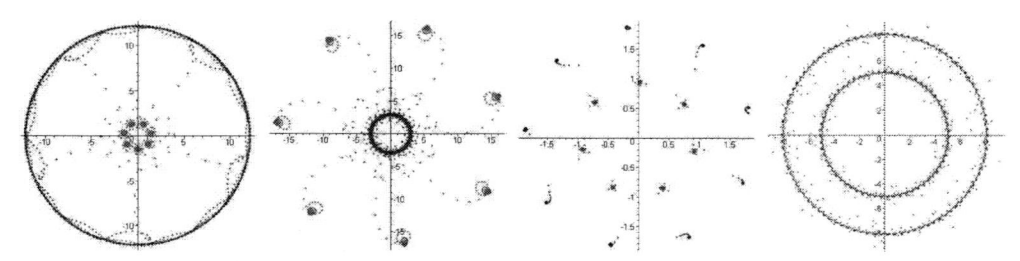

Figure 4. (Ex. 7.): The attractors for various $|a| = |b| = 1$

Carl Friedrich Gauss, who drew (carefully) and computed a great deal, is said to have noted, *I **have** the result, but I do not yet know how to get it.*[32] An excited young Gauss writes: "A new field of analysis has appeared to us, self-evidently, in the study of functions etc." (October 1798, reproduced in [9, Fig. 1.2, p.15]). It had and the consequent proofs pried open the doors of much modern elliptic function and number theory.

My penultimate and more comprehensive example is more sophisticated and I beg the less-expert analyst's indulgence. Please consider its structure and not the details.

8. **A full run.** Consider the *unsolved* **Problem 10738** from the 1999 *American Mathematical Monthly*, [10]:

Problem: For $t > 0$ let
$$m_n(t) = \sum_{k=0}^{\infty} k^n \exp(-t) \frac{t^k}{k!}$$
be the *nth* moment of a *Poisson distribution* with parameter t. Let $c_n(t) = m_n(t)/n!$. Show

a) $\{m_n(t)\}_{n=0}^{\infty}$ is log-convex[33] for all $t > 0$.

b) $\{c_n(t)\}_{n=0}^{\infty}$ is not log-concave for $t < 1$.

c*) $\{c_n(t)\}_{n=0}^{\infty}$ is log-concave for $t \geq 1$.

Solution. (a) Neglecting the factor of $\exp(-t)$ as we may, this reduces to
$$\sum_{k,j \geq 0} \frac{(jk)^{n+1} t^{k+j}}{k! j!} \leq \sum_{k,j \geq 0} \frac{(jk)^n t^{k+j}}{k! j!} k^2 = \sum_{k,j \geq 0} \frac{(jk)^n t^{k+j}}{k! j!} \frac{k^2 + j^2}{2},$$
and this now follows from $2jk \leq k^2 + j^2$.

(b) As
$$m_{n+1}(t) = t \sum_{k=0}^{\infty} (k+1)^n \exp(-t) \frac{t^k}{k!},$$

---

[32] Like so many attributions, the quote has so far escaped exact isolation!

[33] A sequence $\{a_n\}$ is *log-convex* if $a_{n+1} a_{n-1} \geq a_n^2$, for $n \geq 1$ and log-concave when the inequality is reversed.

on applying the binomial theorem to $(k+1)^n$, we see that $m_n(t)$ satisfies the recurrence

$$m_{n+1}(t) = t \sum_{k=0}^{n} \binom{n}{k} m_k(t), \qquad m_0(t) = 1.$$

In particular for $t = 1$, we computationally obtain as many terms of the sequence

$$1, 1, 2, 5, 15, 52, 203, 877, 4140 \ldots$$

as we wish. These are the *Bell numbers* as was discovered again by consulting *Sloane's Encyclopedia* which can also tell us that, for $t = 2$, we have the *generalized Bell numbers*, and gives the exponential generating functions.[34] Inter alia, an explicit computation shows that

$$t \frac{1+t}{2} = c_0(t)\, c_2(t) \leq c_1(t)^2 = t^2$$

exactly if $t \geq 1$, which completes (b).

Also, preparatory to the next part, a simple calculation shows that

$$\sum_{n \geq 0} c_n u^n = \exp\left(t(e^u - 1)\right). \tag{7}$$

(c*)[35] We appeal to a recent theorem, [10, p. 42], due to E. Rodney Canfield which proves the lovely and quite difficult result below. A self-contained proof would be very fine.

Theorem 1 *If a sequence $1, b_1, b_2, \cdots$ is non-negative and log-concave then so is the sequence $1, c_1, c_2, \cdots$ determined by the generating function equation*

$$\sum_{n \geq 0} c_n u^n = \exp\left(\sum_{j \geq 1} b_j \frac{u^j}{j}\right).$$

Using equation (7) above, we apply this to the sequence $\mathbf{b_j = t/(j-1)!}$ which is log-concave exactly for $t \geq 1$. ∎

A search in 2001 on *MathSciNet* for "Bell numbers" since 1995 turned up 18 items. Canfield's paper showed up as number 10. Later, *Google* found it immediately!

Quite unusually, the given solution to (c) was the only one received by the *Monthly*. The reason might well be that it relied on the following sequence of steps:

---

[34] Bell numbers were known earlier to Ramanujan—an example of *Stigler's Law of Eponymy*, [10, p. 60]. Combinatorially they count the number of nonempty subsets of a finite set.

[35] The '*' indicates this was the unsolved component.

$$\boxed{\text{A (\textbf{Question Posed})} \Rightarrow \text{Computer Algebra System} \Rightarrow \text{Interface} \Rightarrow}$$
$$\boxed{\text{Search Engine} \Rightarrow \text{Digital Library} \Rightarrow \text{Hard New Paper} \Rightarrow \textbf{(Answer)}}$$

Without going into detail, we have visited most of the points elaborated in Section 4.1. Now if only we could already automate this process!

Jacques Hadamard, describes the role of proof as well as anyone—and most persuasively given that his 1896 proof of the Prime number theorem is an inarguable apex of rigorous analysis.

> *"The object of mathematical rigor is to sanction and legitimize the conquests of intuition, and there was never any other object for it."* (Jacques Hadamard[36])

Of the eight uses of computers instanced above, let me reiterate the central importance of heuristic methods for determining what is true and whether it merits proof. I tentatively offer the following surprising example which is very very likely to be true, offers no suggestion of a proof and indeed may have no reasonable proof.

9. **Conjecture.** *Consider*

$$x_n = \left\{ 16x_{n-1} + \frac{120n^2 - 89n + 16}{512n^4 - 1024n^3 + 712n^2 - 206n + 21} \right\} \qquad (8)$$

The sequence $\beta_n = (\lfloor 16x_n \rfloor)$, where (x_n) is the sequence of iterates defined in equation (8), precisely generates the hexadecimal expansion of $\pi - 3$.

(Here $\{\cdot\}$ denotes the fractional part and $(\lfloor \cdot \rfloor)$ denotes the integer part.) In fact, we know from [9, Chapter 4] that the first million iterates are correct and in consequence:

$$\sum_{n=1}^{\infty} \|x_n - \{16^n \pi\}\| \leq 1.46 \times 10^{-8} \ldots \qquad (9)$$

where $\|a\| = \min(a, 1-a)$. By the first Borel-Cantelli lemma this shows that the hexadecimal expansion of π only finitely differs from (β_n). Heuristically, the probability of any error is very low. ∎

6 Conclusions

To summarize, I do argue that reimposing the primacy of mathematical knowledge over proof is appropriate. So I return to the matter of what it takes to persuade an individual to adopt new methods and drop time honoured ones. Aptly, we may start by consulting Kuhn on the matter of paradigm shift:

---

[36] J. Hadamard, in E. Borel, Lecons sur la theorie des fonctions, 3rd ed. 1928, quoted in [49, (2), p. 127]. See also [47].

> *"The issue of paradigm choice can never be unequivocally settled by logic and experiment alone. ··· in these matters neither proof nor error is at issue. The transfer of allegiance from paradigm to paradigm is a conversion experience that cannot be forced."* (Thomas Kuhn[37])

As we have seen, the pragmatist philosopher John Dewey eloquently agrees, while Max Planck, [46], has also famously remarked on the difficulty of such paradigm shifts. This is Kuhn's version[38]:

> *"And Max Planck, surveying his own career in his Scientific Autobiography, sadly remarked that "a new scientific truth does not triumph by convincing its opponents and making them see the light, but rather because its opponents eventually die, and a new generation grows up that is familiar with it."* (Albert Einstein, [37, 46])

This transition is certainly already apparent. It is certainly rarer to find a mathematician under thirty who is unfamiliar with at least one of *Maple*, *Mathematica* or *MatLab*, than it is to one over sixty five who is really fluent. As such fluency becomes ubiquitous, I expect a re-balancing of our community's valuing of deductive proof over inductive knowledge.

In his famous lecture to the Paris International Congress in 1900, Hilbert writes[39]

> *"Moreover a mathematical problem should be difficult in order to entice us, yet not completely inaccessible, lest it mock our efforts. It should be to us a guidepost on the mazy path to hidden truths, and ultimately a reminder of our pleasure in the successful solution."* (David Hilbert, [56])

Note the primacy given by a most exacting researcher to discovery and to truth over proof and rigor. More controversially and most of a century later, Greg Chaitin invites us to be bolder and act more like physicists.

> *"I believe that elementary number theory and the rest of mathematics should be pursued more in the spirit of experimental science, and that you should be willing to adopt new principles...* And the Riemann Hypothesis isn't self-evident either, but it's very useful. A physicist would say that there is ample experimental evidence for the Riemann Hypothesis and would go ahead and take it as a working assumption. ··· We may want to introduce it formally into our mathematical system."* (Greg Chaitin, [9, p. 254])

Ten years later:

---

[37]In [50], *Who Got Einstein's Office?* The answer is Arne Beurling.

[38]Kuhn is quoting Einstein quoting Planck. There are various renderings of this second-hand German quotation.

[39]See the late Ben Yandell's fine account of the twenty-three *"Mathematische Probleme"* lecture, Hilbert Problems and their solvers, [56]. The written lecture (given in [56]) is considerably longer and further ranging that the one delivered in person.

> *"[Chaitin's] "Opinion" article proposes that the Riemann hypothesis (RH) be adopted as a new axiom for mathematics. Normally one could only countenance such a suggestion if one were assured that the RH was undecidable. However, a proof of undecidability is a logical impossibility in this case, since if RH is false it is provably false. Thus, the author contends, one may either wait for a proof, or disproof, of RH—both of which could be impossible—or one may take the bull by the horns and accept the RH as an axiom. He prefers this latter course as the more positive one."* (Roger Heath Brown[40])

Much as I admire the challenge of Greg Chaitin's statements, I am not yet convinced that it is helpful to add axioms as opposed to proving conditional results that start "Assuming the continuum hypothesis" or emphasize that "without assuming the Riemann hypothesis we are able to show ...". Most important is that we lay our cards on the table. We should explicitly and honestly indicate when we believe our tools to be heuristic, we should carefully indicate why we have confidence in our computations—and where our uncertainty lies— and the like.

On that note, Hardy is supposed to have commented—somewhat dismissively—that Landau, a great German number theorist, would never be the first to prove the Riemann Hypothesis, but that if someone else did so then Landau would have the best possible proof shortly after. I certainly hope that a more experimental methodology will better value independent replication and honour the first transparent proof[41] of Fermat's last theorem as much as Andrew Wiles' monumental proof. Hardy also commented that he did his best work past forty. Inductive, accretive, tool-assisted mathematics certainly allows brilliance to be supplemented by experience and—as in my case—stands to further undermine the notion that one necessarily does one's best mathematics young.

6.1 As for Education

The main consequence for me is that a *constructivist educational curriculum*—supported by both good technology and reliable content—is both possible and highly desirable. In a traditional instructivist mathematics classroom there are few opportunities for realistic discovery. The current sophistication of dynamic geometry software such as *Geometer's Sketchpad*, *Cabri* or *Cinderella*, of many fine web-interfaces, and of broad mathematical computation platforms like *Maple* and *Mathematica* has changed this greatly—though in my opinion both *Maple* and *Mathematica* are unsuitable until late in high-school, as they presume too much of both the student and the teacher. A thoughtful and detailed discussion of many of the central issues can be found in J.P. Lagrange's article [41] on teaching functions in such a milieu.

Another important lesson is that we need to teach procedural or *algorithmic thinking*. Although some vague notion of a computer program as a repeated procedure is probably

---
[40] Roger Heath-Brown's *Mathematical Review* of [18], 2004.
[41] Should such exist and as you prefer be discovered or invented.

ubiquitous today, this does not carry much water in practice. For example, five years or so ago, while teaching future elementary school teachers (in their final year), I introduced only one topic not in the text: extraction of roots by Newton's method. I taught this in class, tested it on an assignment and repeated it during the review period. About half of the students participated in both sessions. On the final exam, I asked the students to compute $\sqrt{3}$ using Newton's method starting at $x_0 = 3$ to estimate $\sqrt{3} = \underline{1.732}050808\ldots$ so that the first three digits after the decimal point were correct. I hoped to see $x_1 = 2$, $x_2 = 7/4$ and $x_3 = 97/56 = \underline{1.732}142857\ldots$. I gave the students the exact iteration in the form

$$x_{\text{NEW}} = \frac{x + 3/x_{\text{OLD}}}{2}, \qquad (10)$$

and some other details. The half of the class that had been taught the method had no trouble with the question. The rest almost without exception "guessed and checked". They tried $x_{\text{OLD}} = 3$ and then rather randomly substituted many other values in (10). If they were lucky they found some x_{OLD} such that x_{NEW} did the job.

My own recent experiences with technology-mediated curriculum are described in Jen Chang's 2006 MPub, [19]. There is a concurrent commercial implementation of such a middle-school *Interactive School Mathematics* currently being completed by *MathResources*.[42] Many of the examples I have given, or similar ones more tailored to school [7], are easily introduced into the curriculum, but only if the teacher is not left alone to do so. Technology also allows the same teacher to provide enriched material (say, on fractions, binomials, irrationality, fractals or chaos) to the brightest in the class while allowing more practice for those still struggling with the basics. That said, successful mathematical education relies on active participation of the learner and the teacher and my own goal has been to produce technological resources to support not supplant this process; and I hope to make learning or teaching mathematics more rewarding and often more fun.

6.2 Last Words

To reprise, I hope to have made convincing arguments that the traditional deductive accounting of Mathematics is a largely ahistorical caricature—Euclid's millennial sway not withstanding.[43] Above all, mathematics is primarily about *secure knowledge* not proof, and that while the aesthetic is central, we must put much more emphasis on notions of supporting evidence and attend more closely to the reliability of witnesses.

Proofs are often out of reach—but understanding, even certainty, is not. Clearly, computer packages can make concepts more accessible. A short list includes linear relation algorithms, Galois theory, Groebner bases, etc. While progress is made *"one funeral at a time,"*[44] in Thomas Wolfe's words *"you can't go home again"* and as the co-inventor of the

---

[42]See http://www.mathresources.com/products/ism/index.html. I am a co-founder of this ten-year old company. Such a venture is very expensive and thus relies on commercial underpinning.

[43]Most of the cited quotations are stored at jborwein/quotations.html

[44]This grim version of Planck's comment is sometimes attributed to Niels Bohr but this seems specious. It is also spuriously attributed on the web to Michael Milken, and I imagine many others

Fast Fourier transform properly observed, in [53][45]

> *"Far better an approximate answer to the right question, which is often vague, than the exact answer to the wrong question, which can always be made precise."*
> (J. W. Tukey, 1962)

References

[1] D.H. Bailey and J.M. Borwein, "Experimental Mathematics: Recent Developments and Future Outlook," pp. 51–66 in Vol. I of *Mathematics Unlimited—2001 and Beyond,* B. Engquist & W. Schmid (Eds.), Springer-Verlag, 2000.

[2] D. Bailey, J. Borwein, N. Calkin, R. Girgensohn, R. Luke, and V. Moll, *Experimental Mathematics in Action,* A.K. Peters, 2007.

[3] D.H. Bailey and J.M. Borwein, "Experimental Mathematics: Examples, Methods and Implications," *Notices Amer. Math. Soc.*, **52** No. 5 (2005), 502–514.

[4] David Berlinski, "Ground Zero: A Review of The Pleasures of Counting, by T. W. Koerner," by David Berlinski. *The Sciences*, July/August 1997, 37–41.

[5] F. Bornemann, D. Laurie, S. Wagon, and J. Waldvogel, *The SIAM 100 Digit Challenge: A Study in High-Accuracy Numerical Computing*, SIAM, Philadelphia, 2004.

[6] E.J. Borowski and J.M. Borwein, *Dictionary of Mathematics*, Smithsonian/Collins Edition, 2006.

[7] J.M. Borwein "The Experimental Mathematician: The Pleasure of Discovery and the Role of Proof," *International Journal of Computers for Mathematical Learning*, **10** (2005), 75–108.

[8] J.M. Borwein, "The 100 Digit Challenge: an Extended Review," *Math Intelligencer*, **27** (4) (2005), 40–48. Available at http://users.cs.dal.ca/~jborwein/digits.pdf.

[9] J.M. Borwein and D.H. Bailey, *Mathematics by Experiment: Plausible Reasoning in the 21st Century,* AK Peters Ltd, 2003.

[10] J.M. Borwein, D.H. Bailey and R. Girgensohn, *Experimentation in Mathematics: Computational Paths to Discovery,* AK Peters Ltd, 2004.

[11] J.M. Borwein and P.B. Borwein, *Pi and the AGM,* CMS Monographs and Advanced Texts, John Wiley, 1987.

[12] J.M. Borwein and P.B. Borwein, "Challenges for Mathematical Computing," *Computing in Science & Engineering,* **3** (2001), 48–53.

---
[45]Ironically, despite often being cited as in that article, I can not locate it!

[13] J.M. Borwein, P.B. Borwein, and D.A. Bailey, "Ramanujan, modular equations and pi or how to compute a billion digits of pi," *MAA Monthly*, **96** (1989), 201–219. Reprint ed in *Organic Mathematics Proceedings*, (http://www.cecm.sfu.ca/organics), April 12, 1996. Print version: *CMS/AMS Conference Proceedings*, **20** (1997), ISSN: 0731-1036.

[14] Jonathan Borwein and Marc Chamberland, "Integer powers of Arcsin," *Int. J. Math. & Math. Sci.*, in press 2007. [D-drive preprint 288].

[15] Jonathan M. Borwein and Robert Corless, "Emerging Tools for Experimental Mathematics," *MAA Monthly*, **106** (1999), 889–909.

[16] J.M. Borwein and T.S. Stanway, "Knowledge and Community in Mathematics," *The Mathematical Intelligencer*, **27** (2005), 7–16.

[17] Julie K. Brown, "Solid Tools for Visualizing Science," *Science*, November 19, 2004, 1136–37.

[18] G.J. Chaitin, "Thoughts on the Riemann hypothesis," *Math. Intelligencer*, **26** (2004), no. 1, 4–7. (MR2034034)

[19] Jen Chang, "The SAMPLE Experience: the Development of a Rich Media Online Mathematics Learning Environment," MPub Project Report, Simon Fraser University, 2006. Available at http://locutus.cs.dal.ca:8088/archive/00000327/.

[20] John Dewey, *Influence of Darwin on Philosophy and Other Essays*, Prometheus Books, 1997.

[21] Keith Devlin, *Mathematics the Science of Patterns*, Owl Books, 1996.

[22] J. Dongarra, F. Sullivan, "The top 10 algorithms," *Computing in Science & Engineering*, **2** (2000), 22–23. (See www.cecm.sfu.ca/personal/jborwein/algorithms.html.)

[23] Michael Dummett, *Frege: Philosophy of Language*, Harvard University Press, 1973.

[24] Freeman Dyson, Review of *Nature's Numbers* by Ian Stewart (Basic Books, 1995). *American Mathematical Monthly*, August-September 1996, p. 612.

[25] "Proof and Beauty," *The Economist*, March 31, 2005. (See www.economist.com/science/displayStory.cfm?story_id=3809661.)

[26] Paul Ernest, "Social Constructivism As a Philosophy of Mathematics. Radical Constructivism Rehabilitated?" A 'historical paper' available at www.people.ex.ac.uk/PErnest/.

[27] Paul Ernest, *Social Constructivism As a Philosophy of Mathematics*, State University of New York Press, 1998.

[28] Kurt Gödel, "Some Basic Theorems on the Foundations," p. 313 in *Collected Works, Vol. III. Unpublished essays and lectures.* Oxford University Press, New York, 1995.

[29] Judith Grabiner, "Newton, Maclaurin, and the Authority of Mathematics," *MAA Monthly*, December 2004, 841–852.

[30] Gila Hanna, "The Influence of Lakatos," preprint, 2006.

[31] Reuben Hersch, "Fresh Breezes in the Philosophy of Mathematics", *MAA Monthly*, August 1995, 589–594.

[32] Reuben Hersh, *What is Mathematics Really?* Oxford University Press, 1999.

[33] Felix Klein, *Development of Mathematics in the 19th Century,* 1928, Trans Math. Sci. Press, R. Hermann Ed. (Brookline, MA, 1979).

[34] Gina Kolata, "In Math, Computers Don't Lie. Or Do They?" *NY Times*, April 6th, 2004.

[35] Christopher Koch, "Thinking About the Conscious Mind," a review of John R. Searle's *Mind. A Brief Introduction,* Oxford University Press, 2004. *Science*, November 5, 2004, 979–980.

[36] Donald Knuth, *American Mathematical Monthly Problem* 10832, November 2002.

[37] T.S. Kuhn, *The Structure of Scientific Revolutions,* 3rd ed., U. of Chicago Press, 1996.

[38] Imre Lakatos, *Proofs and Refutations,* Cambridge Univ. Press, 1976.

[39] by George Lakoff and Rafael E. Nunez, *Where Mathematics Comes From: How the Embodied Mind Brings Mathematics into Being,* Basic Books, 2001.

[40] Clement W.H. Lam, "The Search for a Finite Projective Plane of Order 10," *Amer. Math. Monthly* **98** (1991), 305–318.

[41] J.B. Lagrange, "Curriculum, Classroom Practices, and Tool Design in the Learning of Functions through Technology-aided Experimental Approaches," *International Journal of Computers in Math Learning,* **10** (2005), 143–189.

[42] R. C. Lewontin, "In the Beginning Was the Word," (*Human Genome Issue*), *Science*, February 16, 2001, 1263–1264.

[43] John Mason, *Learning and Doing Mathematics,* QED Press; 2nd revised ed.b, 2006.

[44] Peter B. Medawar, *Advice to a Young Scientist,* HarperCollins, 1979.

[45] Andrew Odlyzko, "The 10^{22}-nd zero of the Riemann zeta function. Dynamical, spectral, and arithmetic zeta functions," *Contemp. Math.*, **290** (2001), 139–144.

[46] Max Planck, *Scientific Autobiography and Other Papers,* trans. F. Gaynor (New York, 1949), pp. 33-34.

[47] Henri Poincaré, "Mathematical Invention," pp. 20–30 in *Musing's of the Masters,* Raymond Ayouib editor, MAA, 2004.

[48] Henri Poincaré, *Mathematical Definitions in Education,* (1904).

[49] George Polya, *Mathematical Discovery: On Understanding, Learning, and Teaching Problem Solving* (Combined Edition), New York, Wiley & Sons, 1981.

[50] Ed Regis, *Who got Einstein's office?* Addison Wesley, 1988.

[51] Kerry Rowe et al., *Engines of Discovery: The 21st Century Revolution.* The Long Range Plan for HPC in Canada, NRC Press, Ottawa, 2005.

[52] D'Arcy Thompson, *On Growth and Form,* Dover Publications, 1992.

[53] J.W. Tukey, "The future of data analysis," *Ann. Math. Statist.* **33**, (1962), 1–67.

[54] *Random House Webster's Unabridged Dictionary,* Random House, 1999.

[55] Eric W. Weisstein, "Lucky Number of Euler," from *MathWorld–A Wolfram Web Resource.* http://mathworld.wolfram.com/LuckyNumberofEuler.html.

[56] Benjamin Yandell, *The Honors Class,* AK Peters, 2002.

Acknowledgements. My gratitude is due to many colleagues who offered thoughtful and challenging comments during the completion of this work, and especially to David Bailey, Neil Calkin and Nathalie Sinclair. Equal gratitude is owed to the editors, Bonnie Gold and Roger Simons, for their careful and appropriately critical readings of several earlier drafts of this chapter.

13. Exploratory experimentation and computation

Discussion

This article revisits the theme of several earlier ones, especially Chapters 7 and 10. Over the past thirty years, our views on the subject have evolved and, we hope, matured.

In the intervening period, the human genome has been decoded, neurobiology has become a science, fiber-optics and the internet have rewired the earth, chess has been beaten, Fermat's last theorem has been proven, the Poincaré conjecture is resolved, and the iron curtain is gone. Technically much that was once hard is now easy.

Yet in mathematics much more has remained the same than has changed dramatically. We echo Martin Raff's sentiment that understanding will always remain a few steps away:

> It is tempting to think that the main principles of neural development will have been discovered by the end of the century [20th] and that the cellular and molecular basis of the mind will be the main challenge for the next. An alternative view is that this feeling that understanding is just a few steps away is a recurring and necessary delusion that keeps scientists from dwelling on the complexity the face and how much more remains to be discovered.[1]

Source

D.H. Bailey and J. M. Borwein, "Exploratory Experimentation and Computation," *Notices of the AMS*. Accepted February 2010.

---

[1] "Neural Development: Mysterious No More?" Editorial on page 1063 of *Science* November 15, 1996.

EXPLORATORY EXPERIMENTATION AND COMPUTATION

DAVID H. BAILEY AND JONATHAN M. BORWEIN

ABSTRACT. We believe the mathematical research community is facing a great challenge to re-evaluate the role of proof in light of recent developments. On one hand, the growing power of current computer systems, of modern mathematical computing packages, and of the growing capacity to data-mine on the Internet, has provided marvelous resources to the research mathematician. On the other hand, the enormous complexity of many modern capstone results such as the Poincaré conjecture, Fermat's last theorem, and the classification of finite simple groups has raised questions as to how we can better ensure the integrity of modern mathematics. Yet as the need and prospects for inductive mathematics blossom, the requirement to ensure the role of proof is properly founded remains undiminished.

1. EXPLORATORY EXPERIMENTATION

The authors' thesis—once controversial, but now a commonplace—is that computers can be a useful, even essential, aid to mathematical research.—Jeff Shallit

Jeff Shallit wrote this in his recent review MR2427663 of [9]. As we hope to make clear, Shallit was entirely right in that many, if not most, research mathematicians now use the computer in a variety of ways to draw pictures, inspect numerical data, manipulate expressions symbolically, and run simulations. However, it seems to us that there has not yet been substantial and intellectually rigorous progress in the way mathematics is presented in research papers, textbooks and

---

Date: August 14, 2010.

Bailey: Lawrence Berkeley National Laboratory, Berkeley, CA 94720, USA. Email: dhbailey@lbl.gov. This work was supported by the Director, Office of Computational and Technology Research, Division of Mathematical, Information, and Computational Sciences of the U.S. Department of Energy, under contract number DE-AC02-05CH11231. Borwein: Centre for Computer Assisted Research Mathematics and its Applications (CARMA), University of Newcastle, Callaghan, NSW 2308, Australia. Email: jonathan.borwein@newcastle.edu.au.

classroom instruction, or in how the mathematical discovery process is organized.

1.1. Mathematicians are humans. We share with George Pólya (1887-1985) the view [24, 2 p. 128] that while learned,

> intuition comes to us much earlier and with much less outside influence than formal arguments.

Pólya went on to reaffirm, nonetheless, that proof should certainly be taught in school.

We turn to observations, many of which have been fleshed out in coauthored books such *Mathematics by Experiment* [9], and *Experimental Mathematics in Action* [3], where we have noted the changing nature of mathematical knowledge and in consequence ask questions such as "How do we teach what and why to students?", "How do we come to believe and trust pieces of mathematics?", and "Why do we wish to prove things?" An answer to the last question is "That depends." Sometimes we wish insight and sometimes, especially with subsidiary results, we are more than happy with a certificate. The computer has significant capacities to assist with both.

Smail [26, p. 113] writes:

> the large human brain evolved over the past 1.7 million years to allow individuals to negotiate the growing complexities posed by human social living.

As a result, humans find various modes of argument more palatable than others, and are more prone to make certain kinds of errors than others. Likewise, the well-known evolutionary psychologist Steve Pinker observes that language [23, p. 83] is founded on

> ...the ethereal notions of space, time, causation, possession, and goals that appear to make up a language of thought.

This remains so within mathematics. The computer offers scaffolding both to enhance mathematical reasoning, as with the recent computation of the Lie group E_8, see `http://www.aimath.org/E8/computerdetails.html`, and to restrain mathematical error.

1.2. Experimental mathodology. Justice Potter Stewart's famous 1964 comment, *"I know it when I see it"* is the quote with which *The Computer as Crucible* [12] starts. A bit less informally, by *experimental mathematics* we intend [9]:

(a) Gaining insight and *intuition*;
(b) *Visualizing* math principles;
(c) *Discovering* new relationships;
(d) *Testing* and especially *falsifying* conjectures;

(e) *Exploring* a possible result to see if it *merits* formal proof;
(f) *Suggesting* approaches for formal proof;
(g) *Computing* replacing lengthy hand derivations;
(h) *Confirming* analytically derived results.

Of these items (a) through (e) play a central role, and (f) also plays a significant role for us, but connotes computer-assisted or computer-directed proof and thus is quite distinct from *formal proof* as the topic of a special issue of these *Notices* in December 2008; see, e.g., [19].

1.2.1. *Digital integrity, I.* For us (g) has become ubiquitous, and we have found (h) to be particularly effective in ensuring the integrity of published mathematics. For example, we frequently check and correct identities in mathematical manuscripts by computing particular values on the LHS and RHS to high precision and comparing results—and then if necessary use software to repair defects.

As a first example, in a current study of "character sums" we wished to use the following result derived in [13]:

$$\sum_{m=1}^{\infty}\sum_{n=1}^{\infty} \frac{(-1)^{m+n-1}}{(2m-1)(m+n-1)^3} \tag{1.1}$$

$$\stackrel{?}{=} 4\operatorname{Li}_4\left(\frac{1}{2}\right) - \frac{51}{2880}\pi^4 - \frac{1}{6}\pi^2\log^2(2) + \frac{1}{6}\log^4(2) + \frac{7}{2}\log(2)\zeta(3).$$

Here $\operatorname{Li}_4(1/2)$ is a polylogarithmic value. However, a subsequent computation to check results disclosed that whereas the LHS evaluates to $-0.872929289\ldots$, the RHS evaluates to $2.509330815\ldots$. Puzzled, we computed the sum, as well as each of the terms on the RHS (sans their coefficients), to 500-digit precision, then applied the "PSLQ" algorithm, which searches for integer relations among a set of constants [15]. PSLQ quickly found the following:

$$\sum_{m=1}^{\infty}\sum_{n=1}^{\infty} \frac{(-1)^{m+n-1}}{(2m-1)(m+n-1)^3} \tag{1.2}$$

$$= 4\operatorname{Li}_4\left(\frac{1}{2}\right) - \frac{151}{2880}\pi^4 - \frac{1}{6}\pi^2\log^2(2) + \frac{1}{6}\log^4(2) + \frac{7}{2}\log(2)\zeta(3).$$

In other words, in the process of transcribing (1.1) into the original manuscript, "151" had become "51." It is quite possible that this error would have gone undetected and uncorrected had we not been able to computationally check and correct such results. This may not always matter, but it can be crucial.

With a current Research Assistant, Alex Kaiser at Berkeley, we have started to design software to refine and automate this process and to run it before submission of any equation-rich paper. This semi-automated integrity checking becomes pressing when verifiable output from a symbolic manipulation might be the length of a Salinger novel. For instance, recently while studying expected radii of points in a hypercube [11], it was necessary to show the existence of a "closed form" for

$$J(t) := \int_{[0,1]^2} \frac{\log(t + x^2 + y^2)}{(1 + x^2)(1 + y^2)} \, dx \, dy. \qquad (1.3)$$

The computer verification of [11, Thm. 5.1] quickly returned a 100000-character "answer" that could be numerically validated very rapidly to hundreds of places. A highly interactive process stunningly reduced a basic instance of this expression to the concise formula

$$J(2) = \frac{\pi^2}{8} \log 2 - \frac{7}{48} \zeta(3) + \frac{11}{24} \pi \, \text{Cl}_2\left(\frac{\pi}{6}\right) - \frac{29}{24} \pi \, \text{Cl}_2\left(\frac{5\pi}{6}\right), \qquad (1.4)$$

where Cl_2 is the *Clausen function* $\text{Cl}_2(\theta) := \sum_{n \geq 1} \sin(n\theta)/n^2$ (Cl_2 is the simplest non-elementary Fourier series). Automating such reductions will require a sophisticated simplification scheme with a very large and extensible knowledge base.

1.3. **Discovering a truth.** Giaquinto's [17, p. 50] attractive encapsulation

> In short, discovering a truth is coming to believe it in an independent, reliable, and rational way.

has the satisfactory consequence that a student can legitimately discover things already "known" to the teacher. Nor is it necessary to demand that each dissertation be absolutely original—only that it be independently discovered. For instance, a differential equation thesis is no less meritorious if the main results are subsequently found to have been accepted, unbeknown to the student, in a control theory journal a month earlier—provided they were independently discovered. Near-simultaneous independent discovery has occurred frequently in science, and such instances are likely to occur more and more frequently as the earth's "new nervous system" (Hillary Clinton's term in a recent policy address) continues to pervade research.

Despite the conventional identification of mathematics with deductive reasoning, Kurt Gödel (1906-1978) in his 1951 Gibbs Lecture said:

> If mathematics describes an objective world just like physics, there is no reason why inductive methods should not be applied in mathematics just the same as in physics.

He held this view until the end of his life despite—or perhaps because of—the epochal deductive achievement of his incompleteness results.

Also, we emphasize that many great mathematicians from Archimedes and Galileo—who reputedly said *"All truths are easy to understand once they are discovered; the point is to discover them."*—to Gauss, Poincaré, and Carleson have emphasized how much it helps to "know" the answer beforehand. Two millennia ago, Archimedes wrote, in the Introduction to his long-lost and recently reconstituted *Method* manuscript,

> For it is easier to supply the proof when we have previously acquired, by the method, some knowledge of the questions than it is to find it without any previous knowledge.

Archimedes' *Method* can be thought of as an uber-precursor to today's interactive geometry software, with the caveat that, for example, *Cinderella* actually does provide proof certificates for much of Euclidean geometry.

As 2006 Abel Prize winner Lennart Carleson describes in his 1966 ICM speech on his positive resolution of Luzin's 1913 conjecture (that the Fourier series of square-summable functions converge pointwise a.e. to the function), after many years of seeking a counterexample, he finally decided none could exist. He expressed the importance of this confidence as follows:

> The most important aspect in solving a mathematical problem is the conviction of what is the true result. Then it took 2 or 3 years using the techniques that had been developed during the past 20 years or so.

1.4. Digital Assistance.

By *digital assistance*, we mean the use of:

(a) *Integrated mathematical software* such as *Maple* and *Mathematica*, or indeed MATLAB and their open source variants.
(b) *Specialized packages* such as CPLEX, PARI, SnapPea, Cinderella and MAGMA.
(c) *General-purpose programming languages* such as C, C++, and Fortran-2000.

(d) *Internet-based applications* such as: Sloane's Encyclopedia of Integer Sequences, the Inverse Symbolic Calculator,[1] Fractal Explorer, Jeff Weeks' Topological Games, or Euclid in Java.[2]

(e) *Internet databases and facilities* including Google, MathSciNet, arXiv, Wikipedia, MathWorld, MacTutor, Amazon, Amazon Kindle, and many more that are not always so viewed.

All entail data-mining in various forms. The capacity to consult the Oxford dictionary and Wikipedia instantly within Kindle dramatically changes the nature of the reading process. Franklin [16] argues that Steinle's "exploratory experimentation" facilitated by "widening technology" and "wide instrumentation," as routinely done in fields such as pharmacology, astrophysics, medicine, and biotechnology, is leading to a reassessment of what legitimates experiment; in that a "local model" is not now a prerequisite. Thus, a pharmaceutical company can rapidly examine and discard tens of thousands of potentially active agents, and then focus resources on the ones that survive, rather than needing to determine in advance which are likely to work well. Similarly, aeronautical engineers can, by means of computer simulations, discard thousands of potential designs, and submit only the best prospects to full-fledged development and testing.

Hendrik Sørenson [27] concisely asserts that experimental mathematics —as defined above—is following similar tracks with software such as *Mathematica*, *Maple* and MATLAB playing the role of wide instrumentation.

> These aspects of exploratory experimentation and wide instrumentation originate from the philosophy of (natural) science and have not been much developed in the context of experimental mathematics. However, I claim that e.g. the importance of wide instrumentation for an exploratory approach to experiments that includes concept formation also pertain to mathematics.

In consequence, boundaries between mathematics and the natural sciences and between inductive and deductive reasoning are blurred and becoming more so. (See also [2].) This convergence also promises some

---

[1]Most of the functionality of the ISC, which is now housed at http://carma-lx1.newcastle.edu.au:8087, is now built into the "identify" function of *Maple* starting with version 9.5. For example, the *Maple* command `identify(4.45033263602792)` returns $\sqrt{3} + e$, meaning that the decimal value given is simply approximated by $\sqrt{3} + e$.

[2]A cross-section of Internet-based mathematical resources is available at http://ddrive.cs.dal.ca/~isc/portal/ and http://www.experimentalmath.info.

relief from the frustration many mathematicians experience when attempting to describe their proposed methodology on grant applications to the satisfaction of traditional hard scientists. We leave unanswered the philosophically-vexing if mathematically-minor question as to whether genuine mathematical experiments (as discussed in [9]) truly exist, even if one embraces a fully idealist notion of mathematical existence. It surely seems to us that they do.

2. Pi, Partitions and Primes

The present authors cannot now imagine doing mathematics without a computer nearby. For example, characteristic and minimal polynomials, which were entirely abstract for us as students, now are members of a rapidly growing box of concrete symbolic tools. One's eyes may glaze over trying to determine structure in an infinite family of matrices including

$$M_4 = \begin{bmatrix} 2 & -21 & 63 & -105 \\ 1 & -12 & 36 & -55 \\ 1 & -8 & 20 & -25 \\ 1 & -5 & 9 & -8 \end{bmatrix} \quad M_6 = \begin{bmatrix} 2 & -33 & 165 & -495 & 990 & -1386 \\ 1 & -20 & 100 & -285 & 540 & -714 \\ 1 & -16 & 72 & -177 & 288 & -336 \\ 1 & -13 & 53 & -112 & 148 & -140 \\ 1 & -10 & 36 & -66 & 70 & -49 \\ 1 & -7 & 20 & -30 & 25 & -12 \end{bmatrix}$$

but a command-line instruction in a computer algebra system will reveal that both $M_4^3 - 3M_4 - 2I = 0$ and $M_6^3 - 3M_6 - 2I = 0$. Likewise, more and more matrix manipulations are profitably, even necessarily, viewed graphically. As is now well known in numerical linear algebra, graphical tools are essential when trying to discern qualitative information such as the block structure of very large matrices. See, for instance, Figure 1.

Equally accessible are many matrix decompositions, the use of Groebner bases, Risch's decision algorithm (to decide when an elementary function has an elementary indefinite integral), graph and group catalogues, and others. Many algorithmic components of a *computer algebra system* are today extraordinarily effective compared with two decades ago, when they were more like toys. This is equally true of extreme-precision calculation—a prerequisite for much of our own work [7, 10, 8]. As we will illustrate, during the three decades that we have seriously tried to integrate computational experiments into research, we have experienced at least 12 Moore's law doublings of computer power and memory capacity [9, 12], which when combined with the utilization of highly parallel clusters (with thousands of processing cores) and

8   DAVID H. BAILEY AND JONATHAN M. BORWEIN

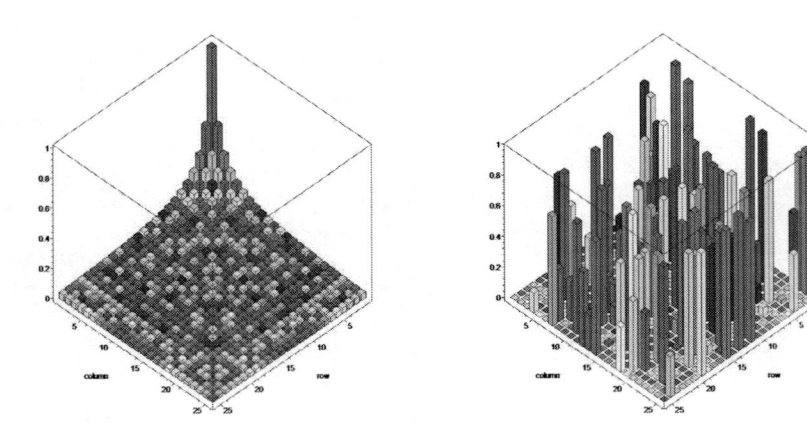

FIGURE 1. Plots of a 25×25 Hilbert matrix (L) and a matrix with 50% sparsity and random $[0, 1]$ entries (R).

fiber-optic networking, has resulted in six to seven orders of magnitude speedup for many operations.

2.1. **The partition function.** Consider the number of additive partitions, $p(n)$, of a natural number, where we ignore order and zeroes. For instance, $5 = 4+1 = 3+2 = 3+1+1 = 2+2+1 = 2+1+1+1 = 1+1+1+1+1$, so $p(5) = 7$. The ordinary generating function (2.1) discovered by Euler is

$$\sum_{n=0}^{\infty} p(n) q^n = \prod_{k=1}^{\infty} \left(1 - q^k\right)^{-1}. \tag{2.1}$$

(This can be proven by using the geometric formula for $1/(1 - q^k)$ to expand each term and observing how powers of q^n occur.)

The famous computation by MacMahon of $p(200) = 3972999029388$ at the beginning of the 20th century, done symbolically and entirely naively from (2.1) on a reasonable laptop, took 20 minutes in 1991 but only 0.17 seconds today, while the many times more demanding computation

$p(2000) = 4720819175619413888601432406799959512200344166$

took just two minutes in 2009. Moreover, in December 2008, Crandall was able to calculate $p(10^9)$ in three seconds on his laptop, using the Hardy-Ramanujan-Rademacher 'finite' series for $p(n)$ along with FFT methods. Using these techniques, Crandall was also able to calculate the probable primes $p(1000046356)$ and $p(1000007396)$, each of which has roughly 35000 decimal digits.

EXPLORATORY EXPERIMENTATION AND COMPUTATION

Such results make one wonder when easy access to computation discourages innovation: Would Hardy and Ramanujan have still discovered their marvelous formula for $p(n)$ if they had powerful computers at hand?

2.2. Quartic algorithm for π. Likewise, the record for computation of π has gone from 29.37 *million* decimal digits in 1986, to 5 *trillion* digits in 2010. Since the algorithm below was used as part of each computation, it is interesting to compare the performance in each case: Set $a_0 := 6 - 4\sqrt{2}$ and $y_0 := \sqrt{2} - 1$, then iterate

$$y_{k+1} = \frac{1 - (1 - y_k^4)^{1/4}}{1 + (1 - y_k^4)^{1/4}},$$
$$a_{k+1} = a_k(1 + y_{k+1})^4 - 2^{2k+3} y_{k+1}(1 + y_{k+1} + y_{k+1}^2). \qquad (2.2)$$

Then a_k converges *quartically* to $1/\pi$—each iteration approximately quadruples the number of correct digits. Twenty-one full-precision iterations of (2.2), which was discovered on a 16K Radio Shack portable in 1983, produce an algebraic number that coincides with π to well more than six trillion places. This scheme and the 1976 Salamin–Brent scheme [9, Ch. 3] have been employed frequently over the past quarter century. Here is a highly abbreviated chronology (based on http://en.wikipedia.org/wiki/Chronology_of_computation_of_pi).

- 1986: One of the present authors used (2.2) to compute 29.4 million digits of π. This required 28 hours on one CPU of the new Cray-2 at NASA Ames Research Center. Confirmation using the Salamin-Brent scheme took another 40 hours. This computation uncovered hardware and software errors on the Cray-2.
- Jan. 2009: Takahashi used (2.2) to compute 1.649 trillion digits (nearly 60,000 times the 1986 computation), requiring 73.5 hours on 1024 cores (and 6.348 Tbyte memory) of a Appro Xtreme-X3 system. Confirmation via the Salamin-Brent scheme took 64.2 hours and 6.732 Tbyte of main memory.
- Apr. 2009: Takahashi computed 2.576 trillion digits.
- Dec. 2009: Bellard computed nearly 2.7 trillion decimal digits (first in binary), using the Chudnovsky series given below in (2.10). This took 131 days on a single four-core workstation with lots of disk storage.
- Aug. 2010: Alexander Yee and Shigeru Kondo used the Chudnovsky formula to compute 5 trillion digits of π over a 90-day period, mostly on a two-core Intel Xeon system with 96 Gbyte of memory. They confirmed the result in two ways, using the BBP formula (see below), which required 66 hours, and a

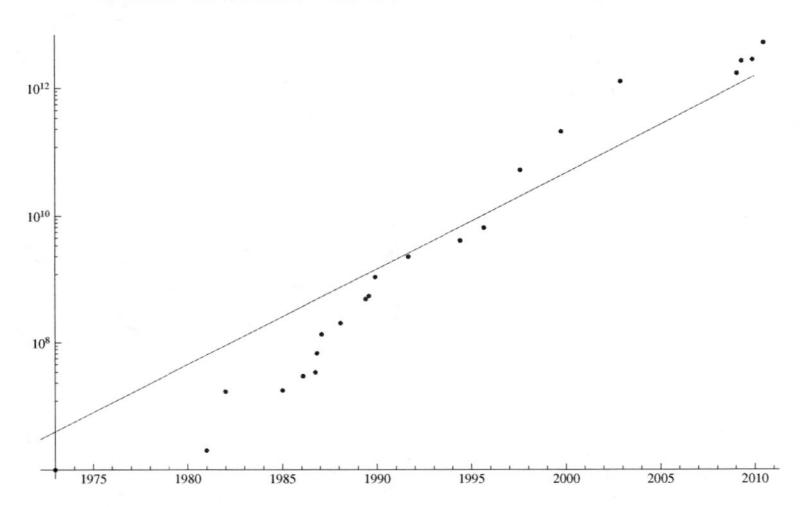

FIGURE 2. Plot of π calculations, in digits (dots), compared with the long-term slope of Moore's Law (line).

variant of the BBP formula due to Bellard, which required 64 hours. Changing from binary to decimal required 8 days. Full details are available at http://www.numberworld.org/misc_runs/pi-5t/details.html.

Daniel Shanks, who in 1961 computed π to over 100,000 digits, once told Phil Davis that a billion-digit computation would be "forever impossible." But both Kanada and the Chudnovskys achieved that in 1989. Similarly, the intuitionists Brouwer and Heyting asserted the "impossibility" of ever knowing whether the sequence 0123456789 appears in the decimal expansion of π, yet it was found in 1997 by Kanada, beginning at position 17387594880. As late as 1989, Roger Penrose ventured, in the first edition of his book *The Emperor's New Mind*, that we likely will never know if a string of ten consecutive sevens occurs in the decimal expansion of π. This string was found in 1997 by Kanada, beginning at position 22869046249.

Figure 2— shows the progress of π calculations since 1970, superimposed with a line that charts the long-term trend of Moore's Law. It is worth noting that whereas progress in computing π exceeded Moore's Law in the 1990s, it has lagged a bit in the past decade.

2.2.1. *Digital integrity, II.* There are many possible sources of errors in these and other large-scale computations:
- The underlying formulas used might conceivably be in error.
- Computer programs implementing these algorithms, which employ sophisticated algorithms such as fast Fourier transforms

EXPLORATORY EXPERIMENTATION AND COMPUTATION 11

to accelerate multiplication, are prone to human programming errors.
- These computations usually are performed on highly parallel computer systems, which require error-prone programming constructs to control parallel processing.
- Hardware errors may occur—this was a factor in the 1986 computation of π, as noted above.

So why would anyone believe the results of such calculations? The answer is that such calculations are always double-checked with an independent calculation done using some other algorithm, sometimes in more than one way. For instance, Kanada's 2002 computation of π to 1.3 trillion decimal digits involved first computing slightly over one trillion hexadecimal (base-16) digits. He found that the 20 hex digits of π beginning at position $10^{12} + 1$ are B4466E8D21 5388C4E014.

Kanada then calculated these hex digits using the "BBP" algorithm [6]. The BBP algorithm for π is based on the formula

$$\pi = \sum_{i=0}^{\infty} \frac{1}{16^i} \left(\frac{4}{8i+1} - \frac{2}{8i+4} - \frac{1}{8i+5} - \frac{1}{8i+6} \right), \quad (2.3)$$

which was discovered using the "PSLQ" integer relation algorithm [15]. Integer relation methods find or exclude potential rational relations between vectors of real numbers. At the start of this millennium, they were named one of the top ten algorithms of the twentieth century by *Computing in Science and Engineering*. The best-known is Helaman Ferguson's PSLQ algorithm [9, 3].

Eventually PSLQ produced the formula

$$\pi = 4\,_2F_1\left(\begin{array}{c} 1, \frac{1}{4} \\ \frac{5}{4} \end{array} \bigg| -\frac{1}{4} \right) + 2\tan^{-1}\left(\frac{1}{2}\right) - \log 5, \quad (2.4)$$

where $_2F_1\left(\begin{array}{c} 1, \frac{1}{4} \\ \frac{5}{4} \end{array} \big| -\frac{1}{4} \right) = 0.955933837\ldots$ is a Gaussian hypergeometric function.

From (2.4), the series (2.3) almost immediately follows. The BBP algorithm, which is based on (2.3), permits one to calculate binary or hexadecimal digits of π beginning at an arbitrary starting point, without needing to calculate any of the preceding digits, by means of a simple scheme that does not require very high precision arithmetic.

The result of the BBP calculation was B4466E8D21 5388C4E014. Needless to say, in spite of the many potential sources of error in both computations, the final results dramatically agree, thus confirming (in a convincing but heuristic sense) that both results are almost certainly correct. Although one cannot rigorously assign a "probability" to this event, note that the chances that two random strings of 20 hex digits perfectly agree is one in $16^{20} \approx 1.2089 \times 10^{24}$.

This raises the following question: What is more securely established, the assertion that the hex digits of π in positions $10^{12} + 1$ through $10^{12} + 20$ are B4466E8D21 5388C4E014, or the final result of some very difficult work of mathematics that required hundreds or thousands of pages, that relied on many results quoted from other sources, and that (as is frequently the case) only a relative handful of mathematicians besides the author can or have carefully read in detail?

In the most recent computation using the BBP formula, Tse-Wo Zse of Yahoo! Cloud Computing calculated 256 binary digits of π starting at the *two quadrillionth* bit [29]. He then checked his result using the following variant of the BBP formula due to Bellard:

$$\pi = \frac{1}{64} \sum_{k=0}^{\infty} \frac{(-1)^k}{1024^k} \left(\frac{256}{10k+1} + \frac{1}{10k+1} - \frac{64}{10k+3} - \frac{4}{10k+5} \right.$$
$$\left. - \frac{4}{10k+7} - \frac{32}{4k+1} - \frac{1}{4k+3} \right) \qquad (2.5)$$

In this case, both computations verified that the 24 hex digits beginning immediately after the 500 trillionth hex digit (i.e., after the two quadrillionth binary bit) are: E6C1294A ED40403F 56D2D764.

2.3. Euler's totient function ϕ.
As another measure of what changes over time and what doesn't, consider two conjectures regarding $\phi(n)$, which counts the number of positive numbers less than and relatively prime to n:

2.3.1. Giuga's conjecture (1950).
An integer $n > 1$, is a prime if and only if $\mathcal{G}_n := \sum_{k=1}^{n-1} k^{n-1} \equiv n - 1 \mod n$.

Counterexamples are necessarily *Carmichael numbers*—rare birds only proven infinite in 1994—and much more. In [10, pp. 227] we exploited the fact that if a number $n = p_1 \cdots p_m$ with $m > 1$ prime factors p_i is a counterexample to Giuga's conjecture (that is, satisfies $s_n \equiv n - 1 \mod n$), then for $i \neq j$ we have $p_i \neq p_j$,

$$\sum_{i=1}^{m} \frac{1}{p_i} > 1,$$

and the p_i form a *normal sequence*: $p_i \not\equiv 1 \mod p_j$ for $i \neq j$. Thus, the presence of '3' excludes $7, 13, 19, 31, 37, \ldots$, and of '5' excludes $11, 31, 41, \ldots$.

This theorem yielded enough structure, using some predictive experimentally discovered heuristics, to build an efficient algorithm to show—over several months in 1995—that any counterexample had at least 3459 prime factors and so exceeded 10^{13886}, extended a few years later to 10^{14164} in a five-day desktop computation. The heuristic is self-validating every time that the programme runs successfully. But this method necessarily fails after 8135 primes; someday we hope to exhaust its use.

While writing this piece, one of us was able to obtain almost as good a bound of 3050 primes in under 110 minutes on a laptop computer, and a bound of 3486 primes and 14,000 digits in less than 14 hours; this was extended to 3,678 primes and 17,168 digits in 93 CPU-hours on a Macintosh Pro, using *Maple* rather than C++, which is often orders-of-magnitude faster but requires much more arduous coding.

An equally hard related conjecture for which much less progress can be recorded is:

2.3.2. Lehmer's conjecture (1932). $\phi(n)|(n-1)$ *if and only if n is prime.* He called this *"as hard as the existence of odd perfect numbers."*

Again, prime factors of counterexamples form a normal sequence, but now there is little extra structure. In a 1997 Simon Fraser M.Sc. thesis, Erick Wong verified the conjecture for 14 primes, using normality and a mix of PARI, C++ and *Maple* to press the bounds of the 'curse of exponentiality.' This very clever computation subsumed the entire scattered literature in one computation but could only extend the prior bound from 13 primes to 14.

For Lehmer's related 1932 question: *when does $\phi(n) \mid (n+1)$?*, Wong showed there are eight solutions with no more than seven factors (six-factor solutions are due to Lehmer). Let

$$\mathcal{L}_m := \prod_{k=0}^{m-1} F_k$$

with $F_n := 2^{2^n} + 1$ denoting the *Fermat primes*. The solutions are

$$2, \mathcal{L}_1, \mathcal{L}_2, \ldots, \mathcal{L}_5,$$

and the rogue pair 4919055 and 6992962672132095, but analyzing just eight factors seems out of sight. Thus, in 70 years the computer only allowed the exclusion bound to grow by one prime.

In 1932 Lehmer couldn't factor 6992962672132097. If it had been prime, a ninth solution would exist: since $\phi(n)|(n+1)$ with $n+2$ prime implies that $N := n(n+2)$ satisfies $\phi(N)|(N+1)$. We say *couldn't* because the number is divisible by 73; which Lehmer—a father of much factorization literature–could certainly have discovered had he anticipated a small factor. Today discovering that

$$6992962672132097 = 73 \cdot 95794009207289$$

is nearly instantaneous, while fully resolving Lehmer's original question remains as hard as ever.

2.4. Inverse computation and Apéry-like series.

Three intriguing formulae for the Riemann zeta function are

$$(a)\ \zeta(2) = 3\sum_{k=1}^{\infty} \frac{1}{k^2 \binom{2k}{k}}, \quad (b)\ \zeta(3) = \frac{5}{2}\sum_{k=1}^{\infty} \frac{(-1)^{k+1}}{k^3 \binom{2k}{k}}, \quad (2.6)$$

$$(c)\ \zeta(4) = \frac{36}{17}\sum_{k=1}^{\infty} \frac{1}{k^4 \binom{2k}{k}}.$$

Binomial identity (2.6)(a) has been known for two centuries, while (b)—exploited by Apéry in his 1978 proof of the irrationality of $\zeta(3)$—was discovered as early as 1890 by Markov, and (c) was noted by Comtet [3].

Using integer relation algorithms, bootstrapping, and the "Pade" function (*Mathematica* and *Maple* both produce rational approximations well), in 1996 David Bradley and one of us [3, 10] found the following unanticipated generating function for $\zeta(4n+3)$:

$$\sum_{k=0}^{\infty} \zeta(4k+3)\, x^{4k} = \frac{5}{2}\sum_{k=1}^{\infty} \frac{(-1)^{k+1}}{k^3 \binom{2k}{k}(1-x^4/k^4)} \prod_{m=1}^{k-1} \left(\frac{1+4x^4/m^4}{1-x^4/m^4}\right). \quad (2.7)$$

Note that this formula permits one to read off an infinity of formulas for $\zeta(4n+3)$, $n > 0$, beginning with (2.6)(b), by comparing coefficients of x^{4k} on the LHS and the RHS.

A decade later, following a quite analogous but much more deliberate experimental procedure, as detailed in [3], we were able to discover a similar general formula for $\zeta(2n+2)$ that is pleasingly parallel to (2.7):

$$\sum_{k=0}^{\infty} \zeta(2k+2)\, x^{2k} = 3\sum_{k=1}^{\infty} \frac{1}{k^2 \binom{2k}{k}(1-x^2/k^2)} \prod_{m=1}^{k-1} \left(\frac{1-4x^2/m^2}{1-x^2/m^2}\right). \quad (2.8)$$

As with (2.7), one can now read off an infinity of formulas, beginning with (2.6)(a). In 1996, the authors could reduce (2.7) to a finite form that they could not prove, but Almquist and Granville did a year later. A decade later, the Wilf-Zeilberger algorithm [28, 22]—for which the inventors were awarded the Steele Prize—directly (as implemented in *Maple*) certified (2.8) [9, 3]. In other words, (2.8) was both discovered and proven by computer.

We found a comparable generating function for $\zeta(2n+4)$, giving (2.6) (c) when $x = 0$, but one for $\zeta(4n+1)$ still eludes us.

2.5. Reciprocal series for π. Truly novel series for $1/\pi$, based on elliptic integrals, were discovered by Ramanujan around 1910 [3, 9, 30]. One is:

$$\frac{1}{\pi} = \frac{2\sqrt{2}}{9801} \sum_{k=0}^{\infty} \frac{(4k)!\,(1103 + 26390k)}{(k!)^4 396^{4k}}. \qquad (2.9)$$

Each term of (2.9) adds eight correct digits. Gosper used (2.9) for the computation of a then-record 17 million digits of π in 1985—thereby completing the first proof of (2.9) [9, Ch. 3]. Shortly thereafter, David and Gregory Chudnovsky found the following variant, which lies in the quadratic number field $Q(\sqrt{-163})$ rather than $Q(\sqrt{58})$:

$$\frac{1}{\pi} = 12 \sum_{k=0}^{\infty} \frac{(-1)^k\,(6k)!\,(13591409 + 545140134k)}{(3k)!\,(k!)^3\,640320^{3k+3/2}}. \qquad (2.10)$$

Each term of (2.10) adds 14 correct digits. The brothers used this formula several times, culminating in a 1994 calculation of π to over four billion decimal digits. Their remarkable story was told in a prizewinning *New Yorker* article [25]. Remarkably, as we already noted earlier, (2.10) was used again in 2010 for the current record computation of π.

2.5.1. Wilf-Zeilberger at work. A few years ago Jésus Guillera found various Ramanujan-like identities for π, using integer relation methods. The three most basic—and entirely rational—identities are:

$$\frac{4}{\pi^2} = \sum_{n=0}^{\infty} (-1)^n r(n)^5 (13 + 180n + 820n^2) \left(\frac{1}{32}\right)^{2n+1} \qquad (2.11)$$

$$\frac{2}{\pi^2} = \sum_{n=0}^{\infty} (-1)^n r(n)^5 (1 + 8n + 20n^2) \left(\frac{1}{2}\right)^{2n+1} \qquad (2.12)$$

$$\frac{4}{\pi^3} \stackrel{?}{=} \sum_{n=0}^{\infty} r(n)^7 (1 + 14n + 76n^2 + 168n^3) \left(\frac{1}{8}\right)^{2n+1}, \qquad (2.13)$$

where $r(n) := (1/2 \cdot 3/2 \cdot \cdots \cdot (2n-1)/2)/n!$.

Guillera proved (2.11) and (2.12) in tandem, by very ingeniously using the Wilf-Zeilberger algorithm [28, 22] for formally proving hypergeometric-like identities [9, 3, 18, 30]. No other proof is known, and there seem to be no like formulae for $1/\pi^N$ with $N \geq 4$. The third, (2.13), is almost certainly true. Guillera ascribes (2.13) to Gourevich, who used integer relation methods to find it.

We were able to "discover" (2.13) using 30-digit arithmetic, and we checked it to 500 digits in 10 seconds, to 1200 digits in 6.25 minutes, and to 1500 digits in 25 minutes, all with naive command-line instructions in *Maple*. But it has no proof, nor does anyone have an inkling of how to prove it; especially, as experiment suggests, since it has no 'mate' in analogy to (2.11) and (2.12) [3]. Our intuition is that if a proof exists, it is more a verification than an explication and so we stopped looking. We are happy just to "know" that the beautiful identity is true (although it would be more remarkable were it eventually to fail). It may be true for no good reason—it might just have no proof and be a very concrete Gödel-like statement.

In 2008 Guillera [18] produced another lovely pair of third-millennium identities—discovered with integer relation methods and proved with creative telescoping—this time for π^2 rather than its reciprocal. They are

$$\sum_{n=0}^{\infty} \frac{1}{2^{2n}} \frac{\left(x+\frac{1}{2}\right)_n^3}{(x+1)_n^3} (6(n+x)+1) = 8x \sum_{n=0}^{\infty} \frac{\left(\frac{1}{2}\right)_n^2}{(x+1)_n^2}, \qquad (2.14)$$

and

$$\sum_{n=0}^{\infty} \frac{1}{2^{6n}} \frac{\left(x+\frac{1}{2}\right)_n^3}{(x+1)_n^3} (42(n+x)+5) = 32x \sum_{n=0}^{\infty} \frac{\left(x+\frac{1}{2}\right)_n^2}{(2x+1)_n^2}. \qquad (2.15)$$

Here $(a)_n = a(a+1)\cdots(a+n-1)$ is the *rising factorial*. Substituting $x = 1/2$ in (2.14) and (2.15), he obtained respectively the formulae

$$\sum_{n=0}^{\infty} \frac{1}{2^{2n}} \frac{(1)_n^3}{\left(\frac{3}{2}\right)_n^3} (3n+2) = \frac{\pi^2}{4}, \qquad \sum_{n=0}^{\infty} \frac{1}{2^{6n}} \frac{(1)_n^3}{\left(\frac{3}{2}\right)_n^3} (21n+13) = 4\frac{\pi^2}{3}.$$

3. Formal Verification of Proof

In 1611, Kepler described the stacking of equal-sized spheres into the familiar arrangement we see for oranges in the grocery store. He asserted that this packing is the tightest possible. This assertion is now known as the Kepler conjecture, and has persisted for centuries without rigorous proof. Hilbert implicitly included the irregular case of the Kepler conjecture in problem 18 of his famous list of unsolved problems in 1900: *whether there exist non-regular space-filling polyhedra?* the regular case having been disposed of by Gauss in 1831.

In 1994, Thomas Hales, now at the University of Pittsburgh, proposed a five-step program that would result in a proof: (a) treat maps that only have triangular faces; (b) show that the face-centered cubic and hexagonal-close packings are local maxima in the strong sense that

EXPLORATORY EXPERIMENTATION AND COMPUTATION

they have a higher score than any Delaunay star with the same graph; (c) treat maps that contain only triangular and quadrilateral faces (except the pentagonal prism); (d) treat maps that contain something other than a triangular or quadrilateral face; and (e) treat pentagonal prisms.

In 1998, Hales announced that the program was now complete, with Samuel Ferguson (son of mathematician-sculptor Helaman Ferguson) completing the crucial fifth step. This project involved extensive computation, using an interval arithmetic package, a graph generator, and *Mathematica*. The computer files containing the source code and computational results occupy more than three Gbytes of disk space. Additional details, including papers, are available at `http://www.math.pitt.edu/~thales/kepler98`. For a mixture of reasons—some more defensible than others—the *Annals of Mathematics* initially decided to publish Hales' paper with a cautionary note, but this disclaimer was deleted before final publication.

Hales [19] has now embarked on a multi-year program to certify the proof by means of computer-based formal methods, a project he has named the "Flyspeck" project. As these techniques become better understood, we can envision a large number of mathematical results eventually being confirmed by computer, as instanced by other articles in the same issue of the *Notices* as Hales' article.

4. Limits of Computation

A remarkable example is the following:

$$\int_0^\infty \cos(2x) \prod_{n=1}^\infty \cos(x/n)\,\mathrm{d}x = \qquad (4.1)$$
$$0.39269908169872415480783042290993786052464\underline{5434187231595926}\ldots$$

The computation of this integral to high precision can be performed using a scheme described in [5]. When we first did this computation, we thought that the result was $\pi/8$, but upon careful checking with the numerical value

$$0.39269908169872415480783042290993786052464\underline{6174921888227621}\ldots,$$

it is clear that the two values disagree beginning with the 43rd digit!

Richard Crandall [14, §7.3] later explained this mystery. Via a physically motivated analysis of *running out of fuel* random walks, he showed that $\pi/8$ is given by the following very rapidly convergent series expansion, of which formula (4.1) above is merely the first term:

$$\frac{\pi}{8} = \sum_{m=0}^{\infty} \int_0^\infty \cos[2(2m+1)x] \prod_{n=1}^{\infty} \cos(x/n) \, \mathrm{d}x. \qquad (4.2)$$

Two terms of the series above suffice for 500-digit agreement.

As a final sobering example, we offer the following "sophomore's dream" identity

$$\begin{aligned}
\sigma_{29} &:= \sum_{n=-\infty}^{\infty} \operatorname{sinc}(n) \operatorname{sinc}(n/3) \operatorname{sinc}(n/5) \cdots \operatorname{sinc}(n/23) \operatorname{sinc}(n/29) \\
&= \int_{-\infty}^{\infty} \operatorname{sinc}(x) \operatorname{sinc}(x/3) \operatorname{sinc}(x/5) \cdots \operatorname{sinc}(x/23) \operatorname{sinc}(x/29) \, dx,
\end{aligned} \qquad (4.3)$$

where the denominators range over the odd primes, which was first discovered empirically. More generally, consider

$$\begin{aligned}
\sigma_p &:= \sum_{n=-\infty}^{\infty} \operatorname{sinc}(n) \operatorname{sinc}(n/3) \operatorname{sinc}(n/5) \operatorname{sinc}(n/7) \cdots \operatorname{sinc}(n/p) \\
&\stackrel{?}{=} \int_{-\infty}^{\infty} \operatorname{sinc}(x) \operatorname{sinc}(x/3) \operatorname{sinc}(x/5) \operatorname{sinc}(x/7) \cdots \operatorname{sinc}(x/p) \, dx.
\end{aligned} \qquad (4.4)$$

Provably, the following is true: The "sum equals integral" identity, for σ_p remains valid at least for p among the first 10176 primes; but stops holding after some larger prime, and thereafter the "sum less the integral" is strictly positive, but *they always differ by much less than one part in a googolplex* $= 10^{100}$. An even stronger estimate is possible assuming the Generalized Riemann Hypothesis (see [14, §7] and [7]).

5. Concluding Remarks

The central issues of how to view experimentally discovered results have been discussed before. In 1993, Arthur Jaffe and Frank Quinn warned of the proliferation of not-fully-rigorous mathematical results and proposed a framework for a "healthy and positive" role for "speculative" mathematics [20]. Numerous well-known mathematicians responded [1]. Morris Hirsch, for instance, countered that even Gauss published incomplete proofs, and the 15,000 combined pages of the proof of the classification of finite groups raises questions as to when we should certify a result. He suggested that we attach a label to each proof – e.g., "computer-aided," "mass collaboration," "constructive," etc. Saunders Mac Lane quipped that "we are not saved by faith alone, but by faith and works," meaning that we need both intuitive work and precision.

At the same time, computational tools now offer remarkable facilities to confirm analytically established results, as in the tools in development to check identities in equation-rich manuscripts, and in Hales' project to establish the Kepler conjecture by formal methods.

The flood of information and tools in our information-soaked world is unlikely to abate. We have to learn and teach judgment when it comes to using what is possible digitally. This means mastering the sorts of techniques we have illustrated and having some idea why a software system does what it does. It requires knowing when a computation is or can—in principle or practice—be made into a rigorous proof and when it is only compelling evidence, or is entirely misleading. For instance, even the best commercial linear programming packages of the sort used by Hales will not certify any solution though the codes are almost assuredly correct. It requires rearranging hierarchies of what we view as hard and as easy.

It also requires developing a curriculum that carefully teaches experimental computer-assisted mathematics. Some efforts along this line are already underway by individuals including Marc Chamberland at Grinnell (`http://www.math.grin.edu/~chamberl/courses/MAT444/syllabus.html`), Victor Moll at Tulane, Jan de Gier in Melbourne, and Ole Warnaar at University of Queensland.

Judith Grabner has noted that a large impetus for the development of modern rigor in mathematics came with the Napoleonic introduction of regular courses: lectures and textbooks force a precision and a codification that apprenticeship obviates. But it will never be the case that quasi-inductive mathematics supplants proof. We need to find a new equilibrium. That said, we are only beginning to tap new ways to enrich mathematics. As Jacques Hadamard said [24]:

> The object of mathematical rigor is to sanction and legitimize the conquests of intuition, and there was never any other object for it.

Never have we had such a cornucopia of ways to generate intuition. The challenge is to learn how to harness them, how to develop and how to transmit the necessary theory and practice. The Priority Research Centre for Computer Assisted Research Mathematics and its Applications (CARMA), `http://www.newcastle.edu.au/research/centres/carmacentre.html`, which one of us directs, hopes to play a lead role in this endeavor: an endeavor which in our view encompasses an exciting mix of exploratory experimentation and rigorous proof.

References

[1] M. Atiyah, et al (1994). "Responses to 'Theoretical Mathematics: Toward a Cultural Synthesis of Mathematics and Theoretical Physics,' by A. Jaffe and F. Quinn," *Bulletin of the American Mathematical Society*, vol. 30, no. 2 (Apr 1994), pp. 178-207.

[2] J. Avigad (2008). "Computers in mathematical inquiry," in *The Philosophy of Mathematical Practice*, P. Mancuso ed., Oxford University Press, pp. 302–316.

[3] D. Bailey, J. Borwein, N. Calkin, R. Girgensohn, R. Luke and V. Moll (2007). *Experimental Mathematics in Action*, A K Peters, Natick, MA, 2007.

[4] D. H. Bailey and J. M. Borwein (2008). "Computer-assisted discovery and proof." *Tapas in Experimental Mathematics*, 21–52, in *Contemporary Mathematics*, vol. 457, American Mathematical Society, Providence, RI, 2008.

[5] D. H. Bailey, J. M. Borwein, V. Kapoor and E. Weisstein. "Ten Problems in Experimental Mathematics," *American Mathematical Monthly*, vol. 113, no. 6 (Jun 2006), pp. 481–409.

[6] D. H. Bailey, P. B. Borwein and S. Plouffe, (1997). "On the Rapid Computation of Various Polylogarithmic Constants," *Mathematics of Computation*, vol. 66, no. 218 (Apr 1997), pp. 903-913.

[7] R. Baillie, D. Borwein, and J. Borwein (2008). "Some sinc sums and integrals," *American Math. Monthly*, vol. 115 (2008), no. 10, pp. 888–901.

[8] J. M. Borwein (2005). "The SIAM 100 Digits Challenge," Extended review in the *Mathematical Intelligencer*, vol. 27 (2005), pp. 40–48.

[9] J. M. Borwein and D. H. Bailey (2008). *Mathematics by Experiment: Plausible Reasoning in the 21st Century*, extended second edition, A K Peters, Natick, MA, 2008.

[10] J. M. Borwein, D. H. Bailey and R. Girgensohn (2004). *Experimentation in Mathematics: Computational Roads to Discovery*, A K Peters, Natick, MA, 2004.

[11] J. M. Borwein, O-Yeat Chan and R. E. Crandall, "Higher-dimensional box integrals." Submitted *Experimental Mathematics*, January 2010.

[12] J. M. Borwein and K. Devlin (2008). *The Computer as Crucible*, A K Peters, Natick, MA, 2008.

[13] J. M. Borwein, I. J. Zucker and J. Boersma, "The evaluation of character Euler double sums," *Ramanujan Journal*, **15** (2008), 377–405.

[14] R. E. Crandall (2007). "Theory of ROOF Walks," 2007, available at http://www.perfscipress.com/papers/ROOF11_psipress.pdf.

[15] H. R. P. Ferguson, D. H. Bailey and S. Arno (1999). "Analysis of PSLQ, An Integer Relation Finding Algorithm," *Mathematics of Computation*, vol. 68, no. 225 (Jan 1999), pp. 351–369.

[16] L. R. Franklin (2005). "Exploratory Experiments," *Philosophy of Science*, **72**, pp. 888–899.

[17] M. Giaquinto (2007). *Visual Thinking in Mathematics. An Epistemological Study*, Oxford University Press, New York, 2007.

[18] J. Guillera (2008). "Hypergeometric identities for 10 extended Ramanujan-type series," *Ramanujan Journal*, vol. 15 (2008), pp. 219–234.

[19] T. C. Hales, "Formal Proof," *Notices of the AMS*, vol. 55, no. 11 (Dec. 2008), pp. 1370–1380.

[20] A. Jaffe and F. Quinn (1993). "'Theoretical Mathematics': Toward a Cultural synthesis of Mathematics and Theoretical Physics," *Bulletin of the American Mathematical Society*, vol. 29, no. 1 (Jul 1993), pp. 1–13.

[21] M. Livio (2009). *Is God a Mathematician?*, Simon and Schuster, New York, 2009.

[22] M. Petkovsek, H. S. Wilf, D. Zeilberger, $A = B$, A K Peters, Natick, MA, 1996.

[23] S. Pinker (2007). *The Stuff of Thought: Language as a Window into Human Nature*, Allen Lane, New York, 2007.

[24] G. Pólya (1981). *Mathematical discovery: On understanding, learning, and teaching problem solving*, (Combined Edition), New York, John Wiley and Sons, New York, 1981.

[25] R. Preston (1992) "The Mountains of Pi," *New Yorker*, 2 Mar 1992, http://www.newyorker.com/archive/content/articles/050411fr_archive01.

[26] D. L. Smail (2008). *On Deep History and the Brain*, Caravan Books, University of California Press, Berkeley, CA, 2008.

[27] H. K. Sørenson (2009). "Exploratory experimentation in experimental mathematics: A glimpse at the PSLQ algorithm," *Philosophy of Mathematics: Sociological Aspects and Mathematical Practice*. In press.

[28] H. S. Wilf and D. Zeilberger, "Rational Functions Certify Combinatorial Identities," *Journal of the American Mathematical Society*, vol. 3 (1990), pp. 147–158.

[29] Tse-Wo Zse, personal communication to the authors, July 2010.

[30] W. Zudilin (2008). "Ramanujan-type formulae for $1/\pi$: A second wind," 19 May 2008, available at http://arxiv.org/abs/0712.1332.

14. Closed forms:
What they are and why they matter

Discussion

This article considers "what is a good mathematical answer?" Such concerns are very old and yet always new. As Bertrand Russell famously put it:

> Thus, mathematics may be defined as the subject in which we never know what we are talking about, nor whether what we are saying is true. People who have been puzzled by the beginnings of mathematics will, I hope, find comfort in this definition, and will probably agree that it is accurate.[1]

There is a great purpose to realistically attempting the impossible, as indicated by the history of the classic Greek paradoxes, the non-solvability of the quintic, or indeed Hilbert's program—all of which have been mentioned in earlier articles in this collection.

Whatever the future of computationally assisted mathematics, some fundamental mathematical matters will remain imprecise, and it is fitting that Hermann Weyl, writing about David Hilbert, should have the final word:

> The question of the ultimate foundations and the ultimate meaning of mathematics remains open: We do not know in what direction it will find its final solution or even whether a final objective answer can be expected at all. 'Mathematizing' may well be a creative activity of man, like language or music, of primary originality, whose Historical decisions defy complete objective rationalisation."—Hermann Weyl.[2]

> The problems of mathematics are not problems in a vacuum. There pulses in them the life of ideas which realize themselves in concreto through our [or throughout] human endeavors in our historical existence, but forming an indissoluble whole transcending any particular science."[3]

Source

J.M. Borwein and R.E. Crandall, "Closed forms: what they are and why we care," *Notices Amer. Math. Soc.* Submitted September 2010.

---

[1] From "Recent Work on the Principles of Mathematics in International Monthly," **4** (July, 1901), 83–101. (Collected Papers, v3, p.366; revised version in Newman's World of Mathematics, v3, p. 1577.)

[2] In "Obituary: David Hilbert 1862 - 1943," *RSBIOS,* **4**, 1944, 547–553; and American Philosophical Society Year Book, 1944, 387–395, p. 392.

[3] In "David Hilbert and his mathematical work," *Bull. Am. Math. Soc.,* **50** (1944), p. 615.

CLOSED FORMS: WHAT THEY ARE AND WHY THEY MATTER

JONATHAN M. BORWEIN AND RICHARD E. CRANDALL

ABSTRACT. The term "closed form" is one of those mathematical notions that is commonplace, yet virtually devoid of rigor. And, there is disagreement even on the intuitive side; for example, most everyone would say that $\pi + \log 2$ is a closed form, but some of us would think that the Euler constant γ is not closed. Like others before us, we shall try to supply some missing rigor to the notion of closed forms and also to give examples from modern research where the question of closure looms both important and elusive.

1. CLOSED FORMS: WHAT THEY ARE

Mathematics abounds in terms which are in frequent use yet which are rarely made precise. Two such are *rigorous proof* and *closed form* (absent the technical use within differential algebra). If a rigorous proof is "that which 'convinces' the appropriate audience" then a closed form is "that which looks 'fundamental' to the requisite consumer." In both cases, this is a community-varying and epoch-dependent notion. What was a compelling proof in 1810 may well not be now; what is a fine closed form in 2010 may have been anathema a century ago. In the article we are intentionally informal as befits a topic that intrinsically has no one "right" answer.

Let us begin by sampling the Web for various approaches to informal definitions of "closed form."

1.0.1. First approach to a definition of closed form. The first comes from MathWorld [55] and so may well be the first and last definition a student or other seeker-after-easy-truth finds.

> An equation is said to be a closed-form solution if it solves a given problem in terms of functions and mathematical operations from a given generally accepted set. For example, an infinite sum would generally not be considered closed-form. However, the choice of what to call closed-form and what not is rather arbitrary since a new "closed-form" function could simply be defined in terms of the infinite sum.—Eric Weisstein

There is not much to disagree with in this but it is far from rigorous.

Date: November 30, 2010.
Borwein: Centre for Computer Assisted Research Mathematics and its Applications (CARMA), University of Newcastle, Callaghan, NSW 2308, Australia. Email: jonathan.borwein@newcastle.edu.au.
Crandall: Center for Advanced Computation, Reed College, Portland OR, USA. Email: crandall@reed.edu.

1.0.2. Second approach.
The next attempt follows a 16 September 1997 question to the long operating "Dr. Math." site[1] and is a good model of what interested students are likely to be told.

> Subject: Closed form solutions
> Dear Dr. Math, What is the exact mathematical definition of a closed form solution? Is a solution in "closed form" simply if an expression relating all of the variables can be derived for a problem solution, as opposed to some higher-level problems where there is either no solution, or the problem can only be solved incrementally or numerically?
> Sincerely, ...

The answer followed on 22 Sept:

> This is a very good question! This matter has been debated by mathematicians for some time, but without a good resolution.
>
> Some formulas are agreed by all to be "in closed form." Those are the ones which contain only a finite number of symbols, and include only the operators $+, -, *, /,$ and a small list of commonly occurring functions such as n-th roots, exponentials, logarithms, trigonometric functions, inverse trigonometric functions, greatest integer functions, factorials, and the like.
>
> More controversial would be formulas that include infinite summations or products, or more exotic functions, such as the Riemann zeta function, functions expressed as integrals of other functions that cannot be performed symbolically, functions that are solutions of differential equations (such as Bessel functions or hypergeometric functions), or some functions defined recursively. Some functions whose values are impossible to compute at some specific points would probably be agreed not to be in closed form (example: $f(x) = 0$ if x is an algebraic number, but $f(x) = 1$ if x is transcendental. For most numbers, we do not know if they are transcendental or not). I hope this is what you wanted.

No more formal, but representative of many dictionary definitions is:

1.0.3. Third approach.
A coauthor of the current article is at least in part responsible for the following brief definition from a recent mathematics dictionary [16]:

> **closed form** n. an expression for a given function or quantity, especially an integral, in terms of known and well understood quantities, such as the evaluation of
> $$\int_{-\infty}^{\infty} \exp(-x^2)\, dx$$
> as $\sqrt{\pi}$.—Collins Dictionary

And of course one cares more for a closed form when the object under study is important, such as when it engages the normal distribution as above.

With that selection recorded, let us turn to some more formal proposals.

---
[1]Available at http://mathforum.org/dr/math/.

1.0.4. Fourth approach.
Various notions of elementary numbers have been proposed.

Definition [30]. A subfield F of $\mathbb{C}$ is *closed under* exp and log if (1) $\exp(x) \in F$ for all $x \in F$ and (2) $\log(x) \in F$ for all nonzero $x \in F$, where log is the branch of the natural logarithm function such that $-\pi < \text{Im}(\log x) \leq \pi$ for all x. The field $\mathbb{E}$ of EL numbers is the intersection of all subfields of $\mathbb{C}$ that are closed under exp and log.—Tim Chow

Tim Chow explains nicely why he eschews capturing all algebraic numbers in his definition; why he wishes only elementary quantities to have closed forms; whence he prefers $\mathbb{E}$ to Ritt's 1948 definition of *elementary numbers* as the smallest algebraically closed subfield $\mathbb{L}$ of $\mathbb{C}$ that is closed under exp and log. His reasons include that:

> Intuitively, "closed-form" implies "explicit," and most algebraic functions have no simple explicit expression.

Assuming *Shanuel's conjecture* that *given n complex numbers $z_1, ..., z_n$ which are linearly independent over the rational numbers, the extension field*

$$\mathbb{Q}(z_1, ..., z_n, e^{z_1}, ..., e^{z_n})$$

has transcendence degree of at least n over the rationals, then the algebraic members of $\mathbb{E}$ are exactly those solvable in radicals [30]. We may thence think of Chow's class as the smallest plausible class of closed forms. Only a mad version of Markov would want to further circumscribe the class.

1.1. Special functions.
In an increasingly computational world, an explicit/implicit dichotomy is occasionally useful; but not very frequently. Often we will prefer computationally the numerical implicit value of an algebraic number to its explicit tower of radicals; and it seems increasingly perverse to distinguish the root of $2x^5 - 10x + 5$ from that of $2x^4 - 10x + 5$ or to prefer $\arctan(\pi/7)$ to $\arctan(1)$. We illustrate these issues further in Example 3.1, 3.3 and 4.3.

We would prefer to view all values of classical *special functions* of mathematical physics [53] at algebraic arguments as being closed forms. Again there is no generally accepted class of special functions, but most practitioners would agree that the solutions to the classical second-order algebraic differential equations (linear or say Painlevé) are included. But so are various *hypertranscendental functions* such as Γ, B and ζ which do not arise in that way.[2]

Hence, we do not wish to accept any definition of special function which relies on the underlying functions satisfying natural differential equations. The class must be extensible, new special functions are always being discovered.

A recent *American Mathematical Monthly* review[3] of [46] says:

> There's no rigorous definition of special functions, but the following definition is in line with the general consensus: functions that are commonly used in applications, have many nice properties, and are not typically available on a calculator. Obviously

---

[2]Of course a value of an hypertranscendental function at algebraic argument may be very well behaved, see Example 1.4.

[3]Available at http://www.maa.org/maa%20reviews/4221.html.

this is a sloppy definition, and yet it works fairly well in practice. Most people would agree, for example, that the Gamma function is included in the list of special functions and that trigonometric functions are not.

Once again, there is much to agree with, and much to quibble about, in this reprise. That said, most serious books on the topic are little more specific. For us, special functions are non-elementary functions about which a significant literature has developed because of their importance in either mathematical theory or in practice. We certainly prefer that this literature includes the existence of excellent algorithms for their computation. This is all consonant with—if somewhat more ecumenical than—Temme's description in the preface of his lovely book [53, Preface p. xi]:

> [W]e call a function "special" when the function, just like the logarithm, the exponential and trigonometric functions (the elementary transcendental functions), belongs to the toolbox of the applied mathematician, the physicist or engineer.

Even more economically, Andrews, Askey and Roy start the preface to their important book *Special functions* [1] by writing:

> Paul Turan once remarked that special functions would be more appropriately labeled "useful functions."

With little more ado, they then start to work on the Gamma and Beta functions; indeed the term "special function" is not in their index. Near the end of their preface, they also write

> [W]e suggest that the day of formulas may be experiencing a new dawn.

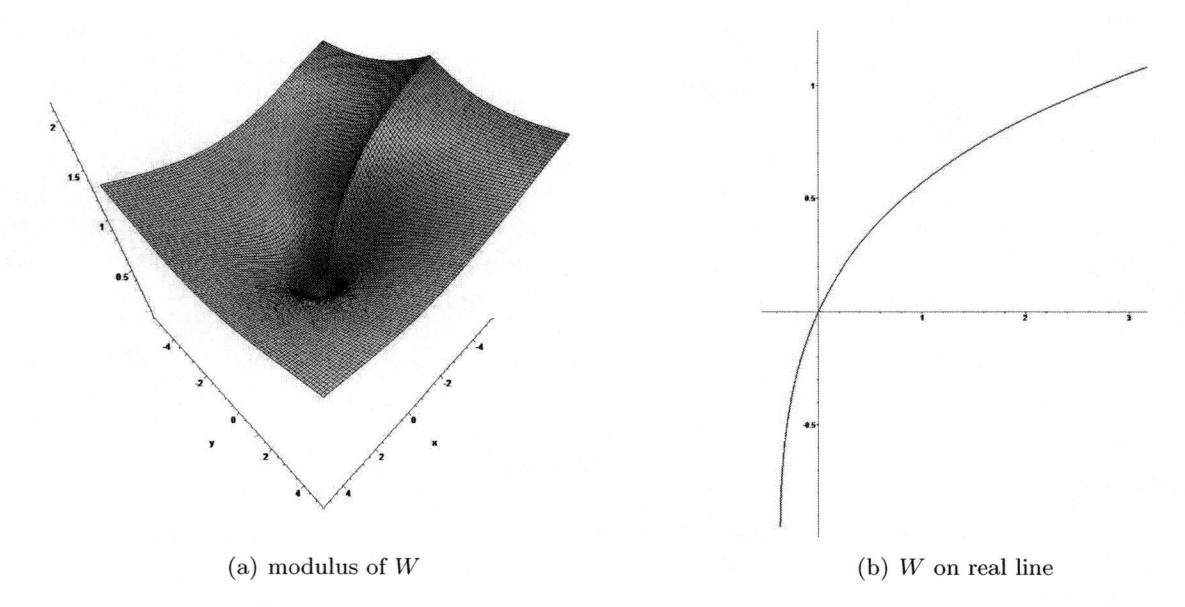

(a) modulus of W (b) W on real line

FIGURE 1. The Lambert W function.

Example 1.1 (Lambert's W). The *Lambert W* function, $W(x)$, is defined by appropriate solution of $y \cdot e^y = x$ [20, pp. 277–279]. This function has been implemented in computer algebra systems (CAS); and has many uses despite being unknown to most scientists and only relatively recently named [40]. It is now embedded as a primitive in *Maple* and *Mathematica* with the same status as any other well studied special or elementary function. (See for example the tome [25].) The CAS know its power series and much more. For instance in *Maple* entering:
```
> fsolve(exp(x)*x=1);identify(%);
```
returns
```
      0.5671432904, LambertW(1)
```

We consider this to be a splendid closed form even though, again assuming Shanuel's conjecture, $W(1) \notin \mathbb{E}$ [30]. Additionally, it is only recently rigorously proven that W is not an elementary function in Liouville's precise sense [25]. We also note that successful simplification in a modern CAS [28] requires a great deal of knowledge of special functions. □

1.2. Further approaches.

1.2.1. Fifth approach.
PlanetMath's offering, as of 15 February 2010[4], is certainly in the elementary number corner.

> **expressible in closed form** (Definition) An expression is expressible in a closed form, if it can be converted (simplified) into an expression containing only elementary functions, combined by a finite amount of rational operations and compositions.— Planet Math

This reflects both much of what is best and what is worst about 'the mathematical wisdom of crowds'. For the reasons adduced above, we wish to distinguish—but admit both—those closed forms which give analytic insight from those which are sufficient and prerequisite for effective computation. Our own current preferred class [7] is described next.

1.2.2. Sixth approach.
We wish to establish a set $\mathbb{X}$ of *generalized hypergeometric* evaluations; see [7] for an initial, rudimentary definition which we shall refine presently. First, we want any convergent sum

$$x = \sum_{n \geq 0} c_n z^n \tag{1.1}$$

to be an element of our set $\mathbb{X}$, where z is algebraic, c_0 is rational, and for $n > 0$,

$$c_n = \frac{A(n)}{B(n)} c_{n-1}$$

for integer polynomials A, B with $\deg A \leq \deg B$. Under these conditions the expansion for x converges absolutely on the open disk $|z| < 1$. However, we also allow x to be any finite analytic-continuation value of such a series; moreover, when z lies on a branch

---
[4]Available at http://planetmath.org/encyclopedia/ClosedForm4.html.

cut we presume both branch limits to be elements of $\mathbb{X}$. (See ensuing examples for some clarification.) It is important to note that our set $\mathbb{X}$ is closed under rational multiplication, due to freedom of choice for c_0.

Example 1.2 (First members of $\mathbb{X}$). The *generalized hypergeometric function* evaluation
$$_{p+1}F_p\left(\begin{array}{c}a_1,\ldots a_{p+1}\\ b_1,\ldots,b_p\end{array}\bigg|z\right)$$
for rational a_i, b_j with all b_j positive has branch cut $z \in (0, \infty)$, and the evaluation is an element of $\mathbb{X}$ for complex z not on the cut (and the evaluation on each side of said cut is also in $\mathbb{X}$).

The *trilogarithm* $\text{Li}_3(z) := \sum_{n \geq 1} z^n/n^3$ offers a canonical instance. Formally,
$$\frac{1}{z}\text{Li}_3(z) = {}_4F_3\left(\begin{array}{c}1,1,1,1\\ 2,2,2\end{array}\bigg|z\right).$$
and for $z = 1/2$ the hypergeometric series converges absolutely, with
$$\text{Li}_3\left(\frac{1}{2}\right) = \frac{7}{8}\zeta(3) + \frac{1}{6}\log^3 2 - \frac{\pi^2}{12}\log 2.$$
Continuation values at $z = 2$ on the branch cut can be inferred as
$$\lim_{\epsilon \to 0^+} \text{Li}_3(2 \pm i\epsilon) = \frac{7}{16}\zeta(3) + \frac{\pi^2}{8}\log 2 \pm i\frac{\pi}{4}\log 2,$$
so both complex numbers on the right here are elements of $\mathbb{X}$. The *quadralogarithmic* value $\text{Li}_4\left(\frac{1}{2}\right)$ is thought not to be similarly decomposable but likewise belongs to $\mathbb{X}$. □

Now we are prepared to posit

Definition [7]. The *ring of hyperclosure* $\mathbb{H}$ is the smallest subring of $\mathbb{C}$ containing the set $\mathbb{X}$. Elements of $\mathbb{H}$ are deemed *hyperclosed*.

In other words, the ring $\mathbb{H}$ is generated by all general hypergeometric evaluations, under the $\cdot, +$ operators, all symbolized by
$$\mathbb{H} = \langle\mathbb{X}\rangle_{\cdot, +}.$$
$\mathbb{H}$ will contain a great many interesting closed forms from modern research. Note that $\mathbb{H}$ contains all closed forms in the sense of Wilf and Zeilberger [47, Ch. 8] wherein only finite linear combinations of hypergeometric evaluations are allowed.

So what numbers are in the ring $\mathbb{H}$? First off, *almost no* complex numbers belong to this ring! This is easily seen by noting that the set of general hypergeometric evaluations is countable, so the generated ring must also be countable. Still, a great many fundamental numbers are provably hyperclosed. Examples follow, in which we let ω denote an arbitrary algebraic number and n any positive integer:

$$\omega, \log\omega, e^\omega, \pi$$
the dilogarithmic combination $\text{Li}_2\left(\frac{1}{\sqrt{5}}\right) + (\log 2)(\log 3),$

the elliptic integral $K(\omega)$,

the zeta function values $\zeta(n)$,

special functions such as the Bessel evaluations $J_n(\omega)$.

Incidentally, it occurs in some modern experimental developments that the real or imaginary part of a hypergeometric evaluation is under scrutiny. Generally, $\Re, \Im$ operations preserve hyperclosure, simply because the series (or continuations) at z and z^* can be linearly combined in the ring $\mathbb{H}$. Referring to Example 1.2, we see that for algebraic z, the number $\Re(\text{Li}_3(z))$ is hyperclosed; and even on the cut, $\Re(\text{Li}_3(2)) = \frac{7}{16}\zeta(3) + \frac{\pi^2}{8}\log 2$ is hyperclosed. In general, $\Re\left({}_{p+1}F_p\left(\begin{smallmatrix}\cdots\\\cdots\end{smallmatrix}\Big| z\right)\right)$ is hyperclosed.

We are not claiming that hyperclosure is any kind of final definition for "closed forms." But we do believe that any defining paradigm for closed forms must include this ring of hyperclosure $\mathbb{H}$. One way to reach further is to define a *ring of superclosure* as the closure

$$\mathbb{S} := \langle \mathbb{H}^{\mathbb{H}} \rangle_{\cdot, +}.$$

This ring contains numbers such as

$$e^\pi + \pi^e, \; \frac{1}{\zeta(3)^{\zeta(5)}},$$

and of course a vast collection of numbers that may not belong to $\mathbb{H}$ itself. If we say that an element of $\mathbb{S}$ is *superclosed*, we still preserve the countability of all superclosed numbers. Again, any good definition of "closed form" should incorporate whatever is in the ring $\mathbb{S}$.

1.3. Seventh approach. In a more algebraic topological setting, it might make sense to define closed forms to be those arising as *periods*—that is as integrals of rational functions (with integer parameters) in n variables over domains defined by algebraic equations. These ideas originate in the theory of elliptic and abelian integrals and are deeply studied [41]. Periods form an algebra and certainly capture many constants. They are especially well suited to the study of L-series, multi zeta values, polylogarithms and the like, but again will not capture all that we wish. For example, e is conjectured not to be a period, as is Euler's constant γ (see Section 5). Moreover, while many periods have nice series, it is not clear that all do.

As this takes us well outside our domain of expertise we content ourselves with two examples originating in the study of Mahler measures. We refer to a fundamental paper by Denninger [38] and a very recent paper of Rogers [49] for details.

Example 1.3 (Periods and Mahler Measures [38]). The logarithmic *Mahler measure* of a polynomial P in n-variables can be defined as

$$\mu(P) := \int_0^1 \int_0^1 \cdots \int_0^1 \log\left|P\left(e^{2\pi i \theta_1}, \cdots, e^{2\pi i \theta_n}\right)\right| d\theta_1 \cdots d\theta_n.$$

Then $\mu(P)$ turns out to be an example of a period and its exponential, $M(P) := \exp(\mu(P))$, is a mean of the values of P on the unit n-torus. When $n = 1$ and P has integer coefficients $M(P)$ is always an algebraic integer. An excellent online synopsis can be found in Dave

Boyd's article http://eom.springer.de/m/m120070.htm. Indeed, Boyd has been one of the driving forces in the field. A brief introduction to the univariate case is also given in [21, 358–359].

There is a remarkable series of recent results—many more discovered experimentally than proven—expressing various multi-dimensional $\mu(P)$ as arithmetic quantities. Boyd observes that there appears to be a tight connection to K-theory. An early result due to Boyd is that $\mu(1 + x + y + 1/x + 1/y) = 7\zeta(3)/\pi^2$, a number that is certainly hyperclosed, being as both $\zeta(3)$ and $1/\pi$ are. A more recent result due to Smyth (see [50], also [52]) is that $\mu(1 + x + y) = L_3'(-1)$. Here L_3 is the Dirichlet L-series modulo three. A conjecture of Denninger [38], confirmed to over 50 places, is that

$$\mu(1 + x + y + 1/x + 1/y) \stackrel{?}{=} \frac{15}{\pi^2} L_E(2) \tag{1.2}$$

is an L-series value over an elliptic curve E with conductor 15. Rogers [49] recasts (1.2) as

$$F(3, 5) \stackrel{?}{=} \frac{15}{\pi^2} \sum_{n=0}^{\infty} \binom{2n}{n}^2 \frac{(1/16)^{2n+1}}{2n + 1}, \tag{1.3}$$

where

$$F(b, c) := (1 + b)(1 + c) \sum_{n,m,j,k} \frac{(-1)^{n+m+j+k}}{((6n + 1)^2 + b(6m + 1)^2 + c(6j + 1)^2 + bc(6k + 1)^2)^2}$$

is a four-dimensional lattice sum.

While (1.3) remains a conjecture, Rogers is able to evaluate many values of $F(b, c)$ in terms of Meijer-G or hypergeometric functions. We shall consider the most famous crystal sum, the Madelung constant, in Example 5.1. □

It is striking how beautiful combinatorial games can be when played under the rubric of hyper- or superclosure.

Example 1.4 (Superclosure of Γ at rational). Let us begin with the *Beta function*

$$B(r, s) := \frac{\Gamma(r)\Gamma(s)}{\Gamma(r + s)}.$$

with $\Gamma(s)$ defined if one wishes as $\Gamma(s) := \int_0^\infty t^{s-1} e^{-t}\, dt$. It turns out that for any rationals r, s both B and $1/B$ are hyperclosed. This is immediate from the hypergeometric identities

$$\frac{1}{B(r, s)} = \frac{rs}{r + s} {}_2F_1\left(\begin{matrix}-r, -s \\ 1\end{matrix}\bigg| 1\right)$$

$$B(r, s) = \frac{\pi \sin \pi(r + s)}{\sin \pi r \sin \pi s} \frac{(1 - r)_M (1 - s)_M}{M!(1 - r - s)_M} {}_2F_1\left(\begin{matrix}r, s \\ M + 1\end{matrix}\bigg| 1\right),$$

where M is any integer chosen such that the hypergeometric series converges, say $M = \lceil 1 + r + s \rceil$. (Each of these Beta relations is a variant of the celebrated Gauss evaluation of ${}_2F_1$ at 1 [1, 53] and is also the reason B is a period.)

CLOSED FORMS: WHAT THEY ARE AND WHY THEY MATTER

We did not seize upon the Beta function arbitrarily, for, remarkably, the hyperclosure of $B^{\pm 1}(r,s)$ leads to compelling results on the Gamma function itself. Indeed, consider for example this product of four Beta-function evaluations:

$$\frac{\Gamma(1/5)\Gamma(1/5)}{\Gamma(2/5)} \cdot \frac{\Gamma(2/5)\Gamma(1/5)}{\Gamma(3/5)} \cdot \frac{\Gamma(3/5)\Gamma(1/5)}{\Gamma(4/5)} \cdot \frac{\Gamma(4/5)\Gamma(1/5)}{\Gamma(5/5)}.$$

We know this product is hyperclosed. But upon inspection we see that the product is just $\Gamma^5(1/5)$. Along such lines one can prove that for any positive rational a/b (in lowest terms), we have hyperclosure of powers of the Gamma-function, in the form:

$$\Gamma^{\pm b}(a/b) \in \mathbb{H}.$$

Perforce, we have therefore a superclosure result for any Γ(rational) and its reciprocal:

$$\Gamma^{\pm 1}(a/b) \in \mathbb{S}.$$

Again like calculations show $\Gamma^b(a/b)$ is a period [41]. One fundamental consequence is thus: $\Gamma^{-2}\left(\frac{1}{2}\right) = \frac{1}{\pi}$ is hyperclosed; thus every integer power of π is hyperclosed.

Incidentally, deeper combinatorial analysis shows that—in spite of our $\Gamma^5\left(\frac{1}{5}\right)$ Beta-chain above, it really only takes *logarithmically many* (i.e., $O(\log b)$) hypergeometric evaluations to write Gamma-powers. For example,

$$\Gamma^{-7}\left(\frac{1}{7}\right) = \frac{1}{2^3 7^6} {}_2F_1\left(\begin{array}{c}-\frac{1}{7},-\frac{1}{7}\\1\end{array}\bigg|1\right)^4 {}_2F_1\left(\begin{array}{c}-\frac{2}{7},-\frac{2}{7}\\1\end{array}\bigg|1\right)^2 {}_2F_1\left(\begin{array}{c}-\frac{4}{7},-\frac{4}{7}\\1\end{array}\bigg|1\right).$$

We note also that for $\Gamma(n/24)$ with n integer, elliptic integral algorithms are known which converge as fast as those for π [26, 21]. □

The above remarks on superclosure of $\Gamma(a/b)$ lead to the property of superclosure for special functions such as $J_\nu(\omega)$ for algebraic ω and rational ν; and for many of the mighty Meijer-G functions, as the latter can frequently be written by Slater's theorem [14] as superpositions of hypergeometric evaluations with composite-gamma products as coefficients. (See Example 3.2 below for instances of Meijer-G in current research.)

There is an interesting alternative way to envision hyperclosure, or at least something very close to our above definition. This is an idea of J. Carette [27], to the effect that solutions at algebraic end-points, and algebraic initial points, for *holonomic* ODEs— i.e. differential-equation systems with integer-polynomial coefficients—could be considered closed. One might say *diffeoclosed*. An example of a diffeoclosed number is $J_1(1)$, i.e., from the Bessel differential equation for $J_1(z)$ with $z \in [0,1]$; it suffices without loss of generality to consider topologically clean trajectories of the variable over $[0,1]$. There is a formal ring of diffeoclosure, which ring is very similar to our $\mathbb{H}$; however there is the caution that trajectory solutions can sometimes have nontrivial topology, so precise ring definitions would need to be effected carefully.

It is natural to ask "what is the complexity of hypergeometric evaluations?" Certainly for the converging forms with variable z on the open unit disk, convergence is geometric, requiring $O(D^{1+\epsilon})$ operations to achieve D good digits. However, in very many cases this can be genuinely enhanced to $O(D^{1/2+\epsilon})$ [21].

2. Closed Forms: Why They Matter

> In many optimization problems, *simple, approximate* solutions are more useful than complex exact solutions.—Steve Wright

As Steve Wright observed in a recent lecture on *sparse optimization* it may well be that a complicated analytic solution is practically intractable but a simplifying assumption leads to a very practical closed form approximation (e.g., in compressed sensing). In addition to appealing to Occam's razor, Wright instances that:

(a) the data quality may not justify exactness;
(b) the simple solution may be more robust;
(c) it may be easier to "*explain/ actuate/ implement/ store*";
(d) and it may conform better to prior knowledge.

As mathematical discovery more and more involves extensive computation, the premium on having a closed form increases. The insight provided by discovering a closed form ideally comes at the top of the list but efficiency of computation will run a good second.

Example 2.1 (The amplitude of a pendulum). *Wikipedia*[5] after giving the classical small angle (simple harmonic) approximation

$$p \approx 2\pi \sqrt{\frac{L}{g}}$$

for the period p of a pendulum of length L and amplitude α, develops the exact solution in a form equivalent to

$$p = 4\sqrt{\frac{L}{g}} \, \mathrm{K}\left(\sin \frac{\alpha}{2}\right)$$

and then says:

> This integral cannot be evaluated in terms of elementary functions. It can be rewritten in the form of the elliptic function of the first kind (also see Jacobi's elliptic functions), which gives little advantage since that form is also insoluble.

True, an elliptic-integral solution is not elementary, yet the notion of insolubility is misleading for two reasons: First, it is known that for some special angles α, the pendulum period can be given a closed form. As discussed in [32], one exact solution is, for $\alpha = \pi/2$ (so pendulum is released from horizontal-rod position),

$$p = \left(2\pi \sqrt{\frac{L}{g}}\right) \frac{\sqrt{\pi}}{\Gamma^2(3/4)}.$$

It is readily measurable in even a rudimentary laboratory that the excess factor here, $\sqrt{\pi}\Gamma^{-2}(3/4) \approx 1.18034$ looks just right, i.e., a horizontal-release pendulum takes 18 per cent longer to fall. Moreover, there is an *exact* dynamical solution for the time-dependent angle $\alpha(t)$; namely for a pendulum with $\alpha(\pm\infty) = \pm\pi$ and $\alpha(0) = 0$, i.e. the bob crosses angle zero (hanging straight down) at time zero, but in the limits of time $\to \pm\infty$ the bob

---

[5] Available at http://en.wikipedia.org/wiki/Pendulum_(mathematics).

ends up straight vertical. We have period $p = \infty$, yet the *exact* angle $\alpha(t)$ for given t can be written down in terms of elementary functions!

The second misleading aspect is this: K is—for any α—remarkably tractable in a computational sense. Indeed K admits of a quadratic transformation

$$\mathrm{K}(k) = (1+k_1)\mathrm{K}(k_1), \quad k_1 := \frac{1-\sqrt{1-k^2}}{1+\sqrt{1-k^2}} \tag{2.1}$$

as was known already to Landen, Legendre and Gauss.

In fact all elementary function to very high precision are well computed via K [21]. So the comment was roughly accurate in the world of slide rules or pocket calculators; it is misleading today—if one has access to any computer package. Nevertheless, both deserve to be called closed forms: one exact and the other an elegant approximate closed form (excellent in its domain of applicability, much as with Newtonian mechanics) which is equivalent to

$$\mathrm{K}\left(\sin\frac{\alpha}{2}\right) \approx \frac{\pi}{2}$$

for small initial amplitude α. To compute $K(\pi/6) = 1.699075885\ldots$ to five places requires using (2.1) only twice and then estimating the resultant integral by $\pi/2$. A third step gives the ten-digit precision shown. □

It is now the case that much mathematical computation is *hybrid*: mixing numeric and symbolic computation. Indeed, which is which may not be clear to the user if, say, numeric techniques have been used to return a symbolic answer or if a symbolic closed form has been used to make possible a numerical integration. Moving from classical to modern physics, both understanding and effectiveness frequently demand hybrid computation.

Example 2.2 (Scattering amplitudes [2]). An international team of physicists, in preparation for the Large Hadron Collider (LHC), is computing scattering amplitudes involving quarks, gluons and gauge vector bosons, in order to predict what results could be expected on the LHC. By default, these computations are performed using conventional double precision (64-bit IEEE) arithmetic. Then if a particular phase space point is deemed numerically unstable, it is recomputed with double-double precision. These researchers expect that further optimization of the procedure for identifying unstable points may be required to arrive at an optimal compromise between numerical accuracy and speed of the code. Thus they plan to incorporate arbitrary precision arithmetic, into these calculations. Their objective is to design a procedure where instead of using fixed double or quadruple precision for unstable points, the number of digits in the higher precision calculation is dynamically set according to the instability of the point. Any subroutine which uses a closed form symbolic solution (exact or approximate) is likely to prove much more robust and efficient. □

3. Detailed Examples

We start with three examples originating in [15].

In the January 2002 issue of *SIAM News*, Nick Trefethen presented ten diverse problems used in teaching *modern* graduate numerical analysis students at Oxford University, the answer to each being a certain real number. Readers were challenged to compute ten digits of each answer, with a $100 prize to the best entrant. Trefethen wrote,

"If anyone gets 50 digits in total, I will be impressed."

To his surprise, a total of 94 teams, representing 25 different nations, submitted results. Twenty of these teams received a full 100 points (10 correct digits for each problem). The problems and solutions are dissected most entertainingly in [15]. One of the current authors wrote the following in a review [18] of [15].

> Success in solving these problems required a broad knowledge of mathematics and numerical analysis, together with significant computational effort, to obtain solutions and ensure correctness of the results. As described in [15] the strengths and limitations of *Maple*, *Mathematica*, MATLAB (The *3Ms*), and other software tools such as PARI or GAP, were strikingly revealed in these ventures. Almost all of the solvers relied in large part on one or more of these three packages, and while most solvers attempted to confirm their results, there was no explicit requirement for proofs to be provided.

Example 3.1 (Trefethen problem #2 [15, 18]).

A photon moving at speed 1 in the x-y plane starts at $t = 0$ at $(x, y) = (1/2, 1/10)$ heading due east. Around every integer lattice point (i, j) in the plane, a circular mirror of radius $1/3$ has been erected. How far from the origin is the photon at $t = 10$?

Using *interval arithmetic* with starting intervals of size smaller than 10^{-5000}, one can actually find the position of the particle at time 2000—not just time ten. This makes a fine exercise in very high-precision interval computation, but in absence of any closed form one is driven to such numerical gymnastics to deal with error propagation. □

Example 3.2 (Trefethen's problem #9 [15, 18]).

The integral $I(a) = \int_0^2 [2 + \sin(10\alpha)] x^\alpha \sin(\alpha/(2 - x)) \, dx$ depends on the parameter α. What is the value $\alpha \in [0, 5]$ at which $I(\alpha)$ achieves its maximum?

The maximum parameter is expressible in terms of a *Meijer-G function* which is a special function with a solid history. While knowledge of this function was not common among the contestants, *Mathematica* and *Maple* both will figure this out [14], and then the help files or a web search will quickly inform the scientist.

This is another measure of the changing environment. It is usually a good idea—and not at all immoral—to data-mine. These Meijer-G functions, first introduced in **1936**, also occur in quantum field theory and many other places [8]. For example, the moments of an n-step random walk in the plane are given for $s > 0$ by

$$W_n(s) := \int_{[0,1]^n} \left| \sum_{k=1}^n e^{2\pi x_k i} \right|^s dx. \tag{3.1}$$

CLOSED FORMS: WHAT THEY ARE AND WHY THEY MATTER

It transpires [23, 35] that for all complex s

$$W_3(s) = \frac{\Gamma(1+s/2)}{\Gamma(1/2)\Gamma(-s/2)} G_{3,3}^{2,1}\left(\begin{array}{c}1,1,1\\ \frac{1}{2},-\frac{s}{2},-\frac{s}{2}\end{array}\bigg|\frac{1}{4}\right), \tag{3.2}$$

Moreover, for s not an odd integer, we have

$$W_3(s) = \frac{1}{2^{2s+1}}\tan\left(\frac{\pi s}{2}\right)\left(\frac{s}{\frac{s-1}{2}}\right)^2 {}_3F_2\left(\begin{array}{c}\frac{1}{2},\frac{1}{2},\frac{1}{2}\\ \frac{s+3}{2},\frac{s+3}{2}\end{array}\bigg|\frac{1}{4}\right) + \binom{s}{\frac{s}{2}}{}_3F_2\left(\begin{array}{c}-\frac{s}{2},-\frac{s}{2},-\frac{s}{2}\\ 1,-\frac{s-1}{2}\end{array}\bigg|\frac{1}{4}\right).$$

We have not given the somewhat technical definition of MeijerG, but *Maple*, *Mathematica*, *Google* searches, *Wikipedia*, the *DLMF* or many other tools will.

There are two corresponding formulae for W_4. We thus know, from our "Sixth approach" section previous in regard to superclosure of Γ-evaluations, that *both $W_3(q), W_4(q)$ are superclosed for rational argument q* for q not an odd integer. We illustrate by showing graphs of W_3, W_4 on the real line in Figure 2 and in the complex plane in Figure 3. The later highlights the utility of the Meijer-G representations. Note the poles and removable singularities.

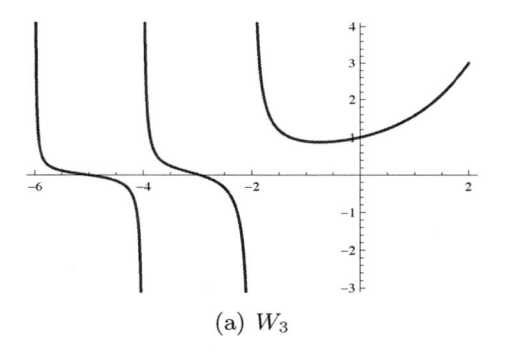

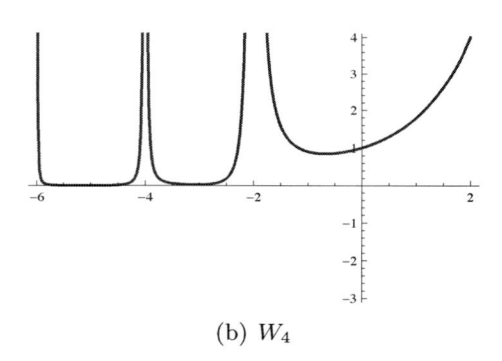

(a) W_3

(b) W_4

FIGURE 2. Moments of n-step walks in the plane. W_3, W_4 analytically continued to the real line.

The Meijer-G functions are now described in the newly completed *Digital Library of Mathematical Functions*[6] and as such are now full, indeed central, members of the family of special functions. □

Example 3.3 (Trefethen's problem #10 [15, 18]).

> A particle at the center of a 10×1 rectangle undergoes Brownian motion (i.e., 2-D random walk with infinitesimal step lengths) till it hits the boundary. What is the probability that it hits at one of the ends rather than at one of the sides?

---

[6]A massive revision of Abramowitz and Stegun—with the now redundant tables removed, it is available at www.dlmf.nist.gov. The hard copy version is also now out [44]. It is not entirely a substitute for the original version as coverage has changed.

14   JONATHAN M. BORWEIN AND RICHARD E. CRANDALL

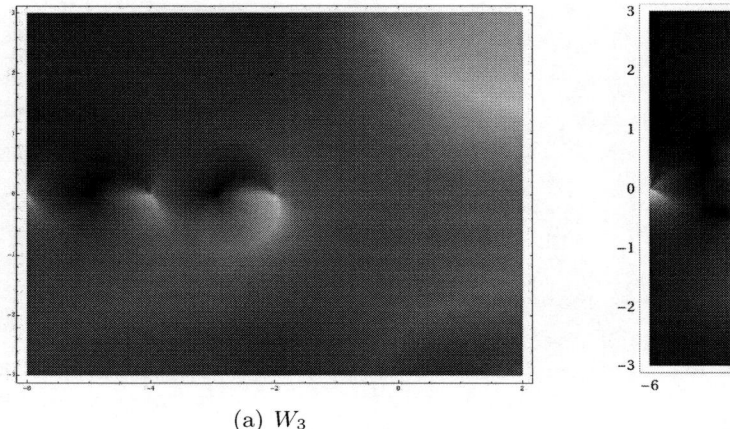

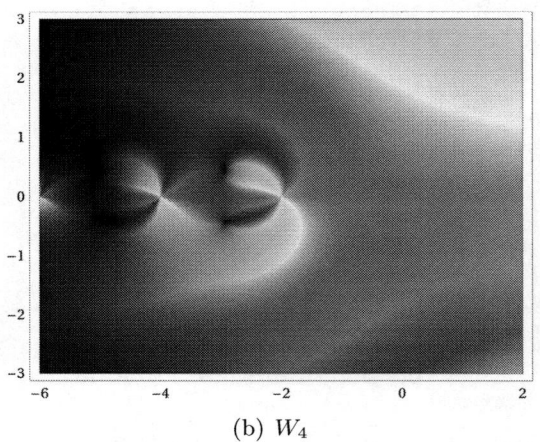

(a) W_3 (b) W_4

FIGURE 3. W_3 via (3.2) and W_4 in the complex plane.

Hitting the Ends. Bornemann [15] starts his remarkable solution by exploring *Monte-Carlo methods*, which are shown to be impracticable. He then reformulates the problem *deterministically* as *the value at the center of a 10×1 rectangle of an appropriate* harmonic measure [56] *of the ends*, arising from a 5-point discretization of *Laplace's equation* with Dirichlet boundary conditions. This is then solved by a well chosen *sparse Cholesky* solver. At this point a reliable numerical value of $3.837587979 \cdot 10^{-7}$ is obtained. And the posed problem is solved *numerically* to the requisite ten places.

This is the warm up. We may proceed to develop two analytic solutions, the first using *separation of variables* on the underlying PDE on a general $2a \times 2b$ rectangle. We learn that

$$p(a,b) = \frac{4}{\pi} \sum_{n=0}^{\infty} \frac{(-1)^n}{2n+1} \operatorname{sech}\left(\frac{\pi(2n+1)}{2}\rho\right) \qquad (3.3)$$

where $\rho := a/b$. A second method using *conformal mappings*, yields

$$\operatorname{arccot} \rho = p(a,b)\frac{\pi}{2} + \arg \mathrm{K}\left(e^{ip(a,b)\pi}\right) \qquad (3.4)$$

where K is again the *complete elliptic integral* of the first kind. It will not be apparent to a reader unfamiliar with inversion of elliptic integrals that (3.3) and (3.4) encode the same solution—though they must as the solution is unique in $(0,1)$—and each can now be used to solve for $\rho = 10$ to arbitrary precision. Bornemann ultimately shows that the answer is

$$p = \frac{2}{\pi} \arcsin(k_{100}), \qquad (3.5)$$

where

$$k_{100} := \left(\left(3 - 2\sqrt{2}\right)\left(2 + \sqrt{5}\right)\left(-3 + \sqrt{10}\right)\left(-\sqrt{2} + \sqrt[4]{5}\right)^2\right)^2.$$

No one (except harmonic analysts perhaps) anticipated a closed form—let alone one like this.

Where does this come from? In fact [21, (3.2.29)] shows that

$$\sum_{n=0}^{\infty} \frac{(-1)^n}{2n+1} \operatorname{sech}\left(\frac{\pi(2n+1)}{2}\rho\right) = \frac{1}{2}\arcsin k, \qquad (3.6)$$

exactly when k_{ρ^2} is parameterized by *theta functions* in terms of the so called *nome*, $q = \exp(-\pi\rho)$, as Jacobi discovered. We have

$$k_{\rho^2} = \frac{\theta_2^2(q)}{\theta_3^2(q)} = \frac{\sum_{n=-\infty}^{\infty} q^{(n+1/2)^2}}{\sum_{n=-\infty}^{\infty} q^{n^2}}, \qquad q := e^{-\pi\rho}. \qquad (3.7)$$

Comparing (3.6) and (3.3) we see that the solution is

$$k_{100} = 6.028069101559710828825407122920\ldots \cdot 10^{-7}$$

as asserted in (3.5).

The explicit form now follows from classical nineteenth century theory as discussed say in [15, 21]. In fact k_{210} is the singular value sent by Ramanujan to Hardy in his famous letter of introduction [20, 21]. If Trefethen had asked for a $\sqrt{210} \times 1$ box, or even better a $\sqrt{15} \times \sqrt{14}$ one, this would have shown up in the answer since in general

$$p(a,b) = \frac{2}{\pi}\arcsin\left(k_{a^2/b^2}\right). \qquad (3.8)$$

Alternatively, armed only with the knowledge that the singular values of rational parameters are always algebraic we may finish entirely computationally as described in [18]. □

FIGURE 4. Typical mirage, due to optical refraction. There appears to be a "lake" in front of the right-hand vehicles.

We finish this section with a quick pair of attractive applied-physics examples—from optics and astrophysics, respectively.

Example 3.4 (Mirages [45]). In [45] the authors, using geometric methods, develop an exact but implicit formula for the path followed by a light ray propagating over the earth, given radial variations in the refractive index. By suitably simplifying they are able to provide an explicit integral as a closed form, then expand asymptotically. This is done with the knowledge that the approximation is good to six or seven places—more than enough to use it on optically realistic scales. Moreover, in the case of quadratic or linear refractive indices these steps may be done *analytically*.

In other words, as advanced by Wright, a tractable and elegant approximate closed form is obtained to replace a problematic exact solution. From these forms interesting qualitative consequences follow. With a quadratic index, images are uniformly magnified in the vertical direction; only with higher order indices can nonuniform vertical distortion occur. This sort of knowledge allows one, for example, to correct distortions of photographic images such as Figure 4, with confidence and efficiently. (Said image taken from free repository `http://upload.wikimedia.org/wikipedia/commons/7/7b`, then processed/refined by the present authors.) □

Example 3.5 (Structure of stars). The celebrated *Lane–Emden equation*, presumed to describe the pressure χ at radius r within a star, can be put in the form:

$$r^{n-1}\frac{d^2\chi}{dr^2} = -\chi^n, \tag{3.9}$$

with boundary conditions $\chi(0) = 0$, $\chi'(0) = 1$, and positive real constant n, all of this giving rise to a unique trajectory $\chi_n(r)$ on $r \in [0, \infty)$. (Some authors invoke the substitution $\chi(r) := r\theta(r)$ to get an equivalent ODE for temperature θ; see [29]).

- The beautiful thing is, where this pressure trajectory crosses zero for positive radius r is supposed to be the *star radius*; call that zero z_n.

Amazingly, the Lane–Emden equation has known exact solutions for $n = 0, 1, 5$. The pressure trajectories for which indices n being respectively

$$\chi_0(r) = -\frac{1}{6}r^3 + r, \tag{3.10}$$

$$\chi_1(r) = \sin r, \tag{3.11}$$

$$\chi_5(r) = \frac{r}{\sqrt{1+r^2/3}}. \tag{3.12}$$

The respective star radii are thus closed forms $z_0 = \sqrt{6}$ and $z_1 = \pi$, while for (3.12) with index $n = 5$ we have infinite star radius (no positive zero for the pressure χ_5).

In the spirit of our previous optics example, the Lane–Emden equation is a simplification of a complicated underlying theory—in this astrophysics case, hydrodynamics—and one is rewarded by some closed-form star radii. But what about, say, index $n = 2$? We do not know a closed-form function for the χ trajectory in any convenient sense. What the present

CLOSED FORMS: WHAT THEY ARE AND WHY THEY MATTER

authors *have* calculated (in 2005) is the $n = 2$ star radius, as a high-precision number

$$z_2 = 4.3528745959461246769735700615261426281123653632130088353021\,51\ldots.$$

If only we could gain a closed form for this special radius, we might be able to guess the nature of the whole trajectory!

$\square$

4. Recent Examples Relating to Our Own Work

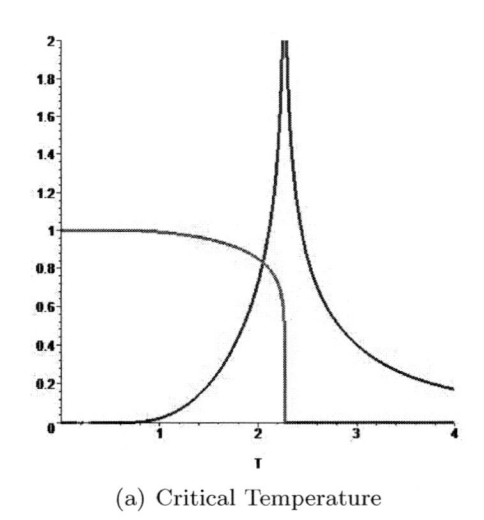

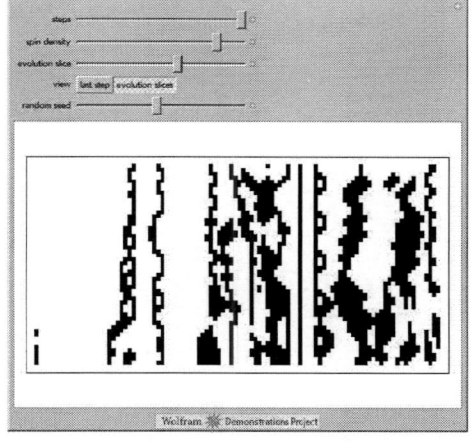

(a) Critical Temperature

(b) *Wolfram Player* Demonstration

FIGURE 5. The 2-dimensional Ising Model of Ferromagnetism, (a) image provided by Jacques Perk, plotting magnetization M (peak) and specific heat C (decay) per site against absolute temperature T [43, p. 91-93, p. 245].

Example 4.1 (Ising integrals [5, 8]). We recently studied the following classes of integrals [5]. The D_n integrals arise in the Ising model of mathematical physics (showing ferromagnetic temperature driven phase shifts see Figure 5 and [31]), and the C_n have tight connections to quantum field theory [8].

$$C_n = \frac{4}{n!} \int_0^\infty \cdots \int_0^\infty \frac{1}{\left(\sum_{j=1}^n (u_j + 1/u_j)\right)^2} \frac{du_1}{u_1} \cdots \frac{du_n}{u_n}$$

$$D_n = \frac{4}{n!} \int_0^\infty \cdots \int_0^\infty \frac{\prod_{i<j} \left(\frac{u_i - u_j}{u_i + u_j}\right)^2}{\left(\sum_{j=1}^n (u_j + 1/u_j)\right)^2} \frac{du_1}{u_1} \cdots \frac{du_n}{u_n}$$

$$E_n = 2 \int_0^1 \cdots \int_0^1 \left(\prod_{1 \leq j < k \leq n} \frac{u_k - u_j}{u_k + u_j} \right)^2 dt_2 \, dt_3 \cdots dt_n,$$

where (in the last line) $u_k = \prod_{i=1}^k t_i$.

Needless to say, evaluating these multidimensional integrals to high precision presents a daunting computational challenge. Fortunately, in the first case, the C_n integrals can be written as one-dimensional integrals:

$$C_n = \frac{2^n}{n!} \int_0^\infty p K_0^n(p) \, dp,$$

where K_0 is the *modified Bessel function*. After computing C_n to 1000-digit accuracy for various n, we were able to identify the first few instances of C_n in terms of well-known constants, e.g.,

$$C_3 = \mathrm{L}_{-3}(2) := \sum_{n \geq 0} \left(\frac{1}{(3n+1)^2} - \frac{1}{(3n+2)^2} \right), \quad C_4 = \frac{7}{12} \zeta(3),$$

where ζ denotes the Riemann zeta function. When we computed C_n for fairly large n, for instance

$$C_{1024} = 0.6304735033743867961220401927108789043545870787127 3234\ldots,$$

we found that these values rather quickly approached a limit. By using the new edition of the *Inverse Symbolic Calculator*[7] this numerical value was identified as

$$\lim_{n \to \infty} C_n = 2e^{-2\gamma},$$

where γ is the *Euler constant*, see Section 5. We later were able to prove this fact—this is merely the first term of an asymptotic expansion—and thus showed that the C_n integrals are fundamental in this context [5].

The integrals D_n and E_n are much more difficult to evaluate, since they are not reducible to one-dimensional integrals (as far as we can tell), but with certain symmetry transformations and symbolic integration we were able to symbolically reduce the dimension in each case by one or two.

In the case of D_5 and E_5, the resulting 3-D integrands are extremely complicated (see Figure 6), but we were nonetheless able to numerically evaluate these to at least 240-digit

---

[7] Available at http://carma.newcastle.edu.au/isc2/.

precision on a highly parallel computer system. This would have been impossible without the symbolic reduction. We give the integral in extenso to show the difference between a humanly accessible answer and one a computer finds useful.

In this way, we produced the following evaluations, all of which except the last we subsequently were able to prove:

$$\begin{aligned} D_2 &= 1/3 \\ D_3 &= 8 + 4\pi^2/3 - 27\,\mathrm{L}_{-3}(2) \\ D_4 &= 4\pi^2/9 - 1/6 - 7\zeta(3)/2 \\ E_2 &= 6 - 8\log 2 \\ E_3 &= 10 - 2\pi^2 - 8\log 2 + 32\log^2 2 \\ E_4 &= 22 - 82\zeta(3) - 24\log 2 + 176\log^2 2 - 256(\log^3 2)/3 + 16\pi^2 \log 2 - 22\pi^2/3 \end{aligned}$$

and

$$\begin{aligned} E_5 \stackrel{?}{=} {}& 42 - 1984\,\mathrm{Li}_4(1/2) + 189\pi^4/10 - 74\zeta(3) - 1272\zeta(3)\log 2 + 40\pi^2 \log^2 2 \\ & - 62\pi^2/3 + 40(\pi^2 \log 2)/3 + 88\log^4 2 + 464\log^2 2 - 40\log 2, \end{aligned} \qquad (4.1)$$

where Li denotes the polylogarithm function.

In the case of D_2, D_3 and D_4, these are confirmations of known results. We tried but failed to recognize D_5 in terms of similar constants (the 500-digit numerical value is accessible[8] if anyone wishes to try to find a closed form; or in the manner of the hard sciences to confirm our data values). The conjectured identity shown here for E_5 was confirmed to 240-digit accuracy, which is 180 digits beyond the level that could reasonably be ascribed to numerical round-off error; thus we are quite confident in this result even though we do not have a formal proof [5].

Note that every one of the D, E forms above, including the conjectured last one, is hyperclosed in the sense of our "Sixth approach" section. □

Example 4.2 (Weakly coupling oscillators [48, 6]). In an important analysis of coupled *Winfree oscillators*, Quinn, Rand, and Strogatz [48] developed a certain N-oscillator scenario whose bifurcation phase offset ϕ is implicitly defined, with a conjectured asymptotic behavior: $\sin \phi \sim 1 - c_1/N$, with experimental estimate $c_1 = 0.605443657\ldots$. In [6] we were able to derive the exact theoretical value of this "QRS constant" c_1 as the unique zero of the Hurwitz zeta $\zeta(1/2, z/2)$ on $z \in (0, 2)$. In so doing were able to prove the conjectured behavior. Moreover, we were able to sketch the higher-order asymptotic behavior; something that would have been impossible without discovery of an analytic formula.

Does this deserve to be called a closed form? In our opinion resoundingly 'yes', unless all inverse functions such as that in Bornemann's (3.8) are to be eschewed. Such constants are especially interesting in light of even more recent work by Steve Strogatz and his collaborators on *chimera*—coupled systems which can self-organize in parts of their domain

---

[8]Available at http://crd.lbl.gov/~dhbailey/dhbpapers/ising-data.pdf.

$$E_5 = \int_0^1 \int_0^1 \int_0^1 \Big[2(1-x)^2(1-y)^2(1-xy)^2(1-z)^2(1-yz)^2(1-xyz)^2$$
$$\Big(- \Big[4(x+1)(xy+1)\log(2) \Big(y^5 z^3 x^7 - y^4 z^2 (4(y+1)z+3)x^6 - y^3 z \Big((y^2+1)z^2 + 4(y+1)z+5 \Big) x^5 + y^2 \Big(4y(y+1)z^3 + 3(y^2+1)z^2 + 4(y+1)z - 1 \Big) x^4 + y \Big(z(z^2+4z+5)y^2 + 4(z^2+1)y + 5z + 4 \Big) x^3 + \Big((-3z^2 - 4z + 1)y^2 - 4zy + 1 \Big) x^2 - (y(5z+4)+4)x - 1 \Big] / \Big[(x-1)^3(xy-1)^3(xyz-1)^3 \Big] + \Big[3(y-1)^2 y^4 (z-1)^2 z^2 (yz-1)^2 x^6 + 2y^3 z \Big(3(z-1)^2 z^3 y^5 + z^2 (5z^3 + 3z^2 + 3z + 5) y^4 + (z-1)^2 z (5z^2 + 16z + 5) y^3 + (3z^5 + 3z^4 - 22z^3 - 22z^2 + 3z + 3) y^2 + 3(-2z^4 + z^3 + 2z^2 + z - 2) y + 3z^3 + 5z^2 + 5z + 3 \Big) x^5 + y^2 \Big(7(z-1)^2 z^4 y^6 - 2z^3 (z^3 + 15z^2 + 15z + 1) y^5 + 2z^2 (-21z^4 + 6z^3 + 14z^2 + 6z - 21) y^4 - 2z (z^5 - 6z^4 - 27z^3 - 27z^2 - 6z + 1) y^3 + (7z^6 - 30z^5 + 28z^4 + 54z^3 + 28z^2 - 30z + 7) y^2 - 2 (7z^5 + 15z^4 - 6z^3 - 6z^2 + 15z + 7) y + 7z^4 - 2z^3 - 42z^2 - 2z + 7 \Big) x^4 - 2y \Big(z^3 (z^3 - 9z^2 - 9z + 1) y^6 + z^2 (7z^4 - 14z^3 - 18z^2 - 14z + 7) y^5 + z (7z^5 + 14z^4 + 3z^3 + 3z^2 + 14z + 7) y^4 + (z^6 - 14z^5 + 3z^4 + 84z^3 + 3z^2 - 14z + 1) y^3 - 3 (3z^5 + 6z^4 - z^3 - z^2 + 6z + 3) y^2 - (9z^4 + 14z^3 - 14z^2 + 14z + 9) y + z^3 + 7z^2 + 7z + 1 \Big) x^3 + \Big(z^2 (11z^4 + 6z^3 - 66z^2 + 6z + 11) y^6 + 2z (5z^5 + 13z^4 - 2z^3 - 2z^2 + 13z + 5) y^5 + (11z^6 + 26z^5 + 44z^4 - 66z^3 + 44z^2 + 26z + 11) y^4 + (6z^5 - 4z^4 - 66z^3 - 66z^2 - 4z + 6) y^3 - 2 (33z^4 + 2z^3 - 22z^2 + 2z + 33) y^2 + (6z^3 + 26z^2 + 26z + 6) y + 11z^2 + 10z + 11 \Big) x^2 - 2 \Big(z^2 (5z^3 + 3z^2 + 3z + 5) y^5 + z (22z^4 + 5z^3 - 22z^2 + 5z + 22) y^4 + (5z^5 + 5z^4 - 26z^3 - 26z^2 + 5z + 5) y^3 + (3z^4 - 22z^3 - 26z^2 - 22z + 3) y^2 + (3z^3 + 5z^2 + 5z + 3) y + 5z^2 + 22z + 5 \Big) x + 15z^2 + 2z$$
$$+ 2y(z-1)^2(z+1) + 2y^3(z-1)^2 z(z+1) + y^4 z^2 (15z^2 + 2z + 15) + y^2 (15z^4 - 2z^3 - 90z^2 - 2z + 15) + 15 \Big] / \Big[(x-1)^2 (y-1)^2 (xy-1)^2 (z-1)^2 (yz-1)^2 (xyz-1)^2 \Big] - \Big[4(x+1)(y+1)(yz+1) \Big(-z^2 y^4 + 4z(z+1)y^3 + (z^2+1)y^2 - 4(z+1)y + 4x (y^2-1)(y^2 z^2 - 1) + x^2 (z^2 y^4 - 4z(z+1)y^3 - (z^2+1)y^2 + 4(z+1)y + 1) - 1 \Big) \log(x+1) \Big] / \Big[(x-1)^3 x (y-1)^3 (yz-1)^3 \Big] - \Big[4(y+1)(xy+1)(z+1) \Big(x^2 (z^2 - 4z - 1) y^4 + 4x(x+1)(z^2-1) y^3 - (x^2+1)(z^2-4z-1) y^2 - 4(x+1)(z^2-1) y + z^2 - 4z - 1 \Big) \log(xy+1) \Big] / \Big[x(y-1)^3 y (xy-1)^3 (z-1)^3 \Big] - \Big[4(z+1)(yz+1) \Big(x^3 y^5 z^7 + x^2 y^4 (4x(y+1)+5) z^6 - xy^3 \Big((y^2+1)x^2 - 4(y+1)x - 3 \Big) z^5 - y^2 \Big(4y(y+1)x^3 + 5(y^2+1)x^2 + 4(y+1)x + 1 \Big) z^4 + y \Big(y^2 x^3 - 4y(y+1)x^2 - 3(y^2+1)x - 4(y+1) \Big) z^3 + (5x^2 y^2 + y^2 + 4x(y+1)y + 1) z^2 + ((3x+4)y + 4)z - 1 \Big) \log(xyz+1) \Big] / \Big[xy(z-1)^3 z (yz-1)^3 (xyz-1)^3 \Big] \Big) \Big]$$
$$/ \Big[(x+1)^2 (y+1)^2 (xy+1)^2 (z+1)^2 (yz+1)^2 (xyz+1)^2 \Big] \, dx \, dy \, dz$$

FIGURE 6. The reduced multidimensional integral for E_5, which integral has led via extreme-precision numerical quadrature and PSLQ to the conjectured closed form given in (4.1).

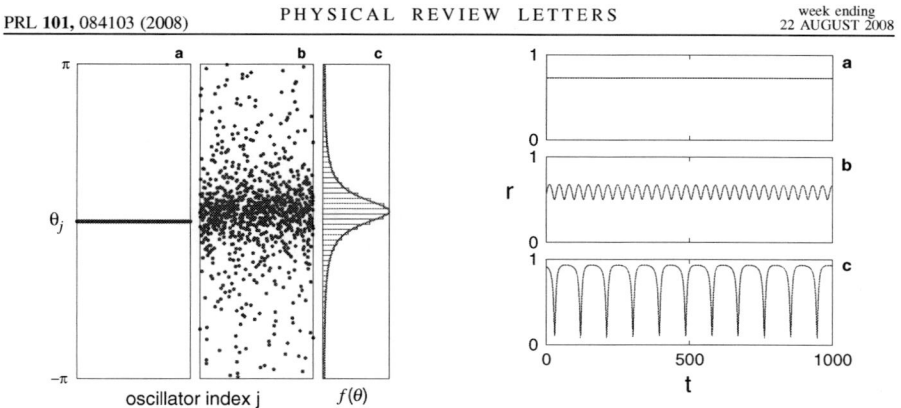

FIG. 1 (color online). Snapshot of a chimera state, obtained by numerical integration of (1) with $\beta = 0.1$, $A = 0.2$, and $N_1 = N_2 = 1024$. (a) Synchronized population. (b) Desynchronized population. (c) Density of desynchronized phases predicted by Eqs. (6) and (12) (smooth curve) agrees with observed histogram.

FIG. 2 (color online). Order parameter r versus time. In all three panels, $N_1 = N_2 = 128$ and $\beta = 0.1$. (a) $A = 0.2$: stable chimera; (b) $A = 0.28$: breathing chimera; (c) $A = 0.35$: long-period breather. Numerical integration began from an initial condition close to the chimera state, and plots shown begin after allowing a transient time of 2000 units.

FIGURE 7. Simulated chimera (figures and parameters from [42]).

and remain disorganized elsewhere, see Figure 7 taken from [42]. In this case observed numerical limits still need to be put in closed form. □

It is a frequent experience of ours that, as in Example 4.2, the need for high accuracy computation drives the development of effective analytic expressions (closed forms?) which in turn typically shed substantial light on the subject being studied.

Example 4.3 (Box integrals [3, 7, 22]). There has been recent research on calculation of expected distance between points inside a hypercube. Such expectations are also called "box integrals" [22]. So for example, the expectation $\langle|\vec{r}|\rangle$ for random $\vec{r} \in [0,1]^3$ has the closed form
$$\frac{1}{4}\sqrt{3} - \frac{1}{24}\pi + \frac{1}{2}\log\left(2 + \sqrt{3}\right).$$

Incidentally, box integrals are not just a mathematician's curiosity—the integrals have been used recently to assess the randomness of brain synapses positioned within a parallelepiped [37]. Indeed, we had cognate results for
$$\Delta_d(s) := \int_{[0,1]^d} \int_{[0,1]^d} \|x - y\|_2^s \, dx dy$$
which gives the moments of the distance between two points in the hypercube.

In a lovely recent paper [51] Stephan Steinerberger has shown that in the limit as the dimension goes to infinity
$$\lim_{d \to \infty} \left(\frac{1}{d}\right)^{s/p} \int_{[0,1]^d} \int_{[0,1]^d} \|x - y\|_p^s \, dx dy = \left(\frac{2}{(p+1)(p+2)}\right)^{s/p} \tag{4.2}$$

for any $s, p > 0$. In particular, with $p = 2$ this gives a first-order answer to our earlier published request for the asymptotic behavior of $\Delta_d(s)$.

A quite recent result is that all box integrals $\langle |\vec{r}|^n \rangle$ for integer n, and dimensions $1, 2, 3, 4, 5$ are *hyperclosed*, in the sense of our "Sixth attempt" section. It turns out that five-dimensional box integrals have been especially difficult, depending on knowledge of a hyperclosed form for a single definite integral $J(3)$, where

$$J(t) := \int_{[0,1]^2} \frac{\log(t + x^2 + y^2)}{(1 + x^2)(1 + y^2)} \, dx \, dy. \tag{4.3}$$

A proof of hyperclosure of $J(t)$ for algebraic $t \geq 0$ is established in [22, Thm. 5.1]. Thus $\langle |\vec{r}|^{-2} \rangle$ for $\vec{r} \in [0, 1]^5$ can be written in explicit hyperclosed form involving a 10^5-character symbolic $J(3)$; the authors of [22] were able to reduce the 5-dimensional box integral down to "only" 10^4 characters. A companion integral $J(2)$ also starts out with about 10^5 characters but reduces stunningly to a only a few dozen characters, namely

$$J(2) = \frac{\pi^2}{8} \log 2 - \frac{7}{48} \zeta(3) + \frac{11}{24} \pi \operatorname{Cl}_2\left(\frac{\pi}{6}\right) - \frac{29}{24} \pi \operatorname{Cl}_2\left(\frac{5\pi}{6}\right), \tag{4.4}$$

where Cl_2 is the *Clausen function* $\operatorname{Cl}_2(\theta) := \sum_{n \geq 1} \sin(n\theta)/n^2$ (Cl_2 is the simplest non-elementary Fourier series).

Automating such reductions will require a sophisticated simplification scheme with a very large and extensible knowledge base. With a current Research Assistant, Alex Kaiser at Berkeley, we have started to design software to refine and automate this process and to run it before submission of any equation-rich paper. This semi-automated integrity checking becomes pressing when—as above—verifiable output from a symbolic manipulation can be the length of a Salinger novella. □

5. Profound curiosities

In our treatment of numbers enjoying hyperclosure or superclosure, we admitted that such numbers are countable, and so almost all complex numbers cannot be given a closed form along such lines. What is stultifying is: *How do we identify an explicit number lying outside of such countable sets?* The situation is tantamount to the modern bind in regard to *normal numbers*—numbers which to some base have statistically random digit-structure in a certain technical sense. The bind is: Though almost all numbers are *absolutely normal* (i.e. normal to every base $2, 3, \ldots$), we do not know a single fundamental constant that is provably absolutely normal. (We do know some "artificial" normal numbers, see [13].)

Here is one possible way out of the dilemma: In the theory of computability, the existence of *noncomputable* real numbers, such as an encoded list of halting Turing machines, is well established. The celebrated *Chaitin constant* Ω is a well-known noncomputable. So a "folk" argument goes: "Since every element of the ring of hyperclosure $\mathbb{H}$ can be computed via converging series, it should be that $\Omega \notin \mathbb{H}$." A good research problem would be to make this heuristic rigorous.

Let us focus on some constants that *might* not be hyperclosed (nor superclosed). One such constant is the celebrated *Euler constant* $\gamma := \lim_{n \to \infty} \sum_{k=1}^{n} 1/k - \log n$. We know of no hypergeometric form for γ; said constant may well lie outside of $\mathbb{H}$ (or even $\mathbb{S}$). There *are* expansions for the Euler constant, such as

$$\gamma = \log \pi - 4 \log \Gamma\left(\frac{3}{4}\right) + \frac{4}{\pi} \sum_{k \geq 1} (-1)^{k+1} \frac{\log(2k+1)}{2k+1},$$

and even more exotic series (see [12]). But in the spirit of the present treatment, we do not want to call the infinite series closed because it is not hypergeometric *per se*. Relatedly, the classical Bessel expansion is

$$K_0(z) = -\left(\ln\left(\frac{z}{2}\right) + \gamma\right) I_0(z) + \sum_{n=1}^{\infty} \frac{\sum_{k=1}^{n-1} \frac{1}{k}}{(n!)^2} \left(\frac{z^2}{4}\right)^n.$$

Now $K_0(z)$ has a (degenerate) Meijer-G representation—so potentially is superclosed for algebraic z—and $I_0(z)$ is accordingly hyperclosed, but the nested-harmonic series on the right is again problematic. Again, γ is conjectured not to be a period [41].

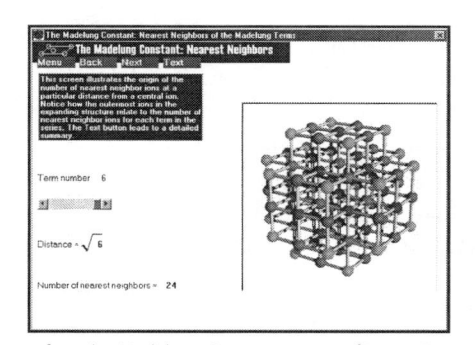

(a) NaCl nearest neighbors

(b) CMS Prize Sculpture

FIGURE 8. Two representations of salt.

Example 5.1 (Madelung constant [21, 36, 57]). Another fascinating number is the *Madelung constant*, $\mathcal{M}$, of chemistry and physics [21, Section 9.3]. This is the potential energy at the origin of an oscillating-charge crystal structure (most often said crystal is NaCl (salt) as illustrated in Figure 8). The right-hand image is of a Helaman Ferguson sculpture based on $\mathcal{M}$ that is awarded biannually by the Canadian Mathematical Society as part of the *David Borwein Career Award*.) and is given by the formal (conditionally convergent [17]) sum

$$\mathcal{M} := \sum_{(x,y,z) \neq (0,0,0)} \frac{(-1)^{x+y+z}}{\sqrt{x^2 + y^2 + z^2}} = -1.747564594633..., \tag{5.1}$$

and has never been given what a reasonable observer would call a closed form. Nature plays an interesting trick here: There are other crystal structures that *are* tractable, yet somehow this exquisitely symmetrical salt structure remains elusive. In general, even dimensional crystal sums are more tractable than odd for the same modular function reasons that the number of representations of a number as the sum of an even number of squares is. But this does not make them easy as illustrated by Example 1.2.

Here we have another example of a constant having no known closed form, yet rapidly calculable. A classical rapid expansion for the Madelung constant is due to Benson:

$$\mathcal{M} = -12\pi \sum_{m,n \geq 0} \text{sech}^2\left(\frac{\pi}{2}\sqrt{(2m+1)^2 + (2n+1)^2}\right), \tag{5.2}$$

in which convergence is exponential. Summing for $m, n \leq 3$ produces $-1.747564594\ldots$, correct to 8 digits. There are great many other such formulae for $\mathcal{M}$ (see [21, 34]).

Through the analytic methods of Buhler, Crandall, Tyagi and Zucker since 1999 (see [34, 36, 54, 57]), we now know approximations such as

$$\mathcal{M} \approx \frac{1}{8} - \frac{\log 2}{4\pi} + \frac{8\pi}{3} + \frac{\Gamma(1/8)\Gamma(3/8)}{\pi^{3/2}\sqrt{2}} + \log\frac{k_4^2}{16 k_4 k_4'},$$

where $k_4 := ((2^{1/4} - 1)/(2^{1/4} + 1))^2$. Two remarkable things: First, this approximation is good to the same 13 decimals we give in the display (5.1); the missing $O(10^{-14})$ error here is a rapidly, exponentially converging—but alas infinite—sum in this modern approximation theory. Second: this 5-term approximation itself is indeed hyperclosed, the only problematic term being the Γ-function part, but we did establish in our "Sixth approach" section that $B(1/8, 3/8)$ and also $1/\pi$ are hyperclosed, which is enough. Moreover, the work of Borwein and Zucker [26] also settles hyperclosure for that term. □

Certainly we have nothing like a proof, or even the beginnings of one, that $\mathcal{M}$ (or γ) lies outside $\mathbb{H}$ (or even $\mathbb{S}$), but we ask on an intuitive basis: Is a constant such as the mighty $\mathcal{M}$ telling us that it is not hyperclosed, in that our toil only seems to bring more "closed-form" terms into play, with no exact resolution in sight?

6. Concluding Remarks and Open Problems

- We have posited several approaches to the elusive notion of "closed form." But what are the intersections and interrelations of said approaches? For example, can our "Fourth approach" be precisely absorbed into the evidently more general "Sixth approach" (hyperclosure and superclosure)?

- How do we find a *single* number that is provably not in the ring of hyperclosure $\mathbb{H}$? (Though no such number is yet known, *almost all* numbers are as noted not in said ring!) The same question persists for the ring of hyperclosure, $\mathbb{S}$. Furthermore, how precisely can one create a *field* out of $\mathbb{H}^{\mathbb{H}}$ via appropriate operator extension?

- Though $\mathbb{H}$ is a subset of $\mathbb{S}$, how might one prove that $\mathbb{H} \neq \mathbb{S}$? (Is the inequality even true?) Likewise, is the set of closed forms in the sense of [47, Ch. 8] (only finite linear combinations of hypergeometric evaluations) properly contained in our $\mathbb{H}$? And what about a construct such as $\mathbb{H}^{\mathbb{H}^{\mathbb{H}}}$? Should such an entity be anything really new? Lest one remark on the folly of such constructions, we observe that most everyone would say π^{π^π} is a closed form!

- Having established the property of hyperclosure for $\Gamma^b(a/b)$, are there any cases where the power b may be brought down? For example, $1/\pi$ is hyperclosed, but what about $1/\sqrt{\pi}$?

- What is a precise connection between the ring of hyperclosure (or superclosure) and the set of periods or of Mahler measures (as in Example 1.3)?

- There is expounded in reference [22] a theory of "expression entropy," whereby some fundamental entropy estimate gives the true complexity of an expression. So for example, an expression having 1000 instances of the `polylog` token Li$_3$ might really involve only about 1000 characters, with that polylogarithm token encoded as a single character, say. (In fact, during the research for [22] it was noted that the entropy of *Maple* and *Mathematica* expressions of the *same* entity often had widely varying text-character counts, but similar entropy assessments.)

  On the other hand, one basic notion of "closed form" is that explicitly infinite sums not be allowed. Can these two concepts be reconciled? Meaning: Can we develop a theory of expression entropy by which an explicit, infinite sum is given infinite entropy? This might be difficult, as for example a sum $\sum_{n=1}^\infty \frac{1}{n^{3/2}}$ only takes a few characters to symbolize, as we just did hereby! If one can succeed, though, in resolving thus the entropy business for expressions, "closed form" might be rephrased as "finite entropy."

In any event, we feel strongly that the value of closed forms increases as the complexity of the objects we manipulate computationally and inspect mathematically grows—and we hope we have illustrated this. Moreover, we belong to the subset of mathematicians which finds fun in finding unanticipated closed forms.

Acknowledgements. Thanks are due to David Bailey and Richard Brent for many relevant conversations and to Armin Straub for the complex plots of W_3 and W_4.

References

[1] G. E. Andrews and R. Askey and R. Roy (1999). *Special Functions*, Cambridge University Press.

[2] D.H. Bailey, R. Barrio, and J.M. Borwein (2010). "High performance computation: mathematical physics and dynamics." Prepared for Joint SIAM/RSME-SCM-SEMA Meeting on *Emerging Topics in Dynamical Systems and Partial Differential Equations* (DSPDEs'10) May 31–June 4, 2010. Available at http://carma.newcastle.edu.au/jon/hpmd.pdf.

[3] D. Bailey, J. Borwein, N. Calkin, R. Girgensohn, R. Luke and V. Moll (2007). *Experimental Mathematics in Action*, A K Peters, Natick, MA.

[4] D. H. Bailey and J. M. Borwein (2008). "Computer-assisted discovery and proof." *Tapas in Experimental Mathematics*, 21–52, in *Contemporary Mathematics*, **457**, American Mathematical Society, Providence, RI, 2008.

[5] D. H. Bailey, D. Borwein, J.M. Borwein and R.E. Crandall (2007). "Hypergeometric forms for Ising-class integrals." *Experimental Mathematics*, **16**, 257–276.

[6] D.H. Bailey, J.M, Borwein and R.E. Crandall (2008) "Resolution of the Quinn–Rand–Strogatz constant of nonlinear physics." *Experimental Mathematics*, **18**, 107–116.

[7] D. H. Bailey, J. M. Borwein and R. E. Crandall (2010). "Advances in the theory of box integral." *Mathematics of Computation*, E-published February 2010.

[8] D.H. Bailey, J.M. Borwein, D.M. Broadhurst and L. Glasser (2008). "Elliptic integral representation of Bessel moments." *J. Phys. A: Math. Theory*, **41** 5203–5231.

[9] D. H. Bailey, J. M. Borwein, V. Kapoor and E. Weisstein (2006). "Ten Problems in Experimental Mathematics." *American Mathematical Monthly*, **113** 481–409.

[10] D. H. Bailey, P. B. Borwein and S. Plouffe (1997). "On the Rapid Computation of Various Polylogarithmic Constants." *Mathematics of Computation*, **66** 903–913.

[11] R. Baillie, D. Borwein, and J. Borwein (2008). "Some sinc sums and integrals." *American Math. Monthly*, **115** 888–901.

[12] D. Bailey, and R. Crandall (2001). "On the random character of fundamental constant expansions." *ExperimentaL Mathematics*, **10** 175–190.

[13] D. Bailey, and R. Crandall (2002)."Random generators and normal number.," *Experimental Mathematics,* **11** 527–547.

[14] Folkmar Bornemann (2005). "How Mathematica and Maple get Meijer's G-function into Problem 9." Preprint. Available at http://www-m3.ma.tum.de/bornemann/Numerikstreifzug/Chapter9/MeijerG.pdf.

[15] F. Bornemann, D. Laurie, S. Wagon, and J. Waldvogel (2004). *The Siam 100-Digit Challenge: A Study In High-accuracy Numerical Computing*, SIAM, Philadelphia.

[16] E. J. Borowski and J. M. Borwein (2006). *Web-linked Dictionary of Mathematics*, Smithsonian/Collins Edition.

[17] D. Borwein, J.M. Borwein and K.F. Taylor (1985), "Convergence of lattice sums and Madelung's constant," *J. Math. Phys.*, **26**, 2999–3009.

[18] J. M. Borwein (2005). "The SIAM 100 Digits Challenge." Extended review in the *Mathematical Intelligencer*, **27** 40–48.

[19] J. M. Borwein and D. H. Bailey (2008). *Mathematics by Experiment: Plausible Reasoning in the 21st Century*, extended second edition, A K Peters, Natick, MA.

[20] J. M. Borwein, D. H. Bailey and R. Girgensohn (2004). *Experimentation in Mathematics: Computational Roads to Discovery*, A K Peters, Natick, MA.

[21] J. M. Borwein and P. B. Borwein (1987). *Pi and the AGM*, John Wiley.

[22] J. M. Borwein, O-Yeat Chan and R. E. Crandall (2010). "Higher-dimensional box integrals." Submitted *Experimental Mathematics*.

[23] (2010). J. Borwein, D. Nuyens, A. Straub, and James Wan, "Random walk integrals." Preprint available at http://carma.newcastle.edu.au/jon/walks.pdf.

[24] J. M. Borwein and K. Devlin (2008). *The Computer as Crucible*, A K Peters, Natick, MA.

[25] M. Bronstein, R. M. Corless, J.H. Davenport, and D. J. Jeffrey (2008). "Algebraic properties of the Lambert W function from a result of Rosenlicht and of Liouville." *Integral Transforms and Special Functions,* **19** 709–712.

[26] J. M. Borwein and I. J. Zucker (1991). "Fast Evaluation of the Gamma Function for Small Rational Fractions Using Complete Elliptic Integrals of the First Kind," *IMA J. Numerical Analysis* **12** 519–526.

[27] J. Carette (2010). Private communication.

[28] J. Carette (2004). "Understanding expression simplification." Proceedings of International Conference on Symbolic and Algebraic Computation, *ACM* 72–79.

[29] S. Chandrasekhar (1967). *An Introduction to the Study of Stellar Structures*, Dover, New York.

[30] T. Y. Chow, (1999). "What is a closed-Form Number?" *American Mathematical Monthly*, **106** 440–448.

[31] B. A. Cipra (1987), "An introduction to the Ising model," *Amer. Math. Monthly*, **94** (10), 937–959.

[32] R. E. Crandall (1996). *Topics in Advanced Scientific Computation*, Springer, New York.

[33] R. E. Crandall (2007). "Theory of ROOF Walks." Available at http://people.reed.edu/~crandall/papers/ROOF.pdf.

[34] R. E. Crandall (1999). "New representations for the Madelung constant." *Experimental Mathematics*, **8** 367–379.

[35] R.E. Crandall (2009). "Analytic representations for circle-jump moments." *PSIpress*, 21 Nov 09, www.perfscipress.com/papers/analyticWn_psipress.pdf

[36] R. E. Crandall and J. E. Buhler (1987). "Elementary expansions for Madelung constants." *J. Phys. A: Math Gen.* **20** 5497–5510.

[37] R. Crandall and T. Mehoke (2010). "On the fractal distribution of brain synapses." Preprint.

[38] C. Deninger (1997) "Deligne periods of mixed motives, theory and the entropy of certain actions." *J. Amer. Math. Soc.*, **10** 259–281.

[39] H. R. P. Ferguson, D. H. Bailey and S. Arno (1999). "Analysis of PSLQ, An Integer Relation Finding Algorithm." *Mathematics of Computation*, vol. 68, no. 225 (Jan 1999), pg. 351–369.

[40] Brian Hayes (2005). "Why W?" *American Scientist*, **93** 1004–1008.

[41] M. Kontsevich and D. Zagier (2001). "Periods." In *Mathematics unlimited—2001 and beyond*, 771–808, Springer-Verlag.

[42] E. A. Martens, C. R. Laing and S. H. Strogatz (2010). "Solvable model of spiral wave chimeras." *Physical Review Letters*, **104** 044101.

[43] B. McCoy and Tai Tsun Wu (1973), *The Two-Dimensional Ising Model*, Harvard Univ Press.

[44] F. W. J. Olver, D. W. Lozier, R. F. Boisvert, and C. W. Clark (2010). *NIST Handbook of Mathematical Functions*, Cambridge University Press.

[45] B. D. Nener, N. Fowkes, and L. Borredon (2003). "Analytical models of optical refraction in the troposphere." *JOSA*, **20** 867–875.

[46] K. Oldham, J. Myland, and J. Spanier (2009). *An Atlas of Functions: With Equator, The Atlas Function Calculator*, Springer-Verlag.

[47] M. Petkovsek, H. Wilf and D. Zeilberger (1996). A K Peters, Ltd. Available at http://www.math.upenn.edu/~wilf/AeqB.html.

[48] D. Quinn, R. Rand, and S. Strogatz (2007). "Singular unlocking transition in the Winfree model of coupled oscillators." *Phys. Rev E*, **75** 036218-1-10.

[49] M. Rogers (2010). "Hypergeometric formulas for lattice sums and Mahler measures." Preprint.

[50] C. J. Smyth (2003). "Explicit Formulas for the Mahler Measures of Families of Multivariable Polynomials." Preprint.

[51] S. Steinerberger (2010). "Extremal uniformly distributed sequences and random chord lengths." *Acta Mathematica Hungarica*. In press.

[52] C. J. Smyth (2008). "The Mahler measure of algebraic numbers: a survey." Conference Proceedings, University of Bristol, 3-7 April 2006. *LMS Lecture Note Series* **352**, Cambridge University Press, Cambridge, pp. 322–349.

[53] N. Temme (1996). *Special Functions, an Introduction to the Classical Functions of Mathematical Physics*, John Wiley & Sons, New York.

[54] S. Tyagi (2003). Private communication.

[55] E. W. Weisstein, "Closed-Form Solution." From MathWorld – A Wolfram Web Resource. http://mathworld.wolfram.com/Closed-FormSolution.html.

[56] B. L. Walden and L. A. Ward (2007). "A harmonic measure interpetation of the arithmetic-geometric mean." *American Mathematical Monthly*, **114** 610–622.

[57] I. J. Zucker (1984). "Some infinite series of exponential and hyperbolic functions." *SIAM J. Math. Anal.*, **15** 406–413.

Index

$P = NP$ problem, 29
π, 30, 31, 49, 55, 61, 64, 87, 91, 92, 99, 119, 168

algorithm
    addition, 24
    algebraic, 58
    BBP, 167, 253
    Borchardt's, 4
    Brent's, 16, 42
    converging, 11
    elliptical integral, 275
    Ferguson-Forcade, 50
    integer relation, 97, 181
    iterative, 41
    Newton's, 5, 6, 10, 25, 52, 53
    PSLQ, 167, 181, 195, 245
    quartic, 251
    Salamin's, 3, 42
    self-correcting, 25, 53
    tanh-sinh quadrature, 200
    Wilf-Zeilberger, 257
approximation
    Padé, 22, 25, 78
    polynomial, 21, 22
    rational, 21, 22
    to π, 40
Archimedes's Method, 38, 51
asymptotic expansion, 27

Baire category, 28
binary splitting, 17

Cauchy Integral Formula, 13
closed form, 267
coefficient
    binomial, 101, 122, 148
    polynomial, 122
        constrained, 126
    integer, 182, 185, 271, 273
compact, 15, 16, 24
complex plane, 22
complexity
    algebraic, 23, 29
    bit, 23, 26, 29, 41, 52
    digit, 23, 28
        of multiplication, 24
    rational, 23, 26, 29
conjugate scale, 6
constant
    Catalan's, 27, 172
    Euler's, 24, 27, 49, 55, 273, 281
    Feigenbaum's delta, 182
    hard hexagon, 99
    Madelung, 185, 289
    multi-Zeta, 170
convergence
    absolute, 271
    exponential, 5, 10, 13, 14
    linear, 4, 5, 11
    quadratic, 6, 11, 26, 42, 251
    quintic, 48
    rate of, 8, 22, 58
    second order, 5
    uniform, 14, 15

differential equations
    Bessel, 273
    coefficients of, 27
    holonomic, 275
    linear, 27
    nonlinear
        algebraic, 29, 32
distribution
    Gauss-Kuzmin, 192
    Poisson, 231

equation

Diophantine, 77
Laplace, 280
modular, 59, 61
    solvable, 58
Parseval's, 170
quintic, 59, 61
exponentially computable, 11

Feynmann diagram, 105
Formula
    reduction, 203
    reflection, 203
    sum, 203
Fourier transform
    discrete, 54
    fast, 24–26, 41, 49, 53, 54
fractions
    continued, 43, 76, 82, 119, 125, 173
        Ramanujan's, 193
        simple, 154
    finite, 76
    infinite, 76
functions
    additive partitioning, 223, 250
    algebraic, 6, 9, 25
    analytic, 29
    Bessel, 175, 206
        modified, 281
    beta, 173, 270
    binary counting, 74
    Clausen, 190, 246
    computable, 29
    elementary, 3, 5, 17, 21, 25, 26
    elliptical, 3, 56, 58
    eta, 59
    gamma, 24, 27, 270
    generating, 135, 137, 203, 224, 232
        ordinary, 250
    hypergeometric, 24, 205
        Gaussian, 253
    inverse hyperbolic, 4

    Lambert W, 157, 271
    measurable, 71
    meromorphic, 57
    modular, 40, 56–58
    ordinary generating, 135
    polynomial, 6
    rational, 6, 22, 25, 26
    Riemann Zeta, 27, 100, 170, 186, 200, 256, 284
        alternating, 72
        multiple, 103, 189, 203
    sinc, 201
    theta, 29, 55, 59, 281
        Jacobian, 56, 173, 186, 194
    totient, 254
    trancendental, 6, 17
    transcendental, 26
    trigonometric, 4, 10

group
    λ, 57
    Galois, 59
    modular, 57
    non-solvable, 59
    solvable, 59

induction, 7
integral
    complementary, 55
    elliptical, 6, 8, 11, 12, 25, 26, 56
        of the first kind, 12, 55, 194, 280
        of the second kind, 11, 12, 55
    indefinite, 145
    iterated, 104
    of the first kind, 280
    Riemann's, 27
    Watson, 206
iteration
    backward, 7
    chaotic, 182
    computation of π, 39

Gaussian arithmetic-geometric mean, 3, 12, 22, 25, 27
    Complex, 15

Legendre's relation, 11, 55, 60
limit
    algebraic, 3
log 2, 100

number
    algebraic, 275
    Bell, 232
    Bernoulli, 175, 199
    Carmichael, 254
    Catalan, 136
    complex, 272
    constructable, 89
    diffeoclosed, 275
    Fibonacci, 84, 104
    irrational, 77
    Liouville, 49, 82
    normal, 288
    pentagonal, 152
    prime, 139, 140, 169, 177
        Fermat, 255
    quasi-rational, 124
    rational, 79, 275
    Skewes, 140
    Stirling, 123
    sublinear, 28, 30, 32
    superclosed, 272
    transcendental, 49, 79

Pascal's triangle, 147
polynomial
    Chebyshev, 122
    Legendre, 123
    roots of, 229

Ramanujan, S., 36, 60, 140, 152
ring
    hyperclosure, 272, 284
    superclosure, 279, 288

self-similarity, 148
series
    converging, 29
    hypergeometric, 272
    Lambert, 80, 187
    Maclaurin, 192
    power, 29
    Taylor, 21, 22, 175, 199
sparse multipliers, 29
sum
    Euler, 191, 203, 206, 228
    Euler-Maclaurin, 171
    Poisson, 186
    Ramanujan's, 27, 61
    Riemann, 173, 194
    telescoping, 192

theorem
    Binomial, 232
    Cauchy-Lindelof, 173, 194
    Parseval's, 176
    Roth's, 74
transformation
    Landen, 12
    Laplace, 198
    linear fraction, 57
    Mellin, 186
    modular, 85
    theta, 186